Universitext

Rudolf Ahlswede · Vladimir Blinovsky

Lectures on Advances in Combinatorics

 Springer

Rudolf Ahlswede
Universität Bielefeld
Fakultät für Mathematik
Universitätsstr. 25
33615 Bielefeld
Germany
ahlswede@math.uni-bielefeld.de

Vladimir Blinovsky
Institute of Information
Transmission Problems
Russian Academy of Sciences
Bol'shoi karetnyi per. 19
127994 Moscow
Russia
vblinovs@yandex.ru

ISBN 978-3-540-78601-6   e-ISBN 978-3-540-78602-3

Library of Congress Control Number: 2008923540

Mathematics Subject Classification (2000): 05-XX, 11-XX, 40-XX, 52-XX, 68-XX, 94-XX

© Springer-Verlag Berlin Heidelberg 2008

Cover design: WMXDesign GmbH, Heidelberg

Printed on acid-free paper

9 8 7 6 5 4 3 2 1

springer.com

Preface

The lectures concentrate on highlights in Combinatorial (Chapters II and III) and Number Theoretical (Chapter IV) Extremal Theory, in particular on the solution of famous problems which were open for many decades.

However, the organization of the lectures in six chapters does neither follow the historic developments nor the connections between ideas in several cases. With the specified auxiliary results in Chapter I on Probability Theory, Graph Theory, etc., all chapters can be read and taught independently of one another.

In addition to the 16 lectures organized in 6 chapters of the main part of the book, there is supplementary material for most of them in the Appendix. In particular, there are applications and further exercises, research problems, conjectures, and even research programs.

The following books and reports [B97], [ACDKPSWZ00], [A01], and [ABCABDM06], mostly of the authors, are frequently cited in this book, especially in the Appendix, and we therefore mark them by short labels as [B], [N], [E], and [G]. We emphasize that there are also "Exercises" in [B], a "Problem Section" with contributions by several authors on pages 1063–1105 of [G], which are often of a combinatorial nature, and "Problems and Conjectures" on pages 172–173 of [E].

The book includes the two well-known results (both in Chapter V), the Ahlswede/Zhang identity, which improves the LYM-inequality, and the Ahlswede/Daykin inequality, which is more general and also sharper than known correlation inequalities in Statistical Physics, Probability Theory, and Combinatorics (cf. the survey by Fishburn/Shepp in [N], 501–516). These inequalities were started in Probability Theory (percolation) around 1960 with Harris, in Combinatorics in 1966 with Kleitman's Lemma, and in physics 1971 with Fortuin/Kasteleyn/Ginibre (FKG). In many books the AD-inequality (also called "4-Function Theorem") is viewed in connection with lattices. We emphasize that a much more general inequality of [AD79b] makes no reference to lattices.

Its essence is a Cartesian product property of sets and therefore there is a wider range of possible applications. Also there is nothing holy about the number of operations and factors on either side of the inequality as long as proper weight-expansiveness is ensured. In the following, we come to another surprise concerning number-theoretical inequalities.

A spectacular series of results started with a lecture in 1992 of Erdös, who raised in 1962 (and repeatedly spoke about) the problem "What is the maximal cardinality of a set of numbers smaller than n with no $k + 1$ of its members being pairwise relatively prime?"

This stimulated Ahlswede and Khachatrian to make a systematic investigation of this and related number-theoretical extremal problems. Its immediate successes are solutions for several well-known conjectures of Erdös and Erdös/Graham (Chapter VI). More importantly, they gained an understanding for the role of the prime number distribution for such problems, which distinguishes them from combinatorial extremal problems. These investigations had another surprising fruit. The AD-inequality implies a number-theoretical correlation inequality for Dirichlet densities of sets of numbers, which implies and is sharper than the classical inequalities in [H37] and [R37], which settled a conjecture of Hasse concerning an identity due to Dirichlet and Behrend, the number theoretical form of FKG! Number Theory came first and AD is a crossroad between Pure and Applied Mathematics (in Chapter V). Also another inequality, seemingly without predecessors, was discovered.

Finally, the analysis led to the discovery of a new "pushing" method with a wide applicability. In particular, it led to the solution of well-known combinatorial problems like the famous $4m$-conjecture (Erdös/Ko/Rado 1938, one of the oldest problems in Combinatorial Extremal Theory) or the diametric problem in Hamming spaces (optimal anticodes).

Actually, the $4m$-conjecture just concerned the first unsolved case of the following much more general problem: A system of sets $\mathcal{A} \subset \binom{[n]}{k}$ is called t-intersecting if $|A_1 \cap A_2| \geq t$ for all $A_1, A_2 \in \mathcal{A}$, and $I(n,k,t)$ denotes the set of all such systems. Determine the function $M(n,k,t) = \max_{\mathcal{A} \in I(n,k,t)} |\mathcal{A}|$ and the structure of maximal systems! Ahlswede and Khachatrian gave the complete solution for every n,k,t. It has a very clear geometrical interpretation (Chapter II).

Most lectures in Chapter III are devoted to combinatorics of multiple packings, which are equivalent to list codes in Information Theory and as such relevant for estimating error probabilities. Fundamental works of Blinovsky [B01a] represented here deliver the solutions of the problems for list codes, which were stated by the classics of information theory in the middle of the last century. These problems give a beautiful example of interplay between Extremal Combinatorics and Information Theory. Covering and packing are classical topics in Geometry. In Chapter III they concern sequence spaces for problems primarily motivated by Information Theory: Data Compression and Shannon's zero error capacity problem of a noisy channel, which is a packing problem for product hypergraphs. A highlight was Lovász' solution of the pentagon case. In general, the progress is rather slow. Here we deal with a related partition problem.

Origins of problems and theories, which developed with them, are only briefly discussed because of limited space. However, we consider it especially important for students to think about Mathematics in connection with other sciences and real world phenomena. In fact, a large part of Chapters IV and V originated that way. Stimuli came mostly from questions in Information Theory. We give a quote from [N], page xvi.

"The deep interplay between several disciplines and a broad philosophical view is a thread through Ahlswede's work. For him, Information Theory deals with *gaining* information (that is, statistics), *transfer* of information without and with secrecy constraints (that is, cryptology), and *storing* information (Memories, Data Compression). Applying ideas from one area to another often led to unexpected and beautiful results and even to new theories.

Let us give an example involving storage. Motivated by the practical problem of storing data using a new laser technique, code models for reusable memories were introduced in Information Theory. It turned out that the analysis was much more efficient when stating the question as a combinatorial extremal problem, which led immediately to the connections with hypergraph coloring, novel iso-diametric problems in sequence spaces and finally to the new class of the so-called "Higher-level extremal problems" in Combinatorics.

Among them are also Sperner-type questions for "clouds" of antichains. These problems are by one degree more complex than those usually considered: sets take the role of elements, families of sets (clouds) take the role of sets, etc. ([N], P.L. Erdös, L.A. Szekely, 117–124).

In another direction, generalizing models for reusable memories, Ahlswede and Zhang introduced write–efficient rewritable memories leading to diametrical problems for sequence spaces.

Imagine a tape with n cells into which we can write letters from an alphabet $\mathcal{X}$. A word $x^n = (x_1, \ldots, x_n)$ stores some messages. When we want to update this record to a message represented by $y^n = (y_1, \ldots, y_n)$ the per letter costs $\varphi(x_t, y_t)$ add up to $\varphi_n(x^n, y^n) = \sum_{t=1}^n \varphi(x_t, y_t)$. When there is a cost constraint D, then we require

$$\varphi_n(x^n, y^n) \leq D.$$

To be able to update many messages we come to the diametric problem to characterize

$$M(\varphi_n, D) = \max\{|C| : \varphi_n(x^n, y^n) \leq D \text{ for all } x^n, y^n \in C\}$$

for the "sum-type" cost φ_n, which can also be a distance function like the Hamming, Taxi or Lee metric, etc. These problems are discussed also in Chapter II (and in the Appendix).

There, in Lecture 6, also two families of sequences with constant mutual distances are considered. This falls into the subject of monochromatic rectangles, which arose in Yao's investigation of communication complexity.

The topic of the last lecture in Chapter II also has its origin in Computer Science and was communicated to us by Mullin in 1990.

Finally, we emphasize that another approach, the study of extremal problems under dimension constraints (in Chapter IV) has found an application in Computer Science. Several questions on databases find answers as immediate consequences.

Another new concept, that of splitting antichains, is treated in Chapter V.

The *lectures* primarily address basic extremal problems and inequalities – two sides of the same coin. Thus, they also prepare to ways of thinking and to methods, which are useful and applicable in a broader mathematical context.

At the end of every chapter, in addition to *exercises*, which also open eyes for new connections, we present several *problems* and sometimes we offer also *conjectures*.

Another important feature of the present book is that several of its concepts and problems arise in response or by remodeling the given questions and the models in sciences like Information Theory, Computer Science, or Statistical Physics. The interdisciplinary character gives Ars Combinatoria a special status in Mathematics. Again with occasional references to the books and reports we hope to create an atmosphere rich of incentives for new discoveries.

We hope that Ars Combinatoria gives light and joy to all minds and hearts striving for understanding and happiness through the world – from their origins to their destinations.

Our next remarks go to the potential readers. Above all, as indicated in the title, the book is meant for lectures dealing with advances in Combinatorics.

They are suited for all Mathematics students at the graduate level and for undergraduate students who very early specialize in certain parts of Combinatorics to combine it, for instance, with Computer Science, especially Complexity Theory or Data Structures, or with Information Theory, especially Coding Theory, Computer Systems Organization, and Communication Networks or Cryptology, or with Biochemistry, especially Sequencing in Genetics, etc.

Chapters of the book can be combined with the books mentioned below.

There seems to be no book on Combinatorics with such a concentration of high-level results, novel concepts, and advanced novel proof techniques. Therefore, the book is especially recommended to experienced researchers.

Concerning content, there is an overlap with the research collection "Sperner Theory" by K. Engel, Cambridge University Press, 1997, which, however, contains no Combinatorial Number Theory. The two books "Combinatorics (Set Systems, Hypergraphs, Families of Vectors, and Combinatorial Probability)", Cambridge University Press, 1986, by B. Bollobás and "A Course in Combinatorics", Cambridge University Press, 1992 by J.H. van Lint and R.M. Wilson treat basic combinatorial concepts and results. Probabilistic proofs can be learned from "Probabilistic Methods in Combinatorics" by N.A. Alon, J. Spencer, and P. Erdös, Akademiai Kiado, Budapest, 1974.

All of them are well-suited for undergraduate and graduate courses and qualify as excellent preparations for chapters of this much more advanced book, which has as a special ingredient also an Appendix with a wealth of research problems and conjectures sometimes constituting even research programs for already established mathematicians, but also for Ph.D. students.

There are also three research perspectives, novel links between Information Theory and Combinatorics:

- A direction in Extremal Theory of Sequences: Creating Order with Simple Machines.
- Information Flows in Networks.
- Information Theory and the Regularity Lemma.

Chapters can also be used to give fresh air to graduate courses. Perhaps the most recent and important are Chapter VI with results on divisors contributing to Elementary, Analytic, and Algebraic Number Theory, Chapter II with the Complete Intersection Theorem, which recently found connections to complexity theory, and Chapter IV with solutions to extremal problems with dimension constraints, which have consequences for Statistical Databases.

For computer scientists, S. Jukna's book "Extremal Combinatorics with Applications in Computer Science", Springer, 2001, finds now substantial additional material in most chapters.

The senior author gratefully acknowledges that he was given the opportunity to head two research projects "Combinatorics on Sequence Spaces" and "Models with Information Exchange" in the Sonderforschungsbereich "Discrete Structures in Mathematics" of the German Science Foundations (DFG) in Bielefeld from 1989 to 2000 and a research project "General Theory of Information Transfer and Combinatorics" at the Center for Interdisciplinary Research (ZiF) in Bielefeld from 2000 to 2004, and that parallel and even afterwards the DFG has continued its very generous support with several projects.

Most of the work reported here is an outgrowth of the cooperation with guests in Bielefeld. Among them, essentially working in Combinatorics, were Noga Alon, Harout Aydinian, Christian Bey, Sergei Bezrukov, Vladimir Blinovsky, Aart Blokhuis, Ning Cai, Konrad Engel, Peter Erdős, Levon Khachatrian, and Zhen Zhang, and in Number Theory, Vladimir Blinovsky, Paul Erdős, Christian Mauduit, Levon Khachatrian, and András Sárközy.

We are indebted to Christian Deppe and Christian Wischmann for diligent proof reading and the reduction of the number of Russian or German styled phrases.

Last but not least, the junior author is grateful to Leonid Bassalygo for introducing him to Combinatorial Coding Theory in the early eighties and to Dan Kleitman for instructive discussions on diametric problems and the senior author is grateful to Gyula Katona and David Daykin for their extreme encouragement in the seventies to look at Extremal Set Theory.

Table of Contents

Chapter I
Conventions and Auxiliary Results

We assume that the reader is familiar with basic concepts in Set Theory, Combinatorics, Probability Theory, Linear Algebra, and Elementary Number Theory. Still textbooks are mentioned if some material needs refreshment.

All chapters can be studied and used for courses, independently of one another.

The following notation is used everywhere in the book.

Denote $\mathbb{N}, \mathbb{Z}, \mathbb{R}$ the set of natural numbers, integers, and real numbers, respectively. For $i < j \in \mathbb{N}$ let $[i, j] = [i, i+1, \ldots, j]$ and $[n] = \{1, 2, \ldots, n\}$. For the arbitrary (finite) set $\mathcal{X}$ denote $2^{\mathcal{X}} = \{A \subset \mathcal{X}\}$ and $\binom{\mathcal{X}}{k} = \{A \subset \mathcal{X} : |A| = k\}$, where $|A|$ is the number of elements in A. We consider the natural bijection between $2^{[n]}$ and set of binary tuples $a^n = (a_1, \ldots, a_n)$, $a_i \in \{0, 1\}$ of length $[n]$. Actually in many places in the text we make no difference between a set $A \in 2^{[n]}$ and the corresponding binary n-tuple a^n and think that it does not make any difficulties in understanding the text. We use capital letters $A, B, \ldots$ to denote sets, we denote n-tuples as $a^n, b^n, \ldots$, and we denote families of sets as $\mathcal{A}, \mathcal{B}, \ldots$. But sometimes when it is convenient we use other notation: for example, S^n for n-tuples from the set S or a, b for the points from some set S.

We assume that the reader is familiar with basic notions from Lattice Theory such as distributive lattice, order isomorphic lattices, isomorphism of a finite distributive lattice to a sublattice of $(2^{[n]}, \subset)$ for some n, upset and downset, complement element, partially ordered set (poset), etc. For the facts about lattices we refer to [R65], [G71], [B73a].

We write $x(n) \overset{n \to \infty}{\to} y$ if the limit of $x(n)$ is y when n tends to infinity. We write $f_n = O(g_n)$ iff $\exists C$ such that $|f_n| < C|g_n|$ for sufficiently large n and $f_n = o(g_n)$ iff $\forall C \ |f_n| < C|g_n|$ for sufficiently large n. Also $f_n = \Omega(g_n)$ iff $g_n = O(f_n)$. At last $f_n \sim g_n$ iff $f_n/g_n \overset{n \to \infty}{\to} 1$. The notation $\overset{\Delta}{=}$ means "equal by definition."

We use Stirling's formula

$$n! = \sqrt{2\pi n} \left(\frac{n}{e}\right)^n (1 + o(1)).$$

More special notions are introduced gradually throughout the book.

Chapter II deals solely with Combinatorics.

In Chapters III–V it is combined with Probability Theory and finally Number Theory comes on stage in Chapter V and becomes then the main topic in Chapter VI.

To make the lectures self-contained, in addition to the prerequisites mentioned we need the following auxiliary results from Probability Theory (in Chapters III and IV) and Graph Theory (in Chapter IV).

The reader should know the following basic notions from Probability Theory, and for this purpose see, for example, [F68].

If we have a series of trials, that is, a sequence of probability spaces $\{(\Omega_n, P_n)\}$ and some property A (common for every Ω_n) is true with probability $P_n(A) \to 1$ as $n \to \infty$, we say that property A is valid with high probability (if the probability $P_n(A^c)$ of the complement event A^c decreases to zero). If $P_n(A^c)$ decreases exponentially with n, we say that property A is true with exponentially high probability.

We need several estimations of the deviations of sums of independent random variables. These estimations are not best possible. All we are worried about is their form, which should be convenient for our applications.

We often use the Chernoff bound: for a sequence of independent identically distributed (i.i.d.) random variables $X, X_1, \ldots, X_n, \ldots$, such that the moment generating function $\mathbb{E}e^{hX} < \infty$, which is the mathematical expectation of exponent e^{hX}, we have

$$\ln P_n(Y_n - n\mathbb{E}(X) \geq n\rho) \leq n \left[\ln \mathbb{E}e^{hX} - h(\mathbb{E}(X) + \rho) \right], \ h, \rho \geq 0, \ Y_n = \sum_{i=1}^{n} X_n,$$

where

$$\frac{\mathbb{E}(Xe^{hX})}{\mathbb{E}e^{hX}} = \mathbb{E}(X) + \rho$$

and

$$\ln P_n(-Y_n + n\mathbb{E}(X) \geq n\rho) \leq n \left[\ln \mathbb{E}e^{-hX} + h(\mathbb{E}(X) - \rho) \right],$$

where

$$\frac{\mathbb{E}(Xe^{-hX})}{\mathbb{E}e^{-hX}} = \mathbb{E}(X) - \rho.$$

We also use the so-called Chebyshev inequality, which we introduce in two forms. First form: for a random variable $\xi \geq 0$ and a number $c > 0$ we have

$$P(\xi > c\mathbb{E}\xi) < \frac{1}{c}.$$

Second form (which is a consequence of the first one): if $\mathbb{E}\xi^2 < \infty$, then

$$P(|\xi - \mathbb{E}(\xi)| > c\sqrt{Var(\xi)}) < \frac{1}{c^2},$$

where

$$Var(\xi) = \mathbb{E}\xi^2 - (\mathbb{E}(\xi))^2$$

is the variance of ξ. The reader can find more about such inequalities in [DS89].

We assume that the reader is familiar with basic notions from Graph Theory, such as simple graph, hypergraph, chromatic number, degree of a vertex, matching number, etc. One can find these and other notions of Graph Theory in [B98].

A graph-theoretic function for the given ground set $[n]$ is a function, which depends on the set of edges of graph G [AS92]. A graph-theoretic function L satisfies the Lipschitz condition if $|L(G) - L(G')| \leq 1$, where G, G' are graphs on the same vertex set which differ by only one edge.

Lemma 1 *Let $Y_1, Y_2, \ldots, Y_m$ be i.i.d. random variables with $P(Y_i = 1) = p$, $P(Y_i = 0) = 1 - p$. Then*

$$P\left(\left| \sum_{i=1}^{m} Y_i - mp \right| > \alpha \sqrt{mp} \right) < 2e^{-\alpha^2/2}, \tag{1}$$

$$P\left(\sum_{i=1}^{m} Y_i \geq [esmp] \right) < 2s^{-[esmp]}, \ s \geq 1. \tag{2}$$

Consider a random graph $G(n, p)$ on n vertices whose edges are chosen independently with probability $p \in (0, 1)$. If L is a graph-theoretic function satisfying the Lipschitz condition, then for any $\lambda > 0$

$$P\left(\left| L(G(n, p)) - \mathbb{E}(L(G(n, p))) \right| > \lambda \sqrt{\binom{n}{2}} \right) < 2e^{-\lambda^2/2}. \tag{3}$$

Proof. The proof of the second relation (2) is quite simple. The LHS of (2) is

$$\sum_{j \geq [esmp]} \binom{m}{j} p^j (1-p)^{m-j} < \sum_{j \geq [esmp]} \binom{m}{j} p^j < 2 \binom{m}{[esmp]} p^{[esmp]}.$$

From these inequalities and the inequalities

$$\binom{m}{t} p^t < \frac{(mpe)^t}{t^t} \leq s^{-t}, \ (t = [esmp])$$

follows (2).

Inequality (1) is a special case of (4) if we consider the martingale $X_k = \sum_{j=1}^{k} Y_j - kp$ (the definition of a martingale comes next). We prove this later.

Instead of proving (3) we prove a little bit more general bound, which is called Azuma–Hoeffding bound ([A67], [H63]). We present it in its usual form. Let Ω be a finite set and let $(\Omega, \mathcal{F}, P)$ be a probability space. Further let $\mathcal{F}_0 = \{\emptyset, \Omega\} \subseteq \mathcal{F}_1 \subset \ldots \subset \mathcal{F}_n = \mathcal{F}$ be an increasing sequence of sub-σ-fields. A sequence of random variables $X_0, X_1, \ldots, X_n$ is called a martingale if for $\mathcal{F}_k = \sigma(X_0, \ldots, X_k)$ ($\mathcal{F}_k$ generated by the first $k+1$ random variables $X_0, \ldots, X_k$), we have $\mathbb{E}(X_{k+1} | \mathcal{F}_k) = X_k$. □

Lemma 2 *Let $(X_0, X_1, \ldots, X_n)$ be a martingale, $X_n = X$, $X_0 = \mathbb{E}(X)$ and suppose there exist constants $c_k > 0$ such that*

$$|X_k - X_{k-1}| \leq c_k.$$

Then for $t > 0$ we have

$$P(|X - \mathbb{E}X| \geq t) \leq 2e^{-\frac{t^2}{2\sum_{k=1}^{n} c_k^2}}. \tag{4}$$

Proof. Actually we prove that

$$P(X \geq \mathbb{E}X + t) \leq e^{-\frac{t^2}{2\sum_{k=1}^{n} c_k^2}}. \tag{5}$$

Then the inequality complementary to (5) has a symmetric proof and (4) follows. Setting $Y_k = X_k - X_{k-1}$, $S_k = \sum_{i=1}^{k} Y_k$ Chernoff's inequality yields

$$P(X \geq \mathbb{E}(X) + t) = P(S_n \geq t) \leq e^{-ht} \mathbb{E}(e^{hS_n}), \quad h \geq 0.$$

Since $|Y_n| \leq c_n$, we conclude that

$$\mathbb{E}(e^{hY_n}|\mathcal{F}_{n-1}) \leq e^{\frac{h^2 c_n^2}{2}}.$$

Indeed, by convexity of e^{hx} we have

$$e^{hY_n} \leq \frac{c_n + Y_n}{2c_n} e^{hc_n} + \frac{c_n - Y_n}{2c_n} e^{-hc_n}.$$

Thus

$$\mathbb{E}(e^{hY_n}|\mathcal{F}_{n-1}) \leq \cosh(hc_n) \leq e^{h^2 c_n^2/2}.$$

Here we used the relation $\mathbb{E}(Y_n|\mathcal{F}_{n-1}) = 0$.

Next we have

$$\mathbb{E}(e^{hS_n}) = \mathbb{E}(e^{hS_{n-1}} \mathbb{E}(e^{hY_n}|\mathcal{F}_{n-1})) \leq e^{h^2 c_n^2/2} \mathbb{E}(e^{hS_{n-1}}).$$

Iterating the last inequality we obtain the relation

$$\mathbb{E}(e^{hS_n}) \leq e^{h^2 \sum_{k=1}^{n} c_k^2/2}$$

and hence

$$P(X \geq \mathbb{E}(X) + t) \leq e^{-ht + h^2 \sum_{i=k}^{n} c_k^2/2} \leq e^{-\frac{t^2}{2\sum_{k=1}^{n} c_k^2}},$$

where we put $h = t/\sum_{k=1}^{n} c_k^2$.

Now if we have a function $f = f(Z_1, \ldots, Z_N)$ of $N = \binom{n}{2}$ variables Z_i, $P(Z_i = 1) = p$, $P(Z_i = 0) = 1 - p$, and $|f(Z_1, \ldots, Z_N) - f(Z_1', \ldots, Z_N')| \leq 1$ if $(Z_1, \ldots, Z_N), (Z_1', \ldots Z_N')$ differ in one coordinate, then we can consider the σ-fields $\mathcal{F}_k$ generated by $Z_1, \ldots, Z_k$ and set $X_k = \mathbb{E}(f(Z_1, \ldots, Z_N)|\mathcal{F}_k)$. It is easy to see that $|X_k - X_{k-1}| \leq 1$ and we can apply the previous lemma with $c_k = 1, t = \lambda \sqrt{N}$ and obtain inequality (3).   □

The next result is called Vizing's Theorem. It was proved by Vizing [V64] and later reproved by Gupta [G66]. Let $\chi(G)$ denote the edge chromatic number of the graph $G = (V(G), E(G))$, that is, the minimal number of colors such that every vertex of the graph G is adjacent to edges with different colors and let $\Delta(G)$ be the maximal degree among the vertices of the graph. Graph G is simple if it does not have loops or multiple edges.

Theorem 1 (Vizing 1964) *For a simple graph G the following inequality holds:*

$$\chi(G) \leq \Delta(G) + 1. \tag{6}$$

To prove this we need the following lemma. Let G be a simple graph. For a given vertex $w \in V(G)$ we say that a sequence of distinct edges $wv_1, wv_2, \ldots, wv_p$ is good if for each $i \geq 2$ we have $c(wv_i) \notin C(v_{i-1})$, where $c(w,v)$ is the color of edge wv and $C(v)$ is the set of colors of the edges adjacent to vertex v. Denote $\deg(w)$ the degree of vertex $w \in V(G)$. An (a,b)-chain is a maximal path or circuit of the subgraph of edges colored by a or b.

Lemma 3 *Let $t \geq \Delta(G)$ and suppose that the graph G is not t-colorable, but for some edge $e = wu$, $G - e$ is t-colorable. Let c be some t-coloring of $G - e$ and let wv_1 be an edge such that $c(wv_1) \notin C(u)$. Let also $wv_1, \ldots, wv_p$ be the longest good sequence starting at wv_1. Then we have the following properties:*

(i) $C(u) \setminus C(w) \subseteq C(v_i)$ for all i,
(ii) $c(wv_j) \in C(v_i)$ for all $j < i$,
(iii) $d(v_p) = \Delta(G) = t$.

Proof of the Theorem. If we suppose that this is valid, then Vizing's Theorem easily follows. Indeed, suppose that $\chi(G) \geq \Delta(G) + 2$. Set $t = \Delta(G) + 1$ and let G' be a minimal subgraph of G that is not t-colorable. Then $G' - e$ is t-colorable for each edge $e \in E(G')$. The lemma implies that $t = \Delta(G') \leq \Delta$, which gives a contradiction. ∎

Proof of the Lemma. First of all note that $C(u) \setminus C(w) \neq \emptyset$. This is because $|C(u) \cup C(w)| = t$ and if $C(u) \subseteq C(w)$, then $d(w) \geq t + 1 > \Delta(G)$. Now if $a \in C(u) \setminus C(w)$ and $a \notin C(v_i)$, then we can color (or recolor) some of the edges of G as follows:

$$\begin{array}{cccc} e, & wv_1, & \ldots, wv_{i-1}, & wv_i \\ c(wv_1), & c(wv_2), & \ldots, c(wv_i), & a \end{array} \tag{7}$$

and obtain a t-coloring of G and this is a contradiction that proves (i).

Now we prove (ii). Let $a \in C(u) \setminus C(w)$ and $\exists j < i$ such that $c(wv_j) \notin C(v_i)$. Let P be the $(a, c(wv_j))$-chain starting from v_i (it can be empty). Next we consider two cases:

(i) Let $j = 1$. Then if P does not end at u, we interchange the colors along P and then (re)color the edges according to table (7). If P does end at u, then we interchange the colors along the $(a, c(wv_1))$-chain starting at wv_1 (this chain does not end at u) and color e with $c(wv_1)$. In either case we come to a contradiction.

(ii) Let now $j \geq 2$. If P does not end at v_{j-1}, then we interchange colors along P and use the (re)coloring (7). Otherwise, interchange colors along P and then (re)color as follows:

$$e, \qquad wv_1, \qquad \ldots, \qquad wv_{j-2}, \qquad wv_{j-1}$$
$$c(wv_1), \quad c(wv_2), \quad \ldots, \qquad c(wv_{j-1}), \quad a.$$

This leads to a contradiction and proves (ii). To prove (iii) note that if some color c' would be missing from v_p, then by (i), (ii), and the relation $|C(u) \cup C(w)| = t$, there would be an edge w of color $c' \notin \{c(wv_1), \ldots, c(wv_s)\}$, and so there would be a longer good sequence. Thus all t colors are present at v_p. $\qquad\qquad \square$

We Turn to Another Fundamental Problem in Graph Theory. We will give a proof of Ramsey's Theorem. Let $\mathcal{H} = \binom{[n]}{k}$ be a complete k-uniform hypergraph $\mathcal{H} = (\mathcal{V}, \mathcal{E})$, $|\mathcal{V}| = n$, $|E| = k$, $E \in \mathcal{E}$. Let L positive numbers $r_1, \ldots, r_L$ be given. Consider a coloring of the edges of $\mathcal{H}$ by the "colors" $1, \ldots, L$, that is, each edge has its own number from $[L]$. The question is what is the minimal n_0 such that when $n > n_0$ for arbitrary coloring of $\mathcal{H}$ there exists at least one color i such that the edges having this color generate a complete subgraph (clique) $\mathcal{H}_i = (\mathcal{V}_i, \mathcal{E}_i)$ with the number of vertices $|\mathcal{V}_i| > r_i$.

The next result from [R30], see also [GRS90], states that such a number n_0 exists.

Theorem 2 (Ramsey 1930) *Let $r_1, \ldots, r_L, L$ be positive numbers. There exists n_0 such that for $n > n_0$ a k-uniform hypergraph with n vertices whose edges are colored by L numbers contains a monochromatic complete subhypergraph with $n' \geq r_i$ vertices.*

Proof. The proof is by induction on k. When $k = 1$, we have the degenerate graph and Ramsey's Theorem in this case reduces to the pigeon-hole principle, which states that among n vertices colored by L colors there exist not less than n/L monochromatic vertices.

Assume now that the statement of the theorem is true for $k - 1$. Let $L_j(n)$ be the lower bound on the number of vertices in a complete monochromatic j-uniform subhypergraph $\hat{\mathcal{H}}$ of the hypergraph $\mathcal{H}$ on n vertices. We have to prove that $L_j(n) \overset{n \to \infty}{\to} \infty$ for arbitrary given j. Consider the vertices of the k-uniform complete hypergraph $\mathcal{H} = (\mathcal{V}, \mathcal{E})$, $|\mathcal{V}| = n$ in some order $v_1, \ldots, v_n$ and fix the vertex v_1. Consider the $k-1$-uniform complete hypergraph $\mathcal{H}'$ on $n - 1$ vertices $v_2, \ldots, v_n$ and the following edge-coloring of it. The edge $\{v_{i_1}, \ldots, v_{i_{k-1}}\}$, $1 < i_1 < \cdots, < i_{k-1} \leq n$ has color i iff the edge $\{v_1, v_{i_1}, \ldots, v_{i_{k-1}}\}$ of the initial graph $\mathcal{H}$ has color i. By induction, hypergraph $\mathcal{H}'$ has a monochromatic complete $k-1$-uniform subhypergraph $C\mathcal{H}' = (\mathcal{V}_1, \mathcal{E}_1)$ on at least $L_{k-1}(n-1)$ vertices $u_1, \ldots, u_{m_1}$, $m_1 \geq L_{k-1}(n-1)$. We expurgate all the other vertices from the set $v_2, \ldots, v_n$ and obtain the ordered vertex set $v_1, u_1, \ldots, u_{m_1}$. Next we fix u_1 and consider the $k-1$-uniform complete hypergraph on vertex set $u_2, \ldots, u_{m_1}$. Then we carry out a similar procedure with this hypergraph and the fixed vertex u_1. After finding a corresponding monochromatic hypergraph and expurgating all vertices that do not belong to the underlying set of this hypergraph, we obtain the set $v_1, u_1, w_1, \ldots, w_{m_2}$ and $m_2 \geq L_{k-1}(L_{k-1}(n-1)-1)$. Also we have two fixed vertices v_1, u_1. Continuing this procedure we stop at the step when the remaining set of nonfixed vertices has less than $k-1$ elements. The number of fixed vertices in this case is lowerbounded by the value

$$M = \min\{j : \underbrace{L_{k-1}(L_{k-1}\ldots L_{k-1}}_{j}(n-1)-1)\cdots -1) < k-1\}.$$

This set of vertices $h_1, \ldots, h_M$ generates the hypergraph $\tilde{\mathcal{H}}$ with the property that the color of an arbitrary edge $\{h_{i_1}, \ldots, h_{i_k}\}$, $i_1 < i_2 < \ldots < i_k$ depends only on i_1. Next, it is easy to see by the pigeon-hole principle that not less than M/L vertices from $\tilde{\mathcal{H}}$ generate the k-uniform complete monochromatic hypergraph. It is easy to show that if $L_{k-1}(n) \overset{n\to\infty}{\to} \infty$, then $L_k(n) = M/L \overset{n\to\infty}{\to} \infty$ and $L_k(n)$ is a lower bound for the number of vertices in the complete monochromatic subhypergraph of that hypergraph $\mathcal{H}$ on n vertices. This proves Theorem 2. ■

Note that from the proof of the theorem it follows that as $n \to \infty$, we can choose the value of L increasing with n but the speed of increase of L should be extremely low in order that the statement of the theorem is still valid.

The theorem gives a lower bound on the numbers $R_k(r_1, \ldots, r_L)$, which are the minimum n such that for an arbitrary k-uniform complete hypergraph on n vertices whose edges are colored with L colors, for at least one $i \in \{1, \ldots, L\}$ there exists a complete subhypergraph with r_i vertices, whose edges all have color i. To find the precise values of $R_k(r_1, \ldots, r_L)$ in general is a difficult problem. Very little is known about tight asymptotic growth of this numbers (they are called Ramsey numbers). For $R_2(3,t)$ the order of magnitude is known. Ajtai, Komlós, and Szemerédi in [AKS80], [AKS81] proved that $R_2(3,t) < ct^2/\ln t$ and then Shearer in [S83] improved this bound to $R_2(3,t) \le (1 + o(1))t^2/\ln t$, where $o(1) \overset{t\to\infty}{\to} 0$. Then using probabilistic arguments Kim proved in [K95] the lower bound $R_2(3,t) \ge c_1 t^2/\ln t$.

In conclusion, we also assume that the reader is familiar with notions from linear algebra, such as rank of a matrix, dimension of linear space, $span(\mathcal{A})$ for the set of vectors $\mathcal{A}$, etc. These notions can be found in [L69].

Chapter II
Intersection and Diametric Problems

In this chapter we first introduce the problem of finding the maximal cardinality of a system (or family) of subsets (in particular from $\binom{[n]}{k}$), such that any two subsets from the system intersect in not less than t elements. We call such a system of subsets t-intersecting family. We also consider the diametric problem in two different spaces. The diametric problem in one of the spaces is closely connected with the intersection problem. This connection is based on a technique that was invented by Ahlswede and Khachatrian. One can understand how it works by following the proof of the Complete Intersection Theorem, which we introduce later. This technique the reader first meets in the proof of Lemma 5. It is quite different from induction or other methods known before. In some sense it is a combination of shifting, with proving of necessity of the symmetry of the family under permutations of a sufficiently large number of components. The whole method becomes clear when the reader goes through the proofs and finds out that this method allows to solve several problems that had been considered hopeless for solving before. The reward for the efforts of the reader going along the lines of rather long proofs will be the satisfaction he attains at the end.

Lecture 1 The Complete Intersection Theorem

We turn to the problem of finding the maximal cardinality of a t-intersecting family of k-subsets (subsets of cardinality k of ground set $[n]$). It is easy to see that when $n \leq 2k - t$, then the whole family $\binom{[n]}{k}$ of k-subsets is t-intersecting. We thus consider the case when $n > 2k - t$.

We come to necessary considerations and definitions. A system of sets $\mathcal{A} \subset 2^{[n]}$ is called t-intersecting if for arbitrary $A_1, A_2 \in \mathcal{A}$, $|A_1 \cap A_2| \geq t$. Denote by $\mathcal{I}(n,t)$ the set of unrestricted t-intersecting systems and

$$\mathcal{I}(n,k,t) = \left\{ \mathcal{A} \in \mathcal{I}(n,t) : \mathcal{A} \subset \binom{[n]}{k} \right\}.$$

Our main goal is to determine the value

$$M(n,k,t) = \max_{\mathcal{A} \in \mathcal{I}(n,k,t)} |\mathcal{A}|.$$

We denote

$$\mathcal{F}(i) = \left\{ F \in \binom{[n]}{k} : |F \cap [t+2i]| \geq t+i \right\}, 0 \leq i \leq k-t.$$

In words: $\mathcal{F}(i)$ is the family of all k-element subsets of $[n]$ containing at least $t+i$ elements in the first $[t+2i]$ positions. Obviously, $\mathcal{F}(i)$ is a t-intersecting system: all pairs of k-element subsets from $\mathcal{F}(i)$ intersect in at least t elements already in the first $[t+2i]$ positions.

Theorem 3 (Complete Intersection Theorem (Ahlswede and Khachatrian 1997))

(i) For $n = 2k$, $t = 1$

$$M(n,k,1) = \binom{n-1}{k-1}.$$

For $1 \leq t \leq k \leq n$, $n > 2k-t$,
(ii) If for some $r \in \{0,1,2\ldots\}$

$$(k-t+1)\left(2 + \frac{t-1}{r+1}\right) < n < (k-t+1)\left(2 + \frac{t-1}{r}\right), \qquad (1)$$

then we have

$$M(n,k,t) = |\mathcal{F}(r)|.$$

Here we set $\frac{t-1}{r} = \infty$ if $r = 0$.
(iii) If for some $r \in \{0,1,2\ldots\}$ and $t > 1$

$$(k-t+1)\left(2 + \frac{t-1}{r+1}\right) = n, \qquad (2)$$

then

$$M(n,k,t) = |\mathcal{F}(r)| = |\mathcal{F}(r+1)|.$$

Moreover, all optimal systems are known. In case (i) one must choose for each $A \in 2^{[n]}$ one set from $A, \bar{A}$. In the other cases, up to permutations on $[n]$, there is uniqueness in case (ii) and there are two systems in case (iii).

From this theorem it follows that if $n \geq (k-t+1)(t+1)$ then

$$M(n,k,t) = \binom{n-t}{k-t}. \qquad (3)$$

In other words in this case the set

$$A(n,k,t) = \left\{ A \in \binom{[n]}{k} : [1,t] \subset A \right\}$$

is a maximal intersecting set.

Most famous in this subject is the $4m$-conjecture, which was stated more than 70 years ago (see [E87]) and says that

$$M(4m,2m,2) = \frac{\binom{4m}{2m} - \binom{2m}{m}^2}{2},$$

which is based on the construction

$$\left\{ F \in \binom{[4m]}{2m} \right\} : |F \cap [1,2m]| \geq m+1 \}.$$

It is the first case ($t = 2$) for which relation (3), which is based on the "naive" construction $\mathcal{A} = \left\{ A \in \binom{[n]}{k} : [1,t] \subset A \right\}$, is not optimal. Indeed, simple calculations give for $m = 1$ still $M(4,2,2) = 1 = \binom{4-2}{2-2}$; however, for $m = 2$ $M(8,4,2) = 17 > 15 = \binom{6}{2}$.

The $4m$-conjecture, which was mentioned in [E90] as the last open problem from [EKR61], attracted the attention of many mathematicians for a long time (see [DF83] and [CF92] for an upper bound). It also made it into the book [CG98].

Proof of Case (i). Obviously $M(2k,k,1) \geq \binom{2k-1}{k-1}$ and since with every A its complement cannot be in an intersecting family,

$$M(2k,k,1) \leq \frac{1}{2}\binom{2k}{k} = \frac{1}{2}\left(\binom{2k-1}{k} + \binom{2k-1}{k-1}\right) = \binom{2k-1}{k-1}.$$

Now we turn to other cases of this theorem. For every $1 \leq i < j \leq n$ define on $2^{[n]}$ the left shifting operator L_{ij} by the equation

$$L_{ij}(A) = \begin{cases} \{i\} \cup (A \setminus \{j\}), & \text{if } i \notin A, \ j \in A, \\ A, & \text{otherwise.} \end{cases}$$

Also for every $1 \leq i < j \leq n$ and set system $\mathcal{A} \in 2^{[n]}$ define an operator on $\mathcal{A}$

$$\mathcal{L}_{ij}(A,\mathcal{A}) = \begin{cases} L_{ij}(A), & \text{if } L_{ij}(A) \notin \mathcal{A}, \\ A, & \text{otherwise.} \end{cases}$$

Also set

$$\mathcal{L}_{ij}(\mathcal{A}) = \{\mathcal{L}_{ij}(A,\mathcal{A}), \ A \in \mathcal{A}\}.$$

We say that the set system $\mathcal{A} \subset 2^{[n]}$ is left-compressed if $\mathcal{A} = \mathcal{L}_{ij}(\mathcal{A})$ for all $1 \leq i < j \leq n$. Let $L\mathcal{I}$ be the set of all left-compressed systems belonging to $\mathcal{I}$. It is easy to see (Exercise 3) that

$$M(n,k,t) = \max_{\mathcal{A} \in L\mathcal{I}(n,k,t)} |\mathcal{A}|. \tag{4}$$

For arbitrary $A \in 2^{[n]}$ we denote by $A_{i,j}$, $1 \leq i, j \leq n$ the set obtained from A by exchanging coordinates i, j. For a set system $\mathcal{A} \subset 2^{[n]}$ and $1 \leq i, j \leq n$ we denote by $\mathcal{A}_{i,j}$ the set system obtained from $\mathcal{A}$ by exchanging the coordinates i, j in every $A \in \mathcal{A}$. Suppose that $\mathcal{A} \in LI(n,k,t)$ and $\mathcal{A}$ is not right-compressed. Let $\ell < n$ be the largest integer such that $\mathcal{A}$ is invariant under exchange operations in $[0, \ell]$, that is,

$$\mathcal{A} = \mathcal{A}_{i,j}, 1 \leq i, j \leq \ell, \text{ but } \mathcal{A} \neq \mathcal{A}_{i,\ell+1} \text{ for some } 1 \leq i \leq \ell. \tag{5}$$

We set

$$\mathcal{A}' = \{A \in \mathcal{A} : A_{i,\ell+1} \notin \mathcal{A} \text{ for some } 1 \leq i \leq \ell\}. \tag{6}$$

Lemma 4 *The following relations are valid (Exercise 4):*

(i) $\ell + 1 \notin A$ *for all* $A \in \mathcal{A}'$.
(ii) *Let* $A \in \mathcal{A}'$ *and* $j \in A$, $1 \leq j \leq \ell$, *then* $A_{j,\ell+1} \notin \mathcal{A}$.
(iii) *Let* $A \in \mathcal{A}'$, $A = B \cup C$, *where* $B = A \cap [\ell]$, $C = A \cap [\ell+1, n]$, *then* $B' \cup C \in \mathcal{A}'$ *for every* $B' \subset [1, \ell]$ *with* $|B'| = |B|$.
(iv) *Let* $A \in \mathcal{A}'$ *and* $D \in \mathcal{A} \setminus \mathcal{A}'$, *then*

$$|A_{i,\ell+1} \cap D| \geq t$$

for all $1 \leq i \leq \ell$.
(v) *Let* $A_1, A_2 \in \mathcal{A}'$, $B_i = A_i \cap [\ell]$; $i = 1, 2$ *and suppose that* $|B_1| + |B_2| \neq \ell + t$, *then* $|A_1 \cap A_2| \geq t + 1$.

Proposition 1 *Let* $\mathcal{B} \subset 2^{[n]}$ *be a set system, such that* $\mathcal{B} = \bar{\mathcal{B}} = \{\bar{B} : B \in \mathcal{B}\}$ *and* $\bar{B} = [n] \setminus B$. *Then every maximal intersecting* $\mathcal{B}' \subset \mathcal{B}$ *has cardinality* $|\mathcal{B}|/2$ *(Exercise 4).*

We will need the following key lemma.

Lemma 5 *Let* $\mathcal{A} \subset LI(n,k,t)$, $|\mathcal{A}| = M(n,k,t)$, $n > 2k - t$ *and*

$$n < (k - t + 1) \left(2 + \frac{t-1}{r} \right). \tag{7}$$

Then $\mathcal{A}_{i,j} = \mathcal{A}$ *for all* $1 \leq i, j \leq t + 2r$.

Proof. We can assume that $t \geq 2$, because in the case $t = 1$ inequalities (7) and $n > 2k - t$ are incomparable ($r \neq 0$). We suppose that the statement of the lemma is not valid and come to a contradiction. Let $\ell < t + 2r$ be such that $\mathcal{A}_{i,j} = \mathcal{A}$, $1 \leq i, j \leq \ell$, but $\mathcal{A}' = \{A \in \mathcal{A} : A_{i,\ell+1} \notin \mathcal{A} \text{ for some } 1 \leq i \leq \ell\} \neq \emptyset$. We will prove that in this case under the assumption (7) there exists a $\mathcal{B} \in \mathcal{I}(n,k,t)$ such that $|\mathcal{B}| > |\mathcal{A}|$, a contradiction.

Let

$$\mathcal{A}' = \bigcup_{i=1}^{\ell} \mathcal{A}(i), \quad \mathcal{A}(i) = \{A \in \mathcal{A}' : |A \cap [1, \ell]| = i\}.$$

From Lemma 4 it follows that $\mathcal{A}(i) = \emptyset$ when $1 \leq i < t$. We will prove that all set systems $\mathcal{A}(i)$ are empty. Suppose that $\mathcal{A}(i) \neq \emptyset$ for some $i : t \leq i \leq \ell$. From (iii) of Lemma 4 it follows that

$$|\mathcal{A}(i)| = \binom{\ell}{i}|\mathcal{A}^*(i)|, \tag{8}$$

where

$$\mathcal{A}^*(i) = \{A \cap [\ell+2, n] : A \in \mathcal{A}(i)\}. \tag{9}$$

From (i) of Lemma 4 it follows that $\ell+1 \notin A$ for all $A \in \mathcal{A}'$. Also note that when $n = \ell+1$, we have $\mathcal{A}^*(i) = \{\emptyset\}$ and hence $|\mathcal{A}^*(i)| = 1$. Let

$$\mathcal{B}(i) = \{B : |B \cap [1, \ell]| = i-1, \ \ell+1 \in B, \ B \cap [\ell+2, n] \in \mathcal{A}^*(i)\}.$$

Then by (ii) of Lemma 4

$$|\mathcal{B}(i)| = \binom{\ell}{i-1}|\mathcal{A}^*(i)|, \ \mathcal{B}(i) \cap \mathcal{A} = \emptyset. \tag{10}$$

Similar to (8) and (10) we have

$$|\mathcal{A}(\ell+t-i)| = \binom{\ell}{\ell+t-i}|\mathcal{A}^*(\ell+t-i)|, \tag{11}$$

$$|\mathcal{B}(\ell+t-i)| = \binom{\ell}{\ell+t-i-1}|\mathcal{A}^*(\ell+t-i)|. \tag{12}$$

Next we consider two subcases: **1.** $i \neq \ell+t-i$ and **2.** $i = \ell+t-i$.

Subcase 1. From (v) of Lemma 4 it follows:

For $B \in \mathcal{B}(i)$, $A \in \mathcal{A}(j)$ with $i+j \neq \ell+t$ we have $|B \cap A| \geq t$. Thus using this inequality, from (iv) of Lemma 4 we obtain

$$\mathcal{H}_1 = ((\mathcal{A} \setminus \mathcal{A}(\ell+t-i)) \cap \mathcal{B}(i)) \in \mathcal{I}(n, k, t),$$
$$\mathcal{H}_2 = ((\mathcal{A} \setminus \mathcal{A}(i)) \cap \mathcal{B}(\ell+t-i)) \in \mathcal{I}(n, k, t).$$

Next we show that in this case

$$\max\{|\mathcal{H}_1|, |\mathcal{H}_2|\} > |\mathcal{A}| = M(n, k, t), \tag{13}$$

which will be a contradiction. If the opposite to (13) is true, then from (8), (10)–(12) the inequalities

$$\binom{\ell}{i-1}|\mathcal{A}^*(i)| \leq \binom{\ell}{\ell+t-i}|\mathcal{A}^*(\ell+t-i)|, \tag{14}$$

$$\binom{\ell}{\ell+t-i-1}|\mathcal{A}^*(\ell+t-i)| \leq \binom{\ell}{i}|\mathcal{A}^*(i)|$$

follow. As $\mathcal{A}(i) \neq \emptyset$, from the first inequality in (14) it follows that $\mathcal{A}(\ell+t-i) \neq \emptyset$.

However, (14) yields the inequality

$$i(\ell+t-i) \le (\ell-i+1)(i+1-t),$$

which could not be true, because $t \ge 2$, and hence

$$i > i+1-t, \quad \ell+t-i > \ell-i+1.$$

Thus $\mathcal{A}(i) = \emptyset$ for all $i \ne \ell+t-i$.

Subcase 2. Here we have $2|(\ell+t)$ and hence $\ell+2 \le n$. Therefore, if $\mathcal{A}\left(\frac{\ell+t}{2}\right) \ne \emptyset$, then $\mathcal{A}^*\left(\frac{\ell+t}{2}\right) \ne \emptyset$.

We have

$$\left|\mathcal{A}\left(\frac{\ell+t}{2}\right)\right| = \binom{\ell}{\frac{\ell+t}{2}}\left|\mathcal{A}^*\left(\frac{\ell+t}{2}\right)\right|$$

and any $A \in \mathcal{A}\left(\frac{\ell+t}{2}\right)$ can be written as $A = B \cup C$ with

$$B = (A \cap [1,\ell]) \in \binom{[\ell]}{\frac{\ell+t}{2}}, \quad C = (S \cap [\ell+2,n]) \in \mathcal{A}^*\left(\frac{\ell+t}{2}\right),$$

where $|C| = k - \frac{\ell+t}{2}$ since $\ell+1 \notin A$.

Using the pigeon-hole principle we establish the existence of an element in $[\ell+2,n]$ and $\mathcal{D} \subset \mathcal{A}^*\left(\frac{\ell+t}{2}\right)$ such that $d \in D$ for all $D \in \mathcal{D}$ and

$$|\mathcal{D}| \ge \left|\mathcal{A}^*\left(\frac{\ell+t}{2}\right)\right| \frac{k - \frac{\ell+t}{2}}{n-\ell-1}. \tag{15}$$

We set

$$\mathcal{A}_1\left(\frac{\ell+t}{2}\right) = \left\{A \in \mathcal{A}\left(\frac{\ell+t}{2}\right) : (A \cap [\ell+2,n]) \in \mathcal{D}\right\},$$

$$\mathcal{A}_2\left(\frac{\ell+t}{2}\right) = \mathcal{A}\left(\frac{\ell+t}{2}\right) \setminus \mathcal{A}_1\left(\frac{\ell+t}{2}\right).$$

Then

$$\mathcal{A}\left(\frac{\ell+t}{2}\right) = \mathcal{A}_1\left(\frac{\ell+t}{2}\right) \cup \mathcal{A}_2\left(\frac{\ell+t}{2}\right).$$

Also set

$$\mathcal{H} = \left(\mathcal{A} \setminus \mathcal{A}_2\left(\frac{\ell+t}{2}\right)\right) \cup \mathcal{G},$$

where

$$\mathcal{G} = \left\{B \in \binom{[n]}{k} : (B \cap [1,\ell]) \in \binom{[\ell]}{\frac{\ell+t}{2}-1}, \ell+1 \in B, B \cap [\ell+2,n] \in \mathcal{D}\right\}.$$

From (ii) of Lemma 4 follows that $\mathcal{G} \cap \mathcal{A} = \emptyset$. Also note that $\mathcal{H} \in \mathcal{I}(n,k,t)$.

Next we show that under the conditions (1) and

$$\ell < t + 2r, \ 2|(\ell + t) \tag{16}$$

the inequality

$$|\mathcal{H}| > |\mathcal{A}| \tag{17}$$

is valid, which will be a contradiction to the assumption about maximality of $\mathcal{A}$.

Note that from inequalities (16) the inequality

$$\ell \leq t + 2r - 2 \tag{18}$$

follows. Since

$$|\mathcal{G}| = \binom{\ell}{\frac{\ell+t}{2}-1}|\mathcal{D}|, \ \left|\mathcal{A}_2\left(\frac{\ell+t}{2}\right)\right| = \binom{\ell}{\frac{\ell+t}{2}}\left(\left|\mathcal{A}^*\left(\frac{\ell+t}{2}\right)\right| - |\mathcal{D}|\right),$$

from (17) we get

$$\binom{\ell}{\frac{\ell+t}{2}-1}|\mathcal{D}| > \binom{\ell}{\frac{\ell+t}{2}}\left(\left|\mathcal{A}^*\left(\frac{\ell+t}{2}\right)\right| - |\mathcal{D}|\right)$$

or

$$\binom{\ell+1}{\frac{\ell+t}{2}}|\mathcal{D}| > \binom{\ell}{\frac{\ell+t}{2}}\left|\mathcal{A}^*\left(\frac{\ell+t}{2}\right)\right|.$$

From (15) it follows that for the validity of the last inequality it is sufficient to set

$$\binom{\ell+1}{\frac{\ell+t}{2}}\frac{k-\frac{\ell+t}{2}}{n-\ell-1} > \binom{\ell}{\frac{\ell+t}{2}}. \tag{19}$$

Inequality (19) is equivalent to $(k-t+1)\left(2+\frac{2(t-1)}{\ell-t+2}\right) > n$. The validity of the last inequality follows from (18) and assumption (7):

$$(k-t+1)\left(2+\frac{2(t-1)}{\ell-t+2}\right) \geq (k-t+1)\left(2+\frac{t-1}{r}\right) > n.$$

This proves Lemma 5. $\qquad\square$

Proof of Theorem 3.
Case (ii): Suppose first that

$$(k-t+1)\left(2+\frac{t-1}{r+1}\right) < n < (k-t+1)\left(2+\frac{t-1}{r}\right) \tag{20}$$

and $\mathcal{A} \in L\mathcal{I}(n,k,t)$, $|\mathcal{A}| = M(n,k,t)$. From Lemma 5 it follows that $\mathcal{A}$ is invariant under the permutations of any positions in $[1, t+2r]$, hence $k \geq t+r$. Also it is easy to see that $\bar{\mathcal{A}}$ is right-compressed, $\bar{\mathcal{A}} \in \mathcal{I}(n, n-k, n-2k+t)$, and

$|A| = |\bar{A}| = M(n,k,t) = M(n,n-k,n-2k+t)$. From (20) the inequalities

$$(k'-t'+1)\left(2+\frac{t'-1}{r'+1}\right) < n < (k'-t'+1)\left(2+\frac{t'-1}{r'}\right)$$

follow, where $k' = n-k$, $t' = n-2k+t$, and $r' = k-t-r$ $(\frac{t'-1}{r'} = \infty$ when $r' = 0)$.

Now it is easy to see that Lemma 5 can be formulated for right-compressed sets with obvious changes in the proof. This proof shows that $\bar{A}$ is invariant under the permutations of the positions in $[n-t'-2r'+1,n] = [t+2r+1,n]$. Hence such invariance is also valid for A and $[t+2r+1,n]$. Since A is left-compressed and $n > 2k-t$, we have

$$|A_1 \cap A_2 \cap [1,t+2r]| \geq t, \; A_1, A_2 \in A. \tag{21}$$

But A is invariant under permutations of the positions from $[1,t+2r]$. Thus the unique maximal set $A \in L\mathcal{I}(n,k,t)$ is $A = \mathcal{F}(r)$.

Case (iii): $n = (k-t+1)\left(2+\frac{t-1}{r+1}\right)$. Similar to the previous case we consider the complement set of A and using the same approach with one exception $n = 2$, $k,t = 1$, we derive an inequality similar to (21):

$$|A_1 \cap A_2 \cap [1,t+2r+2]| \geq t, \; A_1, A_2 \in A. \tag{22}$$

Then (22) and Lemma 5 deliver two optimal sets: either $A = \mathcal{F}(r)$ or $A = \mathcal{F}(r+1)$ and

$$|A| = |\mathcal{F}(r)| = |\mathcal{F}(r+1)|.$$

The answer in the case $n = 2$, $k,t = 0$ is obvious. The theorem is proved. Thus the problem of finding a maximal t-intersecting family is completely solved.

Now we turn our attention to the uniqueness of the optimal families. In case (i) one gets the optimal families by choosing from every set $\{A,\bar{A}\}$, $A \in A$, exactly one element. Up to permutations there is exactly one optimal family in case (ii) and there are exactly two cases in case (iii). This we prove next.

We will need the following:

Lemma 6 *Suppose $A \in \mathcal{I}(n,k,t)$ and A gets transformed by left shifting operations into the set $\mathcal{F}(r)$ for some $0 \leq r \leq (n-t)/2$. Then necessarily A is obtained from $\mathcal{F}(r)$ by permutations of the elements, provided that*

$$n \geq 2k-t+2, \; for \; t \geq 2,$$
$$n = 2k-t+1, \; for \; t \geq 2 \; and \; k = t+r \; or \; k = t+r+1,$$
$$n \geq 2k+1, \; for \; t = 1 \; and \; r = 0 \; or \; r = 1.$$

Proof. W.l.o.g. we assume that

$$\mathcal{L}_{ij}(A) = \mathcal{F}(r). \tag{23}$$

It is clear that if $i,j \in [1,t+2r]$ or $i,j \notin [1,t+2r]$, then $A = \mathcal{F}(r)$.

Suppose now that $i = t + 2r$ and $j = n$. Let

$$\mathcal{A}_1 = \{A \in \mathcal{A} : j \in A, \ i \notin A, \ ((A \setminus \{j\}) \cup \{i\}) \notin \mathcal{A}\},$$
$$\mathcal{A}_2 = \{A \in \mathcal{A} : j \notin A, \ i \in A, \ ((A \setminus \{i\}) \cup \{j\}) \notin \mathcal{A}\}.$$

Clearly, if $\mathcal{A}_1 = \emptyset$, then $\mathcal{A} = \mathcal{F}(r)$ and if $\mathcal{A}_2 = \emptyset$, then $\mathcal{A}$ is obtained from $\mathcal{F}(r)$ by exchanging the coordinates $i = t + 2r$ and $j = n$. Suppose now that $\mathcal{A}_1, \mathcal{A}_2 \neq \emptyset$ and let us show that in this case $\mathcal{A} \notin \mathcal{I}(n,k,t)$. Consider

$$\mathcal{H} = \left\{ H \in \binom{[n] \setminus \{i,j\}}{k-1} : |H \cap [1, t + 2r - 1]| = t + r - 1 \right\}.$$

Observe that from (23) it follows that, for any $B \in \mathcal{A}_1 \cup \mathcal{A}_2$, $|B \cap [1, t + 2r - 1]| = t + r - 1$ holds. Moreover, from the same assumption (23) we have the following: for every $H \in \mathcal{H}$ either $H \cup \{j\} \in \mathcal{A}_1$ or $H \cup \{i\} \in \mathcal{A}_2$.

Now we form a graph $G = (V, E)$ as follows:

$$V = \mathcal{H}, \ e(H_1, H_2) \in E \text{ iff } |H_1 \cap H_2| = t - 1.$$

One can easily verify that graph G is connected iff the conditions of the lemma hold. Hence under these conditions, if $\mathcal{A}_1 \neq \emptyset$ and $\mathcal{A}_2 \neq \emptyset$, then there exist $B_1 \in \mathcal{A}_1$ and $B_2 \in \mathcal{A}_2$ with $|B_1 \cap B_2| = t - 1$, which contradicts $\mathcal{A} \in \mathcal{I}(n,k,t)$.   □

Now we are ready to prove the uniqueness of the optimal set system in the Complete Intersection Theorem. Let $n > 2k - t$, $\mathcal{A} \in \mathcal{I}(n,k,t)$, and $|\mathcal{A}| = M(n,k,t)$, and after finitely many left shifting operations let $\mathcal{A}$ be transformed to the left-compressed set system $\mathcal{A}' \in LI(n,k,t)$, $|\mathcal{A}'| = M(n,k,t)$. We know that $\mathcal{A}' = \mathcal{F}(r)$ for some $r \in \mathbb{N} \cup 0$, where r is defined by the conditions of the theorem. It can be easily verified that these r's satisfy the conditions of the lemma and hence $\mathcal{A}$ is obtained from $\mathcal{F}(r)$ by permutations of the elements.   ■

Now we consider the case when there is no restriction on the cardinality of a set from the t-intersecting family. This case turns out to be much simpler than the previous one. Denote

$$M(n,t) = \max_{\mathcal{A} \in \mathcal{I}(n,t)} |\mathcal{A}|,$$

$$\mathcal{K}(n,t) = \left\{ A \in 2^{[n]} : |A| \geq \frac{n+t}{2} \right\} = \bigcup_{i=\frac{n+t}{2}}^{n} \binom{[n]}{i}, \ \text{if } 2 | (n+t).$$

Theorem 4 (Unrestricted Intersection Theorem (Katona 1964)) *The following identities hold:*

$$M(n,t) = \begin{cases} |\mathcal{K}(n,t)|, & 2 | (n+t), \\ 2|\mathcal{K}(n-1,t)|, & 2 \nmid (n+t). \end{cases} \tag{24}$$

Moreover, in the case $2|(n+t)$, $t > 1$ *the optimal family is unique, while in the case* $2 \nmid (n+t)$, $t > 1$ *it is unique up to permutations of the ground set* $[n]$.

Proof. We will give the simple proof of this theorem, which was presented in [AK05]. It uses only shifting and induction. Consider only the case $2|(n+t)$, the case $2 \nmid (n+t)$ has a similar proof. For $t = 1$ and $t = n$ the theorem is obviously true $M(n,1) = 2^{n-1}$, because if $A \in \mathcal{A}$, then $[n] \setminus A \notin \mathcal{A}$). We can assume, that $\mathcal{A} \in L\mathcal{I}(n,t)$. Let

$$\mathcal{A}_1 = \{A \in \mathcal{A} : 1 \in A\},$$
$$\mathcal{A}_0 = \mathcal{A} \setminus \mathcal{A}_1,$$
$$\mathcal{A}_j^* = \{A \cap [2,n] : A \in \mathcal{A}_j\}, \; j = 0, 1,$$

Then $\mathcal{A}_1^* \in \mathcal{I}(n-1, t-1)$, $\mathcal{A}_0^* \in \mathcal{I}(n-1, t+1)$.

We have by induction

$$|\mathcal{A}| = |\mathcal{A}_0^*| + |\mathcal{A}_1^*| \leq \sum_{i=\frac{n+t}{2}-1}^{n-1} \binom{n-1}{i} + \sum_{i=\frac{n+t}{2}}^{n-1} \binom{n-1}{i}$$
$$= \sum_{i=\frac{n+t}{2}}^{n} \binom{n}{i}.$$

The uniqueness of the family $\mathcal{A}$ for $t > 1$ also follows using induction. For $t = 1$ it is delegated to Exercise 5.   ∎

Lecture 2 The Diametric Problem for Vertices in the Hamming Metric

Next we consider the diametric problem in the Hamming space $\mathcal{H}_q^n$, which is the space of n-tuples with elements from $[0, q-1]$ endowed with the Hamming metric $d_H(a^n, b^n) = n - \sum_{i=1}^n \delta_{a_i, b_i}$. As we will see, the solution of this problem is closely related to the t-intersection problem. The diametric problem is in some sense similar to the intersection problem. In the case, when all n-tuples in the family have exactly w nonzero symbols, these two problems coincide: if two n-tuples from the family intersect in t positions, then the distance between them is $2(w-t)$. Next we consider the nonrestrictive diametric problem: we find the maximal cardinality of a family of n-tuples with prescribed diameter of this family. Note that in the binary case $q = 2$ this problem was solved a long time ago (formula (5) below). For $a^n, b^n \in \mathcal{H}_q^n$ denote

$$int(a^n, b^n) = \left| \{j : a_j = b_j\} \right|.$$

We call $\mathcal{A} \subset \mathcal{H}_q^n$ a $t - \mathcal{H}_q^n$ intersecting family if for all $a^n, b^n \in \mathcal{A}$,

$$int(a^n, b^n) \geq t.$$

Let $I_q(n,t)$ denote the set of all such families. Since

$$d_H(a^n, b^n) = n - int(a^n, b^n),$$

we have that the diameter of a $t - \mathcal{H}_q^n$ intersecting family $\mathcal{A}$ is not greater than $n - t$. Hence the problem of finding a maximal $t - \mathcal{H}_q^n$ intersecting family is equivalent to the problem of finding a maximal family with given diameter. Also note that the notions of $t - \mathcal{H}_2^n$ intersecting family and t-intersecting family in $2^{[n]}$ are different.

We are interested in finding a formula for the volume of a maximal $t - \mathcal{H}_q^n$ intersecting family

$$N_q(n,t) = \max_{\mathcal{A} \in I_q(n,t)} |\mathcal{A}|. \tag{1}$$

Let

$$B(a^n) = \{j : a_j = q - 1\}, \tag{2}$$

$$\mathcal{K}(i) = \{a^n \in \mathcal{H}_q^n : |B(a^n) \cap [1, t + 2i]| \geq t + i\}.$$

Clearly, $\mathcal{K}(i) \in I_q(n,t)$, $i \in \{0, 1, \ldots, (n-t)/2\}$. Obviously $\mathcal{K}(i)$ has diameter $n - t$. Indeed any two n-tuples already in the first $t + 2i$ positions intersect in t positions. The next theorem gives the complete solution of the diametric problem.

Theorem 5 (Ahlswede and Khachatrian 1998) *For $q \geq 2$, $t > 1$ or $q = 2$, $t = 1$ let $r \in \{0, 1, 2 \ldots\}$ be the largest integer such that*

$$t + 2r < \min\left\{n + 1, t + 2\frac{t-1}{q-2}\right\}. \tag{3}$$

Then $N_q(n,t) = |\mathcal{K}(r)|$. We set here $(t-1)/(q-2) = \infty$ if $q = 2$. Also we have

$$N_q(n, 1) = |\mathcal{K}(0)| = q^{n-1}. \tag{4}$$

Uniqueness properties are delegated to the Exercises. In the case $q = 2$, this theorem was proved by Kleitman [K66a] and we write the explicit solution in that case

$$N_2(n,t) = \begin{cases} \sum_{i=0}^{\frac{n-t}{2}} \binom{n}{i}, & 2 | (n-t), \\ 2\sum_{i=0}^{\frac{n-t-1}{2}} \binom{n-1}{i}, & 2 \nmid (n-t). \end{cases} \tag{5}$$

This equality can be proved using the same arguments as in the proof of Theorem 4 (Exercise 7). We will use this equality later when we demonstrate the solution of the diametric problem in the Taxi metric. For $q > 2$ and small values of n, Frankl and Füredi [FF80] proved $N_q(n,t) = q^{n-t}$ iff $t \leq q - 1$ or $t = n, n - 1$. A generalization to $q \geq 2$ was proved in [M82].

Exercise 8 asks the reader to prove equality (4) directly. A natural candidate for the solution of the diametric problem in the case of arbitrary t is

$$N_q(n,t) = q^{n-t}. \tag{6}$$

The maximal family $B_q(n,t)$ in this case can be chosen to be

$$B_q(n,t) = \{B = (q_1,\ldots,q_n) \in \mathcal{H}_q^n : (q_1,\ldots,q_t) = (a_1,\ldots,a_t)\}$$

for some $(a_1,\ldots,a_t)$, $a_i \in [q]$. However, as it follows from Theorem 5, this is not true in the general case. Before in [FF80] it was proved that this is true when $t \geq 15$ and $n \leq t+1$ or $q \geq t+1$.

Also it is interesting to mention one more particular case when

$$n \leq t+1+\log t/\log(q-1).$$

In this case

$$N_q(n,t) = \left| \mathcal{K}\left(\left\lfloor \frac{n-t}{2} \right\rfloor\right) \right|.$$

Proof of the Theorem. One can see that the definitions of the families $\mathcal{F}(i)$ and $\mathcal{K}(i)$ are quite similar. This gives the hint that the proofs of this theorem and Theorem 3 should also have common features. The reader will find in the proof of Theorem 5 a lot of technique from the proof of the Complete Intersection Theorem. Note that in the case $t = 1, q > 2$ inequality (3) is not satisfied for $r = 0, 1, \ldots$. It can be easily seen by following the beginning of the next proof that in this case $N_q(n,1) = |\mathcal{K}(0)| = q^{n-1}$ (see also Exercise 7). The uniqueness of the optimal configuration in this case up to permutations of the components and elements of the alphabet first was proved in [L79a]. In the case $t = 1$, $q = 2$ there are many possibilities of the choice of the optimal configuration (see Exercise 6). Exercises 7 and 9 ask to establish the uniqueness of the optimal configuration in other cases.

Now we turn to the proof of the theorem.

For $\mathcal{A} \subset \mathcal{H}_q^n$, $a^n \in \mathcal{A}$, and $j \in \{1,2,\ldots,n\}$, $i \in \{0,1,\ldots,q-1\}$ we define

$$T_{ji}(a^n) = \begin{cases} b^n = (a_1,a_2,\ldots,a_{j-1},q-1,a_{j+1},\ldots,a_n), & b^n \notin \mathcal{A} \text{ and } a_j = i, \\ a^n, & \text{otherwise.} \end{cases}$$

Also we put

$$T_{ji}(\mathcal{A}) = \{T_{ji}(a^n) : a^n \in \mathcal{A}\}.$$

We say that the set $\mathcal{A} \subset \mathcal{H}_q^n$ is canonical if

$$T_{ji}(\mathcal{A}) = \mathcal{A}$$

for all $j = 1,2,\ldots,n$ and $i = 0,1,\ldots,q-1$. It is easy to see that by a finite number of operations T_{ji} every set $\mathcal{A}$ becomes canonical. Also, each transformation T_{ji} keeps the cardinality and the $t - \mathcal{H}_q^n$-intersection property unchanged, that is, $|T_{ji}(\mathcal{A})| = |\mathcal{A}|$ and $\mathcal{A} \in I_q(n,t) \Rightarrow T_{ji}(\mathcal{A}) \in I_q(n,t)$.

Hence

$$N_q(n,t) = \max_{\mathcal{A} \in CI_q(n,t)} |\mathcal{A}|, \tag{7}$$

where $CI_q(n,t) \subset I_q(n,t)$ is the set of canonical families in $I_q(n,t)$.

With each system $\mathcal{A} \in CI_q(n,t)$ we associate the "image" $\mathcal{B}(\mathcal{A}) = \{B(a^n) : a^n \in \mathcal{A}\}$, where $B(a^n)$ is defined in (2). It is not difficult to see that if $\mathcal{A} \in CI_q(n,t)$, then

$$\mathcal{B}(\mathcal{A}) \in \mathcal{I}(n,t). \tag{8}$$

∎

Directly from the definition follows (Exercise 10)

Proposition 2 *Let* $\mathcal{A} \in CI_q(n,t)$ *be maximal:* $|\mathcal{A}| = N_q(n,t)$ *and let* $\mathcal{B}(\mathcal{A})$ *be the image of* $\mathcal{A}$. *Then*

(i) $\mathcal{B}(\mathcal{A})$ *is an upset.*
(ii)

$$|\mathcal{A}| = \sum_{B \in \mathcal{B}(\mathcal{A})} (q-1)^{n-|B|} = \sum_{i=0}^{n} g(i)(q-1)^{n-i},$$

where

$$g(i) = \left| \mathcal{B}(\mathcal{A}) \cap \binom{[n]}{i} \right|.$$

Denote by $LCI_q(n,t) \subset CI_q(n,t)$ the set of all systems $\mathcal{A}$ from $CI_q(n,t)$ with $\mathcal{B}(\mathcal{A}) \in L\mathcal{I}(n,t)$. From the definitions it follows that

$$N_q(n,t) = \max_{\mathcal{A} \in LCI_q(n,t)} |\mathcal{A}|. \tag{9}$$

For $E \in 2^{[n]}$ denote

$$\mathcal{V}(E) = \{ a^n \in \mathcal{H}_q^n : B(a^n) \in \mathcal{U}(E) \}. \tag{10}$$

Here $\mathcal{U}(E)$ is the upset with one minimal set E. Obviously

$$|\mathcal{V}(E)| = q^{n-|E|}. \tag{11}$$

For $\mathcal{E} \subset 2^{[n]}$ we put

$$\mathcal{V}(\mathcal{E}) = \bigcup_{E \in \mathcal{E}} \mathcal{V}(E).$$

We call $\mathcal{A}$ a q-upset, if

$$\mathcal{A} = \mathcal{V}(\mathcal{B}(\mathcal{A})).$$

For $E = \{e_1, e_2, \ldots, e_{|E|}\} \in 2^{[n]}$, $e_1 < e_2 < \ldots < e_{|E|}$ we set $s^+(E) = e_{|E|}$ and for $\mathcal{E} \subset 2^{[n]}$ we set

$$s^+(\mathcal{E}) = \max_{E \in \mathcal{E}} s^+(E).$$

The next results follow from the definitions.

Proposition 3 *Let* $\mathcal{A} \in LCI_q(n,t)$ *be a q-upset and* $\mathcal{B}(\mathcal{A})$ *be the image of* $\mathcal{A}$. *Let* $\mathcal{M}(\mathcal{A})$ *be the set of minimal elements of* $\mathcal{B}(\mathcal{A})$ *(in the sense of set inclusion). Then*

A is a disjoint union

$$A = \bigcup_{E \in \mathcal{M}(A)} D(E),$$

where

$$D(E) = \left\{ a^n = (a_1, a_2, \ldots, a_n) \in \mathcal{H}_q^n : B(a^n) \cap [1, s^+(E)] = E \right\}. \qquad (12)$$

Proposition 4 *Let $A \in LCI_q(n,t)$ be a q-upset. For $E \in \mathcal{M}(A)$ such that $s^+(E) = s^+(\mathcal{M}(A))$, denote*

$$A_E = V(E) \setminus V(\mathcal{M}(A) \setminus E).$$

This is the set of elements from A which are generated only by E.
   Then

$$A_E = D(E)$$

and

$$|A_E| = (q-1)^{s^+(E) - |E|} q^{n - s^+(E)}. \qquad (13)$$

Proposition 5 *Let $A \in LCI_q(n,t)$ be a q-upset and let $E_1, E_2 \in \mathcal{M}(A)$ have the properties $i \notin E_1 \cup E_2$, $j \in E_1 \cap E_2$ for some $i, j \in [n]$, $i < j$. Then*

$$|E_1 \cap E_2| \geq t + 1.$$

We need the following key result.

Lemma 7 *For $q > 2$ and $A \in LCI_q(n,t)$ with $|A| = N_q(n,t)$ for some $r \in \{0, 1, 2, \ldots\}$ we have*

$$s^+(\mathcal{M}(A)) = t + 2r \leq t + \frac{2(t-1)}{q-2}. \qquad (14)$$

If $(t-1)/(q-2)$ is a positive integer, then there exists an $A' \in LCI_q(n,t)$ with $|A'| = N_q(n,t)$ such that for some $r' \in \{0, 1, 2, \ldots\}$

$$s^+(\mathcal{M}(A')) = t + 2r' < t + \frac{2(t-1)}{q-2}. \qquad (15)$$

Proof. First we prove (14). Suppose the opposite is true:

$$s^+(\mathcal{M}(A)) = \ell > t + \frac{2(t-1)}{q-2} \qquad (16)$$

or $2 \nmid (\ell - t)$ and

$$\ell \leq t + \frac{2(t-1)}{q-2}. \qquad (17)$$

Let us show that in this case there exists $A' \in I_q(n,t)$ such that $|A'| > |A|$. The proof of this fact is quite similar to the proof of the t-intersection theorem and we frequently refer to it. Consider the partition

$$\mathcal{M}(\mathcal{A}) = \mathcal{M}_0(\mathcal{A}) \cup \mathcal{M}_1(\mathcal{A}),$$

where

$$\mathcal{M}_0(\mathcal{A}) = \{E \in \mathcal{M}(\mathcal{A}) : s^+(E) = s^+(\mathcal{M}(\mathcal{A})) = \ell\}$$

and

$$\mathcal{M}_1(\mathcal{A}) = \mathcal{M}(\mathcal{A}) \setminus \mathcal{M}_0(\mathcal{A}).$$

Note that for $E_1 \in \mathcal{M}_0(\mathcal{A})$ and $E_2 \in \mathcal{M}_1(\mathcal{A})$ we have

$$|(E_1 \setminus \{\ell\}) \cap E_2| \geq t.$$

Similar to Lemma 4 (v), using Proposition 5, it can be proved (Exercise 11) that if $E_1, E_2 \in \mathcal{M}_0(\mathcal{A})$ and $|E_1 \cap E_2| = t$, then

$$|E_1| + |E_2| = \ell + t. \tag{18}$$

Like in the proof of the t-intersection theorem, we consider the partition

$$\mathcal{M}_0(\mathcal{A}) = \bigcup_i \mathcal{R}(i),$$

where $\mathcal{R}(i) = \mathcal{M}_0(\mathcal{A}) \cap \binom{[n]}{i}$ and

$$\mathcal{R}'(i) = \{E \subset [1, \ell - 1] : E \cup \{\ell\} \in \mathcal{R}(i)\}.$$

Now we prove that all $\mathcal{R}(i)$ are empty. Suppose that for some i, $\mathcal{R}(i) \neq \emptyset$. Note that from (18) it follows that if $E_1' \in \mathcal{R}'(i)$, $E_2' \in \mathcal{R}'(j)$, and $i + j \neq \ell + t$, then

$$|E_1' \cap E_2'| \geq t. \tag{19}$$

As before, we consider two cases: **a.** $i \neq (\ell + t)/2$ and **b.** $i = (\ell + t)/2$.

Case a. According to (19) the two sets

$$\mathcal{F}_1 = \mathcal{M}_1(\mathcal{A}) \cup (\mathcal{M}_0(\mathcal{A}) \setminus (\mathcal{R}(i) \cup \mathcal{R}(\ell + t - i))) \cup \mathcal{R}'(i),$$
$$\mathcal{F}_2 = \mathcal{M}_1(\mathcal{A}) \cup (\mathcal{M}_0(\mathcal{A}) \setminus (\mathcal{R}(i) \cup \mathcal{R}(\ell + t - i))) \cup \mathcal{R}'(\ell + t - i),$$

have the property $\mathcal{F}_1$, $\mathcal{F}_2 \in \mathcal{I}(n, t)$ and hence $\mathcal{A}_i = \mathcal{V}(\mathcal{F}_i) \in I_q(n, t)$, $i = 1, 2$. We show that under the assumption $\mathcal{R}(i) \neq \emptyset$ we have

$$\max\{|\mathcal{A}_1|, |\mathcal{A}_2|\} > |\mathcal{A}|, \tag{20}$$

which will be a contradiction to the maximality of $\mathcal{A}$.

From the definitions of $\mathcal{F}_1$ and $\mathcal{R}(i)$ it follows that

$$\mathcal{A} \setminus \mathcal{A}_1 = \bigcup_{E \in \mathcal{R}(\ell + t - i)} D(E),$$

and from Proposition 4 we have

$$|\mathcal{A} \setminus \mathcal{A}_1| = |\mathcal{R}(\ell + t - i)|(q-1)^{i-t}q^{n-\ell}. \tag{21}$$

Now we estimate the value $|\mathcal{A}_1 \setminus \mathcal{A}|$. Let $E_1 \in \mathcal{R}'(i)$. Then, denote

$$D'(E_1) = \{a^n \in \mathcal{H}_q^n : B(a^n) \cap [\ell] = E_1\}. \tag{22}$$

We have

$$D'(E_1) \in \mathcal{A}_1 \setminus \mathcal{A}. \tag{23}$$

Since

$$|D'(E_1)| = (q-1)^{\ell-i+1}q^{n-\ell}$$

and

$$D'(E_1) \cap D'(E_2) = \emptyset, \; E_1, E_2 \in \mathcal{R}(i), \; E_1 \neq E_2$$

we obtain

$$|\mathcal{A}_1 \setminus \mathcal{A}| \geq |\mathcal{R}(i)|(q-1)^{\ell-i+1}q^{n-\ell}. \tag{24}$$

In a similar way we show that

$$|\mathcal{A} \setminus \mathcal{A}_2| = |\mathcal{R}(i)|(q-1)^{\ell-i}q^{n-\ell}, \tag{25}$$
$$|\mathcal{A}_2 \setminus \mathcal{A}| \geq |\mathcal{R}(\ell+t-i)|(q-1)^{i-t+1}q^{n-\ell}. \tag{26}$$

It is left to the reader to show that even more, (24) and (26) are equalities!
From (21), (24), (25), and (26) follows that if (20) is not true, then

$$|\mathcal{R}(i)|(q-1)^{\ell-i+1} \leq |\mathcal{R}(\ell+t-i)|(q-1)^{i-t},$$
$$|\mathcal{R}(\ell+t-i)|(q-1)^{i-t+1} \leq |\mathcal{R}(i)|(q-1)^{\ell-i}.$$

If $\mathcal{R}(i) \neq \emptyset$ and $q > 2$, these inequalities are inconsistent. This implies that $\mathcal{R}(i) = \emptyset$ for all $i \neq (\ell+t)/2$. In particular, we prove that if $\mathcal{R}(i) \neq \emptyset$, then $2|(\ell+t)$ and $i = \frac{\ell+t}{2}$.

Case b. By the pigeon-hole principle there exists an $i \in [1, \ell-1]$ and a $\mathcal{G} \subset \mathcal{R}'\left(\frac{t+\ell}{2}\right)$ such that $i \notin E$ for all $E \in \mathcal{G}$ and

$$|\mathcal{G}| \geq \frac{\ell-t}{2(\ell-1)}\left|\mathcal{R}'\left(\frac{\ell+t}{2}\right)\right|. \tag{27}$$

As $|E_1 \cap E_2| \geq t$, $E_1, E_2 \in \mathcal{G}$, and $\mathcal{R}(i) = \emptyset$, $i \neq (\ell+t)/2$, we have

$$\mathcal{F}' = \left(\mathcal{M}(\mathcal{A}) \setminus \mathcal{R}\left(\frac{\ell+t}{2}\right)\right) \cup \mathcal{G} \in \mathcal{I}(n,t).$$

Thus

$$\mathcal{V}(\mathcal{F}') \in I_q(n,t).$$

Next we show that under the condition (14),

$$|\mathcal{V}(\mathcal{F}')| > |\mathcal{A}|, \tag{28}$$

which is a contradiction to the maximality of $\mathcal{A}$. Consider the partition

$$\mathcal{A} = \mathcal{V}(\mathcal{M}(\mathcal{A})) = D_1 \cup D_2,$$

where

$$D_1 = \mathcal{V}\left(\mathcal{M}(\mathcal{A}) \setminus \mathcal{R}\left(\frac{\ell+t}{2}\right)\right),$$

$$D_2 = \mathcal{V}\left(\mathcal{R}\left(\frac{\ell+t}{2}\right)\right) \setminus \mathcal{V}\left(\mathcal{M}(\mathcal{A}) \setminus \mathcal{R}\left(\frac{\ell+t}{2}\right)\right),$$

and

$$\mathcal{V}(\mathcal{F}') = D_1 \cup D_3,$$

where

$$D_3 = \mathcal{V}(\mathcal{G}) \setminus \mathcal{V}\left(\mathcal{M}(\mathcal{A}) \setminus \mathcal{R}\left(\frac{\ell+t}{2}\right)\right).$$

Inequality (28) is equivalent to the inequality

$$|D_3| > |D_2|. \tag{29}$$

From Proposition 4 we have

$$|D_2| = \left|\mathcal{R}\left(\frac{\ell+t}{2}\right)\right|(q-1)^{(\ell-t)/2}q^{n-\ell}. \tag{30}$$

Let $E \in \mathcal{G}$, $E \subset [\ell-1]$, and $|E| = (\ell+t)/2 - 1$. Denote

$$\mathcal{C}(E) = \left\{a^n \in \mathcal{H}_q^n : B(a^n) \cap [\ell-1] = E\right\}.$$

Then $\mathcal{C}(E) \subset D_3$ and we have the partition

$$D_3 = \bigcup_{E \in \mathcal{G}} \mathcal{C}(E)$$

and hence

$$|D_3| = |\mathcal{G}|(q-1)^{(\ell-t)/2}q^{n-\ell+1}. \tag{31}$$

Using inequality (27), from (29), (30), and (31) we get that the following inequality is sufficient for (28) to hold:

$$\frac{\ell-t}{2(\ell-1)}\left|\mathcal{R}\left(\frac{\ell+t}{2}\right)\right|(q-1)^{(\ell-t)/2}q^{n-\ell+1}$$

$$> \left|\mathcal{R}\left(\frac{\ell+t}{2}\right)\right|(q-1)^{(\ell-t)/2}q^{n-\ell}.$$

From inequality (16) it follows that the last inequality is true ($\mathcal{R}\left(\frac{\ell+t}{2}\right) \neq \emptyset$). Hence assumption (16) is false and the first part of Lemma 7 is proved.

To prove the second part of the lemma, suppose that $(t-1)/(q-2)$ is a positive integer and

$$s^+(\mathcal{M}(\mathcal{A})) = \ell = t + 2\frac{t-1}{q-2}. \tag{32}$$

We have already proved that for all $E \in \mathcal{M}(\mathcal{A})$ with $s^+(E) = \ell$ we have $|E| = (\ell+t)/2$. One can repeat the proof of *Case b* and show that instead of (28) under assumption (32) we have the inequality $|\mathcal{V}(\mathcal{F}')| \geq |\mathcal{A}|$. This completes the proof of the lemma. $\square$

In the proof of Theorem 5 we use a lemma that allows to reduce the problem to another one, which we have already solved with Theorem 3. Let $S \subset 2^{2^{[m]}}$ and

$$H(S, \beta_t, \ldots, \beta_m) = \max_{\mathcal{L} \in S} \sum_{i=t}^{m} |\mathcal{L}(i)| \beta_i, \tag{33}$$

where $\mathcal{L}(i) = \mathcal{L} \cap \binom{[m]}{i}$, $t \leq m$, $\mathcal{L} \subset 2^{[m]}$, and $\beta_t, \beta_{t+1}, \ldots, \beta_m \in \mathbb{R}_+$. Suppose that for some $S \subset 2^{2^{[m]}}$ there is an $\mathcal{L}^* \in S$ such that for some $r \in \{1, 2, \ldots\}$, $\mathcal{L}^*(i) = \emptyset$ for $t \leq i < t+r$ and $|\mathcal{L}^*(i)| \geq |\mathcal{L}(i)|$ for $t+r \leq i \leq m$ and all $\mathcal{L} \in S$.

Lemma 8 *Let $\beta_t, \beta_{t+1}, \ldots, \beta_m \in \mathbb{R}_+$ and*

$$\mathcal{L}^* = \arg\max_{\mathcal{L} \in S} \sum_{i=t}^{m} |\mathcal{L}(i)| \beta_i$$

have the properties described above. Then, for any $\gamma_t, \ldots, \gamma_m \in \mathbb{R}_+$ such that

$$\frac{\beta_i}{\beta_{i+1}} \geq \frac{\gamma_i}{\gamma_{i+1}}, \quad i = t, \ldots, m-1, \tag{34}$$

it holds

$$\mathcal{L}^* = \arg\max_{\mathcal{L} \in S} \sum_{i=t}^{m} |\mathcal{L}(i)| \gamma_i.$$

Proof. W.l.o.g. we can assume that $\beta_m = \gamma_m = 1$. We introduce the numbers $\beta_t, \ldots, \beta_m$ and $\gamma_t, \ldots, \gamma_m$ in the form

$$\beta_m = 1, \quad \gamma_m = 1;$$
$$\beta_{m-1} = \delta_{m-1}, \quad \gamma_{m-1} = \varepsilon_{m-1};$$
$$\beta_{m-2} = \delta_{m-1}\delta_{m-2}, \quad \gamma_{m-2} = \varepsilon_{m-1}\varepsilon_{m-2},$$
$$\vdots \quad \vdots$$
$$\beta_i = \delta_{m-1}\delta_{m-2}\ldots\delta_i, \quad \gamma_i = \varepsilon_{m-1}\varepsilon_{m-2}\ldots\varepsilon_i;$$
$$\vdots \quad \vdots$$
$$\beta_t = \delta_{m-1}\delta_{m-2}\ldots\delta_t, \quad \gamma_t = \varepsilon_{m-1}\varepsilon_{m-2}\ldots\varepsilon_t.$$

We have

$$\delta_i \geq \varepsilon_i, \ i = 1, \ldots, m-1. \tag{35}$$

Let $\ell \in \{1, 2, \ldots\}$ be the largest integer such that $\delta_i = \varepsilon_i, \ i \geq m - \ell + 1$.

Introduce the positive numbers $\beta'_t, \ldots, \beta'_m$ satisfying $\beta'_m = \beta_m, \ldots, \beta'_{m-\ell+1} = \beta_{m-\ell+1}$ and $\beta'_i = \beta_i \varepsilon_{m-\ell}/\delta_{m-\ell}, \ t \leq i \leq m - \ell$.

If $m - \ell + 1 \leq t + r$, then

$$\sum_{i=t}^{m} |\mathcal{L}^*(i)| \beta'_i = \sum_{i=t}^{m} |\mathcal{L}^*(i)| \beta_i \geq \sum_{i=t}^{m} |\mathcal{L}(i)| \beta_i \geq \sum_{i=t}^{m} |\mathcal{L}(i)| \beta'_i.$$

If $m - \ell + 1 > t + r$, then the inequality

$$\sum_{i=1}^{m} |\mathcal{L}^*(i)| \beta'_i \geq \sum_{i=1}^{m} |\mathcal{L}(i)| \beta'_i \tag{36}$$

is equivalent to

$$\sum_{i=m-\ell+1}^{m} |\mathcal{L}^*(i)| \beta_i + \sum_{i=t+r}^{m-\ell} |\mathcal{L}^*(i)| \frac{\beta_i \varepsilon_{m-\ell}}{\delta_{n-\ell}}$$

$$\geq \sum_{i=m-\ell+1}^{m} |\mathcal{L}(i)| \beta_i + \sum_{i=t}^{m-\ell} |\mathcal{L}(i)| \frac{\beta_i \varepsilon_{m-\ell}}{\delta_{m-\ell}},$$

or

$$(\delta_{m-\ell} - \varepsilon_{m-\ell}) \sum_{i=m-\ell+1}^{m} (|\mathcal{L}^*(i)| - |\mathcal{L}(i)|) \beta_i$$

$$+ \varepsilon_{m-\ell} \left(\sum_{i=t}^{m} |\mathcal{L}^*(i)| \beta_i - \sum_{i=t}^{m} |\mathcal{L}(i)| \beta_i \right) \geq 0.$$

The last inequality is true since $\delta_{m-\ell} > \varepsilon_{m-\ell}$ and $|\mathcal{L}^*(i)| \geq |\mathcal{L}(i)|$ for $i \geq m - \ell + 1 > t + r$. Continuing this transformation we obtain step by step the coefficients $\gamma_t, \ldots, \gamma_m$ and this proves Lemma 8. □

Now we are ready to prove Theorem 5. It is convenient now to denote the Hamming ball in the space of binary sequences of length $t + 2r$, which has radius r and is centered at $[t + 2r]$ by

$$\mathcal{D}(r, t) = \left\{ D \in 2^{[t+2r]} : |D| \geq t + r \right\}.$$

Let

$$\mathcal{D}(i) = \mathcal{D}(r, t) \cap \binom{[t+2r]}{i}.$$

We have $|\mathcal{D}(i)| = 0$, $i < t + r$, and $|\mathcal{D}(i)| = \binom{t+2r}{i}$, $i \geq t + r$. Also note that $\mathcal{D}(r,t) \in \mathcal{I}(t + 2r, t)$. It is easy to show that the following relations are valid:

$$|\mathcal{F}(r)| = \sum_{j=0}^{r} \binom{2r+t}{t+r+j}\binom{n-2r-t}{k-t-r-j}$$

$$= \sum_{i=0}^{t+2r} |\mathcal{D}(i)|\binom{n-2r-t}{k-i},$$

$$|\mathcal{K}(r)| = \sum_{j=0}^{r} \binom{2r+t}{t+r+j}(q-1)^{r-j}q^{n-2r-t}$$

$$= q^{n-2r-t}\sum_{j=0}^{t+2r} |\mathcal{D}(i)|(q-1)^{2r+t-i}.$$

We can reformulate Theorem 3 as follows. Let for some $r = \{0, 1, 2, \ldots\}$

$$(k-t+1)\left(2+\frac{t-1}{r+1}\right) < m_0 < (k-t+1)\left(2+\frac{t-1}{r}\right), \tag{37}$$

and for $i \geq t$

$$\gamma_i = \binom{m_0 - 2r - t}{k-i}. \tag{38}$$

Then

$$\mathcal{D}(r,t) = \arg \max_{\mathcal{M} \in \mathcal{I}(2r+t,t)} \sum_{i=t}^{t+2r} |\mathcal{M}(i)|\gamma_i,$$

where $\mathcal{M}(i) = \mathcal{M} \cap \binom{[t+2r]}{i}$.

When $q \geq 2$, $t > 1$ or $q = 2$, $t = 1$, let us choose $r \in \{0, 1, 2, \ldots\}$ such that

$$t + 2r < \min\left\{n+1, t+\frac{2(t-1)}{q-2}\right\} \tag{39}$$

and

$$N_q(n,t) = \max_{\mathcal{M} \in \mathcal{I}(t+2r,t)} \sum_{i=t}^{t+2r} |\mathcal{M}(i)|(q-1)^{t+2r-i}q^{n-t-2r}$$

$$= q^{n-t-2r} \max_{\mathcal{M} \in \mathcal{I}(t+2r,t)} \sum_{i=t}^{t+2r} |\mathcal{M}(i)|(q-1)^{t+2r-i}. \tag{40}$$

The possibility of such a choice follows from Lemma 7. Next we apply Lemma 8 for $m = t + 2r$, $S = \mathcal{I}(t+2r,t) \subset 2^{2^{[t+2r]}}$, $\gamma_i = \binom{m_0-2r-t}{k-i}$, $i = t, t+1, \ldots, t+2r$, where m_0 satisfies (37) and $\beta_i = (q-1)^{t+2r-i}$. Also, we take $\mathcal{L}^* = \mathcal{D}(r,t) \in \mathcal{I}(1+2r,t)$. It is easy to see that $\mathcal{D}(r,t)$ enjoys the properties from Lemma 8 for the set $\mathcal{L}^*$.

Now it remains to make a proper choice of the parameters k and m_0. We will show for given r satisfying (39), the existence of m_0 from the interval (37) with condition

$$\frac{\gamma_i}{\gamma_{i+1}} \geq q-1 = \frac{\beta_i}{\beta_{i+1}}, \ i=t,\dots,t+2r-1 \tag{41}$$

from Lemma 8 holding. Therefore

$$k \geq t+2r, \tag{42}$$
$$m_0 \geq q(k-t)+t+2r-1. \tag{43}$$

It remains to prove that there exists $k \in \{1,2,\dots\}$ such that the system

$$\begin{cases} (k-t+1)\left(2+\frac{t-1}{r+1}\right) < m_0 < (k-t+1)\left(2+\frac{t-1}{r}\right), \\ q(k-t)+2r+t-1 \leq m_0 \end{cases} \tag{44}$$

has a solution $m_0 \in \{1,2,\dots\}$ and

$$k \geq t+2r, \ r < \frac{t-1}{q-2}. \tag{45}$$

We rewrite the system (44) in a way to get the following conditions on k:

$$\begin{cases} \frac{rm_0}{2r+t-1}+t-1 < k < \frac{(r+1)m_0}{2r+t+1}+t-1, \\ k \leq \frac{m_0}{q} - \frac{2r+t-1}{q}+t \end{cases} \tag{46}$$

To be able to choose an integer k satisfying the first inequality, it is enough to satisfy the inequality

$$\frac{rm_0}{2r+t-1}+t-1 < \frac{(r+1)m_0}{2r+t+1}+t-2,$$

or

$$m_0 > \frac{(2r+t+1)(2r+t-1)}{t-1}. \tag{47}$$

Consider now the second inequality from (46). For this we impose the condition

$$\frac{rm_0}{2r+t-1}+t-1 < \frac{m_0}{q} - \frac{2r+t-1}{q}+t-1$$

or, since $r < (t-1)/(q-2)$, we have

$$m_0 > \frac{(2r+t-1)^2}{2r+t-qr-1}. \tag{48}$$

We also impose the condition

$$2r+t < \frac{m_0}{2r+t-1}+t-1,$$

or

$$m_0 > \frac{(2r+1)(2r+t-1)}{r}. \tag{49}$$

Finally, we choose m_0 that satisfies (47)–(49) and take k to be the smallest integer such that $k > rm_0/(2r+t-1)+t-1$. For such a choice of m_0, k inequalities (41) hold and hence we can apply Lemma 8. Thus we get

$$N_q(n,t) = |\mathcal{K}(r)| = \alpha^{n-2r-t} \sum_{i=0}^{t+2r} |\mathcal{D}(i)|(q-1)^{2r+t-i}. \tag{50}$$

It is easy to show that the maximum of the RHS of (50) is achieved when r is the maximal number that satisfies (39). This completes the proof of Theorem 5 in the case $q \geq 2$, $t > 1$ and $q = 2$, $t = 1$. When $q > 2, t = 1$ we derive $r = 0$ from Lemma 7 and the theorem follows trivially. ∎

Lecture 3 The Diametric Problem for Vertices in the Taxi Metric

Now we turn to a problem that has considerably different methods of proof. However, there are several connections with the previous material. First of all we once more deal with the diametric problem, but in the Taxi metric (definitions will come next). In the case of binary n-tuples this metric coincides with the Hamming metric and Kleitman's result (5) gives the solution for both metrics.

Consider the diametric problem in a space, which is a direct product of paths. This problem is in some sense easier than its q-ary Hamming space counterpart (the direct product of the complete graphs of given size) and has been solved before the latter one. The metric in the space, which is a direct product of paths, is called the Taxi metric. In other words, we consider the space T^n of sequences $x^n = (x_1, \ldots, x_n)$ with components $x_i \in \mathcal{X}_i$, where the nodes $\mathcal{X}_i = \{x_1, \ldots, x_{|\mathcal{X}_i|}\}$ are nodes of the path $x_1 - x_2 - \ldots - x_{\mathcal{X}_i}$, $|x_i - x_j| = |i - j|$ and the distance between n-tuples is

$$\Delta(x^n, y^n) = \sum_{i=1}^{n} |x_i - y_i|.$$

In the case $|\mathcal{X}_i| > 2$, the structure and the solution of the diametric problem becomes much more difficult in comparison with the binary case. Next we come to the formulation of the problem and the results. For any subset $A \subset \mathcal{X}^n$, the diameter $D(A)$ and the radius $R(A)$ are defined as usual:

$$D(A) = \max_{x^n, y^n \in A} \Delta(x^n, y^n),$$

$$R(A) = \min_{x^n \in A} \max_{y^n \in A} \Delta(x^n, y^n).$$

We are interested in determining the quantity

$$C(d,n) = \max\{|A| : D(A) \leq d\}.$$

We show how to completely solve this problem in some important cases, namely when all $|\mathcal{X}_i|$ are odd or all $|\mathcal{X}_i|$ are even. In the solution of the diametric problem when all $|\mathcal{X}_i|$ are odd and all $|\mathcal{X}_i|$ are even, quite different approaches are used. But note that in both cases the maximal set of diameter d is the ball of radius $d/2$. It is interesting that the center of the ball in the case of all even $|\mathcal{X}_i|$ is not a point in $\prod_{i=1}^{n} \mathcal{X}_i$ but some point with coordinates in the intervals $[\min_{x \in \mathcal{X}_i} x, \max_{x \in \mathcal{X}_i} x]$.

Note the important (probably the main) conclusion here that in all cases the maximal set is a ball in L^1-metric of radius $d/2$ with some specified center, which can vary in different cases.

We start with the case when all $|\mathcal{X}_i|$ are odd. For convenience we write the alphabets in the form

$$\mathcal{X}_i = \{-q_i, \ldots, -1, 0, 1, \ldots, q_i\}, \quad |\mathcal{X}_i| = 2q_i + 1,$$

denote $q^n = (q_1, \ldots, q_n)$, and define for convenience the q^n-space by $\mathcal{B} = \mathcal{X}^n = \mathcal{X}_1 \times \cdots \times \mathcal{X}_n$. Let

$$B(0^n, r) = \{x^n \in \mathcal{B} : ||x^n|| \leq r\}$$

be the Taxi ball of radius r with the center in the origin (here $||x^n|| = \sum_i |x_i|$). Denote $N(r, n) = |B(0^n, r)|$. The next two theorems give the solution of the diametric problem in Taxi metric when all $|\mathcal{X}_i|$ are odd. The first theorem gives the solution for even diameter, and the next one for odd diameter of the set.

Theorem 6 (Ahlswede, Cai, and Zhang 1992) $C(2r, n) = N(r, n)$, if all $|\mathcal{X}_i|$ are odd.

Proof. We define the order $<_c$ on $\mathcal{X}_i$ by arranging its elements in the form $0, 1, -1, \ldots$ and the order $\leq_c$ on $\mathcal{B}$ by setting $x^n \leq_c y^n$ iff $x_i \leq_c y_i$ for $i = 1, \ldots, n$. By means of this order we introduce the "pushing to the center operator" P as follows: for any set $A \subset \mathcal{B}$ and any $x_j^n = (x_1, \ldots, x_{j-1}, x_{j+1}, \ldots, x_n) \in \prod_{1 \leq i \neq j \leq n} \mathcal{X}_i$ we set

$$A(x_j^n) = \{(z_1, \ldots, z_n) \in A : z_i = x_i \text{ for } i \neq j\},$$

let $P_j A(x_j^n) = \{(x_1, \ldots, x_{j-1}, x, x_{j+1}, \ldots, x_n) : x \text{ be one of the } |A(x_j^n)| \text{ } c\text{-smallest elements in } \mathcal{X}_j\}$ and also let $P_j(A) = \bigcup_{x^n} P_j A(x_j^n)$.

If $P_j A = A$ for all j, then we say that A is a c-downset. It is easy to verify that every $A \subset \mathcal{B}$ can be pushed into a c-downset A' such that

$$|A| = |A'|,$$
$$D(A) \geq D(A').$$

One easily verifies the fact (I) that $||x^n|| - ||y^n|| = 0 \pmod 2$ implies $\Delta(x^n, y^n) = 0 \pmod 2$.

We proceed with the proof of the theorem by induction on n. The case $n = 1$ being trivial, let now $q^n = q_1 q^{n-1}$ and let $A \subset \mathcal{B}$ satisfy $D(A) \leq 2r$. We can assume that A is a c-downset. Therefore, we have for $u >_c v$

$$A_u \subset A_v$$

if $A_u = \{x^{n-1} : ux^{n-1} \in A\}$, and for every nonnegative integer $\theta \leq q_1$ we have $A_{-\theta} \subset A_\theta$. Consider now the sets

$$A_\theta^0 = \{x^{n-1} : ||x^{n-1}|| \text{ is odd, } x^{n-1} \in A_\theta \setminus A_{-\theta}\},$$
$$A_\theta^e = \{x^{n-1} : ||x^{n-1}|| \text{ is even, } x^{n-1} \in A_\theta \setminus A_{-\theta}\}$$

and define

$$A_{-\theta}^* = A_{-\theta} \cup A_\theta^0, \; A_\theta^* = A_\theta \setminus A_\theta^0 = A_{-\theta} \cup A_\theta^e.$$

We then have

$$D(A_{-\theta}^*) = \max\{D(A_{-\theta}), D(A_\theta^0), D(A_{-\theta}, A_\theta^0)\}, \qquad (1)$$

where we define

$$D(U,V) = \max_{u \in U, \, v \in V} \Delta(u,v).$$

We shall show next that

$$D(A_{-\theta}^*) \leq 2(r - \theta). \qquad (2)$$

For this, notice that for $a^{n-1}, b^{n-1} \in A_{-\theta} \subset A_\theta$ and $x^{n-1}, y^{n-1} \in A_\theta^0$ the following sequences are in the set A :

$$(-\theta)a^{n-1}, \; (-\theta)b^{n-1}, \; \theta a^{n-1}, \; \theta b^{n-1}, \; \theta x^{n-1}, \; \theta y^{n-1}, \; (-\theta+1)x^{n-1}, \; (-\theta+1)y^{n-1}.$$

From the fact $D(A) \leq 2r$ we obtain therefore the inequalities

$$\Delta(a^{n-1}, b^{n-1}), \quad \Delta(a^{n-1}, x^{n-1}) \leq 2(r - \theta), \qquad (3)$$
$$\Delta(x^{n-1}, y^{n-1}) \leq 2(r - \theta) + 1.$$

However, since $||x^{n-1}||$ and $||y^{n-1}||$ are odd, by (I), $\Delta(x^{n-1}, y^{n-1})$ must be even. This shows that actually

$$\Delta(x^{n-1}, y^{n-1}) \leq 2(r - \theta).$$

This inequality together with (1) and (3) implies (2).
    Similarly one can prove that

$$D(A_\theta^*) \leq 2(r - \theta).$$

By the induction hypothesis we conclude our proof with

$$|A| = \sum_{u=-q_1}^{q_1} |A_u| = \sum_{u=-q_1}^{q_1} |A_u^*| \leq \sum_{u=-q_1}^{q_1} N(r - |u|, q^{n-1}) = N(r, q^n).$$

∎

We address now the case of an odd diameter. Again we present a complete solution for spaces with odd $|\mathcal{X}_i|$.

For this, we introduce the ball $S^*(r,n)$ in L^1-metric with the center in $(1/2,0,\ldots,0)$. For $d = 2r+1$ and $q^n = q^{n-1}q_n$ with $q_1 \geq q_i$, $i = 2,\ldots,n$ we set

$$S^*(r,n) = \{x^n : x_1 \leq 0, ||x^n|| \leq r, \text{ or } x_1 > 0, ||x^n|| \leq r+1\}.$$

Clearly

$$D(S^*(r,n)) = d.$$

Theorem 7 (Ahlswede, Cai, and Zhang 1992) *If we assume w.l.o.g. $q_1 \geq q_i$ for $i = 2,\ldots,n$, then we have $C(2r+1,n) = |S^*(r,n)|$ for $d = 2r+1$, when all $|\mathcal{X}_i|$ are odd.*

For $a^n, b^n \in \mathcal{B}$ denote

$$\overline{\Delta}(a^n,b^n) = \max\{\Delta(a'^n,b'^n) : a'^n \leq_c a^n, \ b'^n \leq_c b^n\}.$$

We introduce a metric $\Delta^* : \mathcal{B} \times \mathcal{B} \to \mathbb{R}_+$ by

$$\Delta^*(a^n,b^n) = \begin{cases} \overline{\Delta}(a^n,b^n), & a^n \neq b^n, \\ 0, & a^n = b^n, \end{cases}$$

and the diameter

$$D^*(A) = \max\{\Delta^*(a^n,b^n) : a^n, b^n \in A\}.$$

The following result can easily be verified.

Proposition 6 *(i) $\Delta^*(a^n,b^n) = ||a^n|| + ||b^n|| - |\{i : a_i > 0, \ b_i > 0\}|$ if $a^n \neq b^n$,*
(ii) Δ^* is a metric,
(iii) $D^*(M_c(A)) = D(A)$ for a c-downset $A \subset \mathcal{B}$, where $M_c(A)$ is the set of c-maximal elements in A.

We assume that $q_1 \geq \ldots \geq q_n$. The operator below is based on the mapping $\varphi : \mathcal{X}_{n-1} \times \mathcal{X}_n \to \mathcal{X}_{n-1} \times \mathcal{X}_n$ defined by

$$\varphi(x,y) = \begin{cases} (-x,-y), & x < 0, y > 0 \\ (-x+1,-y), & x > 0, y > 0 \\ (y,0), & x = 0, y > 0 \\ (x,y), & \text{otherwise.} \end{cases}$$

We will use this function to define for any $A \subset \mathcal{B}$ a mapping $\phi : A \to \mathcal{B}$ by

$$\phi(a^n) = \begin{cases} a^n, & \text{if } a_n > 0, \ a^{n-2}\varphi(a_{n-1},a_n) \in A, \\ a^{n-2}\varphi(a_{n-1},a_n), & \text{otherwise.} \end{cases}$$

We also write $\phi(A) = \{\phi(a^n) : a^n \in A\}$.

For any set $B \subset \mathcal{B}$ we introduce the associated c-downset $\mathcal{D}_c(B) = \{x^n : \exists b^n \in B \text{ such that } x^n \leq_c b^n\}$. Now we define an operator Q by putting

$$Q(A) = \mathcal{D}_c(\phi(A)).$$

Clearly

$$|Q(A)| \geq |\phi(A)| = |A|.$$

We summarize some properties that follow immediately from the definitions.

Proposition 7 *For any set $A \subset B$*

(i) $M_c(Q(A)) = M_c(\phi(A)) \subset \phi(A) \subset Q(A)$,
(ii) $a^{n-2}a_{n-1}a_n \in \phi(A)$ implies $a^{n-2}\varphi(a_{n-1},a_n) \in \phi(A)$.

We need the following:

Lemma 9 *For a c-downset A, $D(Q(A)) \leq D(A)$.*

Proof. By (iii) in Proposition 6,

$$D(Q(A)) = D^*(M_c(Q(A)) = D^*(M_c(\phi(A))) \leq D^*(\phi(A))$$

and, since A is a c-downset, also

$$D(A) = D^*(A).$$

It suffices therefore to show that $D^*(\phi(A)) \leq D^*(A)$ or that

$$\Delta^*(\phi(a^n),\phi(b^n)) \leq D^*(A) \tag{4}$$

for all $a^n, b^n \in A$. In the case $\phi(a^n) = a^n$, $\phi(b^n) = b^n$, which includes the case $a_n \leq 0$, $b_n \leq 0$, this is of course true.

In the case $a_n \leq 0$, $b_n > 0$ we notice that ϕ does not increase $||\cdot||$ and only in the case when $b_{n-1} > 0$, ϕ may decrease $|\{i: a_i > 0, b_i > 0\}|$, but by at most 1. Furthermore, in the case $b_{n-1} > 0$, $b_n > 0$ we have $||\phi(b^n)|| = ||b^n|| - 1$. Therefore, by (i) in Proposition 6, we obtain

$$\Delta^*(\phi(a^n),\phi(b^n)) \leq \Delta^*(a^n,b^n)$$

and thus (4).

The case $a_n > 0, b_n \leq 0$ being symmetrically the same, we are left with the case $a_n > 0$, $b_n > 0$, and (again by symmetry) $\phi(b^n) \neq b^n$. We divide this into two sub-cases:

1. $\phi(a^n) \neq a^n$. We establish (4) by proving $\Delta^*(\phi(a^n),\phi(b^n)) = \Delta^*(a^n,b^n)$. To prove it one should verify that $\Delta^*(a^n,b^n) - \overline{\Delta}(a^{n-2},b^{n-2})$ and $\Delta^*(\phi(a^n),\phi(b^n)) - \overline{\Delta}(a^{n-2},b^{n-2})$ are equal.

2. $\phi(a^n) = a^n$. Here necessarily $\tilde{a}^n = a^{n-2}\varphi(a_{n-1},a_n) \in A$. We can easily prove that $\Delta^*(\phi(a^n),\phi(b^n)) = \Delta^*(\tilde{a}^n,b^n)$ by verifying the validity of the equality $\Delta^*(\tilde{a}^n,b^n) - \overline{\Delta}(a^{n-2},b^{n-2}) = \Delta^*(\phi(a^n,b^n)) - \overline{\Delta}(a^{n-2},b^{n-2})$. □

Now we are able to prove Theorem 7. As before, we proceed by induction on n. The case $n = 1$ is clear. By Proposition 7 and Lemma 9 we can assume that A is a c-downset with the property

$$a^n = a^{n-2}a_{n-1}a_n \in M_c(A), \tag{5}$$

which implies

$$a^{n-2}\varphi(a_{n-1},a_n) \in A.$$

Let $A_x = \{x^{n-1} : x^{n-1}x \in A\}$ and consider for $\theta > 0$ the sets

$$A_\theta^+ = \{x^{n-2}x_{n-1} \in A_\theta \setminus A_{-\theta} : x_{n-1} > 0\}$$
$$A_\theta^- = \{x^{n-2}x_{n-1} \in A_\theta \setminus A_{-\theta} : x_{n-1} \leq 0\}$$
$$A_\theta^* = A_\theta \setminus A_\theta^- = A_{-\theta} \cup A_\theta^+$$
$$A_{-\theta}^* = A_{-\theta} \cup A_\theta^-, \; A_0^* = A_0.$$

Since A is a c-downset, we have $A_\theta \supset A_{-\theta}$. Therefore, for $a^{n-1},b^{n-1} \in A_{-\theta} \subset A_\theta$ and $x^{n-1} \in A_\theta$ we also have $a^{n-1}(-\theta)$, $b^{n-1}\theta$, $x^{n-1}\theta \in A$ and thus

$$\Delta(a^{n-1},b^{n-1}), \; \Delta(a^{n-1},x^{n-1}) \leq d - 2\theta$$

and

$$D(A_{-\theta}), D(A_{-\theta},A_\theta^-), D(A_{-\theta},A_{-\theta}^+) \leq d - 2\theta. \tag{6}$$

Now we are going to prove that also

$$D(A_\theta^-) = D^*(M_c(A_\theta^-)) \leq d - 2\theta \tag{7}$$
$$D(A_\theta^+) = D^*(M_c(A_\theta^+)) \leq d - 2\theta. \tag{8}$$

Suppose (7) is not true, then for some $a^{n-1},b^{n-1} \in M_c(A_\theta^-)$

$$\Delta^*(a^{n-1},b^{n-1}) > d - 2\theta. \tag{9}$$

Since $a^{n-1} \notin A_{-\theta}$ and $a^{n-1}\theta \in M_c(A)$, we have $a^{n-2}\varphi(a_{n-1},\theta) \in A$ by (5). Moreover, since $a_{n-1} \leq 0$ and $\theta > 0$, by our definitions

$$\varphi(a_{n-1},\theta) = \begin{cases} (-a_{n-1},-\theta), & a_{n-1} < 0, \\ (\theta,0), & a_{n-1} = 0. \end{cases}$$

Thus, noticing that $\theta > 0$ and $b_{n-1} \leq 0$, we can conclude that

$$d \geq D(A) \geq \Delta^*(a^{n-2}\varphi(a_{n-1},\theta),b^{n-1}\theta) = \overline{\Delta}(a^{n-2},b^{n-2})$$
$$+ |a_{n-1}| + |b_{n-1}| + 2\theta = \Delta^*(a^{n-1},b^{n-1}) + 2\theta > d.$$

This contradiction proves (7).

Now suppose that (8) is not true, that is, for some $a^{n-1},b^{n-1} \in M_c(A_\theta^+)$ (9) holds. By the reasoning given before $a^{n-2}\varphi(a_{n-1},\theta) \in A$. Now $\varphi(a_{n-1},\theta) = (-(a_{n-1}-1),-\theta)$, because $a_{n-1} > 0$ and $\theta > 0$ in this case. We arrive again at a contradiction

$$d \geq D(A) \geq \Delta^*(a^{n-2}\varphi(a_{n-1},\theta),b^{n-1}\theta)$$
$$= \overline{\Delta}(a^{n-2},b^{n-2}) + \overline{\Delta}(\varphi(a_{n-1},\theta),b_{n-1}\theta)$$
$$= \overline{\Delta}(a^{n-2},b^{n-2}) + |a_{n-1}| + |b_{n-1}| + 2\theta - 1$$
$$= \Delta^*(a^{n-1},b^{n-1}) + 2\theta > d.$$

So (8) holds. From (6), (7), and (8) we conclude that

$$D(A_\ell^*) \leq d - 2|\ell|$$

for all ℓ and by the induction hypothesis

$$|A_\ell^*| \leq |S^* (r - |\ell|, n)|.$$

Note that $A_\ell^* = \emptyset$ when $|\ell| > r$, and $S^*(z,n) = \emptyset$ when $z < 0$. Therefore,

$$|A| \leq \sum_{\ell=-q_n}^{q_n} |S^* (r - |\ell|, n)| = |S^* (r, n)|.$$

This completes the proof of Theorem 7.   ∎

Now we consider the diametric problem in the case, when all $|\mathcal{X}_i|$ are even. As we have already mentioned, the proof that some ball of radius $d/2$ is a maximal set of diameter d in this case is quite different.

First of all, we prove that in this case a maximal set of diameter $d = b(\mathcal{B}) - 1$, where $b(\mathcal{B}) = \sum_{i=1}^n q_i$, contains half of the points from $\mathcal{B}$. Moreover, such a set can be chosen to be a ball of radius $(b(\mathcal{B}) - 1)/2$ with the center depending on the parity of $b(\mathcal{B}) + n$.

Let L^1 be the space of n-tuples of reals with the L^1-metric

$$\Delta(x^n, y^n) = \sum_{i=1}^n |x_i - y_i|, \ x^n = (x_1, \ldots, x_n), \ y^n = (y_1, \ldots, y_n) \in L^1.$$

We consider the set $\mathcal{B}$ imbedded into the space L^1 in such a way that the ith coordinate of $\mathcal{B}$ takes the values from $\mathcal{X}_i = \{-q_i + \frac{1}{2}, \ldots, -\frac{1}{2}, \frac{1}{2}, \ldots, q_i - \frac{1}{2}\}$. Next we show that a set of diameter $b(\mathcal{B})$ cannot contain more than $|\mathcal{B}|/2$ points from $\mathcal{B}$. Indeed, consider the set $\mathcal{B}_j \subset \mathcal{B}$, which belongs to some orthant of L^1 (by orthant we mean a set with a prescribed sign of each component), and couple it with the set $\tilde{\mathcal{B}}_j : \tilde{\mathcal{B}}_j = -\mathcal{B}_j$. To every point $x^n = (x_1, \ldots, x_n) \in \mathcal{B}_j$ there is a corresponding unique point $\tilde{x}^n = (\tilde{x}_1, \ldots, \tilde{x}_n) \in \tilde{\mathcal{B}}, \ \tilde{x}_i = -q_i + x_i$. It is easy to see that this correspondence is a bijection and

$$\Delta(x^n, \tilde{x}^n) = b(\mathcal{B}).$$

Thus only one of the points from a pair $(x^n, \tilde{x}^n)$ can be in a set of diameter $b(\mathcal{B}) - 1$, and a set of diameter $b(\mathcal{B}) - 1$ contains not more than half of the points from $\mathcal{B}$.

The next lemma solves the problem of representing a maximal set of diameter $d = b(\mathcal{B}) - 1$ as a ball of radius $d/2$.

Lemma 10 Let $d = b(\mathcal{B}) - 1$. If $b(\mathcal{B}) + n$ is odd, then the ball $B\left(0^n, \frac{d}{2}\right)$, $0^n = (0, \ldots, 0) \in L^1$ contains half of the points from $\mathcal{B}$. In the case when $b(\mathcal{B})$ is odd, the ball $B\left(z^n, \frac{d}{2}\right)$, $z^n = (1/2, \ldots, 1/2)$ also contains half of the points from $\mathcal{B}$ and in the case of even $b(\mathcal{B})$ the same assertion is true with $z^n = (0, 1/2, \ldots, 1/2)$.

If $b(\mathcal{B}) + n$ is even, the ball $B\left(z^n, \frac{d}{2}\right)$, $z^n = (1/2, 0, \ldots, 0)$ contains half of the points from $\mathcal{B}$. If $b(\mathcal{B})$ is even, the ball $B\left(z^n, \frac{d}{2}\right)$, $z^n = (0, 1/2, \ldots, 1/2)$ also contains half of the points from $\mathcal{B}$. Here in the case of odd $b(\mathcal{B})$ the same assertion is true with $z^n = (1/2, \ldots, 1/2)$.

Proof. Suppose that $b(\mathcal{B}) + n$ is odd. We call the point $0^n = (0, \ldots, 0) \in L^1$ the center of $\mathcal{B}$. Let us show that each orthant intersects the ball $B\left(0^n, \frac{b-1}{2}\right)$ in exactly half of the points. W.l.o.g. we consider the orthant $\mathcal{B}_+$ with all-positive coordinates. Again we consider coupling, now the points being from $\mathcal{B}_+$. To each point $x^n = (x_1, \ldots, x_n) \in \mathcal{B}_+$ we assign in a one-to-one manner the point $\bar{x}^n = (\bar{x}_1, \ldots, \bar{x}_n) \in \mathcal{B}_+$ with $\bar{x}_i = q_i - x_i$. Next we show that the ball $B\left(0^n, \frac{b(\mathcal{B}) - 1}{2}\right)$ contains at least (actually exactly) one point from each pair $(x, \bar{x})$.

Indeed, if

$$||x^n|| > \frac{b(\mathcal{B}) - 1}{2}, \tag{10}$$

$$||\bar{x}^n|| > \frac{b(\mathcal{B}) - 1}{2},$$

or

$$\frac{b(\mathcal{B}) + 1}{2} > \sum_{i=1}^{n} x_i > \frac{b(\mathcal{B}) - 1}{2}.$$

The only possibility for these inequalities to be valid is

$$\sum_{i=1}^{n} x_i = \frac{b(\mathcal{B})}{2}.$$

For some positive integer y_i, $x_i = y_i - \frac{1}{2} = \frac{2y_i - 1}{2}$; thus

$$\sum_{i=1}^{n} (2y_i - 1) = b(\mathcal{B})$$

or

$$2\sum_{i=1}^{n} y_i = n + b(\mathcal{B}). \tag{11}$$

We see that the RHS of (11) is odd and the LHS is even, leading to a contradiction. Thus, when $b(\mathcal{B}) + n$ is odd, the ball $B\left(0^n, \frac{d}{2}\right)$ is maximal and contains exactly half of the points from $\mathcal{B}_+$ and hence, by symmetry, also from $\mathcal{B}$.

Using the same method, it is easy to check that if $2|(b(\mathcal{B}) - 1)$, then the ball $B\left(z^n, \frac{b(\mathcal{B}) - 1}{2}\right)$, $z^n = (1/2, \ldots, 1/2)$ is also a maximal set of diameter $b(\mathcal{B}) - 1$ and contains half of $\mathcal{B}$. To prove this we consider the pairing $(x^n, \bar{x}^n)$, $x^n = (x_1, \ldots, x_n)$, $\bar{x}^n = (\bar{x}_1, \ldots, \bar{x}_n) \in \mathcal{B}$ with

$$\bar{x}_i = \begin{cases} q_i + x_i, & x_i < 1/2, \\ -q_i + x_i, & x_i \geq 1/2. \end{cases} \tag{12}$$

As before, the correspondence $x^n \leftrightarrow \bar{x}^n$ is a bijection and the relations, similar to (10), look as follows:

$$||x^n - z^n|| = \sum_{i=1}^{n} \left| x_i - \frac{1}{2} \right| > \frac{b(\mathcal{B}) - 1}{2},$$

$$||\bar{x}^n - z^n|| = \sum_{x_i < 1/2} \left| q_i + x_i - \frac{1}{2} \right| + \sum_{x_i \geq 1/2} \left| -q_i + x_i - \frac{1}{2} \right| > \frac{b(\mathcal{B}) - 1}{2}$$

or

$$\sum_{i=1}^{n} |x_i| + \frac{\alpha - \beta}{2} > \frac{b(\mathcal{B}) - 1}{2},$$

$$b(\mathcal{B}) - \sum_{i=1}^{n} |x_i| - \frac{\alpha - \beta}{2} > \frac{b(\mathcal{B}) - 1}{2},$$

where

$$\alpha = |\{i : x_i < 1/2\}|, \ \beta = |\{i : x_i \geq 1/2\}|.$$

Hence we have

$$\frac{b(\mathcal{B}) + 1}{2} > \sum_{i=1}^{n} |x_i| + \frac{\alpha - \beta}{2} > \frac{b(\mathcal{B}) - 1}{2}$$

or

$$2 \sum_{i=1}^{n} |x_i| + \alpha - \beta = b(\mathcal{B}).$$

It is easy to see that the LHS of this equality is even, while $b(\mathcal{B})$ is odd, again a contradiction.

If $b(\mathcal{B})$ is even, we prove that the subset $\mathcal{B}' \subset \mathcal{B}$ with the first coordinate being positive intersects $B\left(z^n, \frac{d}{2}\right)$, $z^n = (0, 1/2, \ldots, 1/2)$ in $|\mathcal{B}'|/2$ points, from which by symmetry follows that the ball $B\left(z^n, \frac{d}{2}\right)$ contains half of the points from $\mathcal{B}$. Consider now the coupling defined by the transformation (12) in all but one coordinate (for example, when $i = 2, \ldots, n$) and set

$$\bar{x}_1 = q_1 - x_1.$$

As before, we impose the conditions

$$||x^n - z^n|| = \sum_{i=2}^{n} \left| x_i - \frac{1}{2} \right| + x_1 > \frac{b(\mathcal{B}) - 1}{2},$$

$$||\bar{x}^n - z^n|| = \sum_{x_i < 1/2} \left| q_i + x_i - \frac{1}{2} \right| + \sum_{i>1, \, x_i \geq 1/2} \left| -q_i + x_i - \frac{1}{2} \right| + q_1 - x_1$$

$$> \frac{b(\mathcal{B}) - 1}{2}$$

or

$$\sum_{i=1}^{n} |x_i| + \frac{\alpha - \beta_1}{2} > \frac{b(\mathcal{B}) - 1}{2},$$

$$b(\mathcal{B}) - \sum_{i=1}^{n} |x_i| - \frac{\alpha - \gamma}{2} > \frac{b(\mathcal{B}) - 1}{2},$$

where

$$\gamma = \#\{i > 1 : x_i \geq 1/2\}.$$

Hence we have

$$\frac{b(\mathcal{B}) + 1}{2} > \sum_{i=1}^{n} |x_i| + \frac{\alpha - \gamma}{2} > \frac{b(\mathcal{B}) - 1}{2}$$

or

$$2\sum_{i=1}^{n} |x_i| + \alpha - \gamma = b(\mathcal{B}).$$

But $\alpha - \gamma$ is odd and $b(\mathcal{B})$ is even, a contradiction.

We are done with the case of odd $b(\mathcal{B}) + n$. The case of even $b(\mathcal{B}) + n$ can be settled analogously and we leave it to the reader.   □

Now we are ready to prove the theorem, which says that a ball of radius $d/2$ with center in some specified point in L^1 is a maximal set of diameter d in $\mathcal{B}$.

Theorem 8 *Let us assume that all $|\mathcal{X}_i|$ are even, then there is a ball (in Taxi metric) of radius $d/2$, which is a maximal set of diameter d in $\mathcal{B}$. The center of the ball can be chosen to be $z^n = (1/2,\ldots,1/2)$ if d is even and $d < b(\mathcal{B})$ or $z^n = (0,1/2,\ldots,1/2)$ if d is odd and $d < b(\mathcal{B})$.*

If $d \geq b(\mathcal{B})$, then we can choose $z^n = (0,\ldots,0)$ if $d - n$ is even and $z^n = (1/2,0,\ldots,0)$ if $d - n$ is odd.

Proof. To prove the theorem we use the result (5) of Lecture 2, which solves the problem in the case when all q_i are equal to 1 (the binary $\mathcal{B}$). It is easy to check that the solution of the problem in the binary case is consistent with the general case, formulated in the theorem.

As in the proof for odd values of $|\mathcal{X}_i|$ we can assume that the maximal set is p-compressed according to the p-order on each $\mathcal{X}_i$: $-q_i + 1/2 >_p q_i - 1/2 >_p -q_i + 3/2 >_p q_i - 3/2 >_p \ldots >_p -1/2 >_p 1/2$.

Suppose first that d is odd and $d < b(\mathcal{B})$. Fix some coordinate i with $q_i > 1$. Let $\mathcal{B}_1$ be the set obtained from $\mathcal{B}$ by deleting the extremal points from the set $\mathcal{X}_i$, that is, the points $q_i - 1/2$, $-q_i + 1/2$, and $\mathcal{B}_2 = \mathcal{B} \setminus \mathcal{B}_1$. We shift the ith coordinate of $\mathcal{B}_2$ to zero in the following way: $q_i - 1/2 \rightarrow 1/2$; $-q_i + 1/2 \rightarrow -1/2$. Let S be the maximal p-compressed set in $\mathcal{B}$ and $S_j = S \cap \mathcal{B}_j$, $j = 1,2$ (we also assume that the ith coordinate of S_2 is shifted simultaneously with the ith coordinate of $\mathcal{B}_2$). Let $z^n = (0,1/2,\ldots,1/2)$. If $d_j = d(S_j)$, then

$$d_1 \leq d,$$
$$b(\mathcal{B}_1) = b(\mathcal{B}) - 1,$$

$$d_2 \leq d - 2q_i + 2,$$
$$b(\mathcal{B}_2) = b(\mathcal{B}) - q_i + 1.$$

Here the inequality for d_2 is valid, because the set S is p-compressed.

We deduce that if $d < b(\mathcal{B})$, then $d_1 \leq b(\mathcal{B}_1)$ and $d_2 \leq b(\mathcal{B}_2) - 1$. Assume at first that $d_1 < b(\mathcal{B})$. Note that the RHS of the restrictions for d_j from (13) have the same parities as d. For the set S_2 we can use induction and choose the ball $B_2 = B\left(z^n, \frac{d-2q_i+2}{2}\right)$ in $\mathcal{B}_2$ as a maximal set of diameter $d - 2q_i + 2$. If $d < b(\mathcal{B}_1) - 1$, then we choose the ball $B_1 = B\left(z^n, \frac{d}{2}\right)$ in $\mathcal{B}_1$ as a maximal set of diameter d. Then the ball $B_1 \cup B_2 = B\left(z^n, \frac{d}{2}\right)$ gives a maximal set of diameter d in $\mathcal{B}$. If $d_1 = d = b(\mathcal{B}_1) = b(\mathcal{B}) - 1$, then we can apply Lemma 10 for even $b(\mathcal{B})$ to justify the statement of the theorem. Also in some step it can happen that $\mathcal{B}_1$ and/or $\mathcal{B}_2$ become binary and in this case we use (5) of Lecture 2 and choose the center of the maximal ball in the binary space as needed for induction (make the necessary considerations in this case!).

Since the RHS of restrictions for d_j in (13) has the same parities as d, the proof of the theorem in the case of even $d < b(\mathcal{B}) - 1$ is similar to the case of odd d and we leave it to the reader.

Now consider the case when $d \geq b(\mathcal{B})$. Here we make another splitting of the set $\mathcal{B}$. Again we choose i such that $q_i > 1$ and choose $\mathcal{B}_3 \subset \mathcal{B}$ to be the set of all n-tuples $x^n = (x_1, \ldots, x_n)$ from $\mathcal{B}$ with $x_1 = \pm 1/2$ and $\mathcal{B}_4 = \mathcal{B} \setminus \mathcal{B}_3$. Again we shift the ith coordinate of $\mathcal{B}_4$ by making the transformation

$$x_i \longrightarrow \begin{cases} x_i + 1, & x_i < 0, \\ x_i - 1, & x_i > 0. \end{cases}$$

With the same notation as before we have

$$d_3 \geq \min\{d, t\},$$
$$b(\mathcal{B}_3) = b(\mathcal{B}) - q_i + 1,$$
$$d_4 = d - 2,$$
$$b(\mathcal{B}_4) = b(\mathcal{B}) - 1,$$

where $t = d(\mathcal{B}_3)$. Hence if $d \leq t$, then in any case $d_3 > b(\mathcal{B}_3)$ and if $d_4 \geq b(\mathcal{B}_4)$, then (as the restrictions on d_j have the same parity as d) we can deduce that the cardinality of S_3 or S_4 is upper-bounded by the cardinality of the balls $B\left(z^n, \frac{d}{2}\right)$ in $\mathcal{B}_3$ or $B\left(z^n, \frac{d-2}{2}\right)$ in $\mathcal{B}_4$, respectively, where the common center z^n depends on the parity of d (or, for fixed n, the parity of $d - n$). Thus the cardinality of S does not exceed the cardinality of the ball $B\left(z^n, \frac{d}{2}\right)$ in $\mathcal{B}$.

If $d > t$, then we can choose the center z^n of the ball $B\left(z^n, \frac{d-2}{2}\right)$ in $\mathcal{B}_4$ as the maximal set of diameter $d - 2$ in $\mathcal{B}_4$ and we have $\mathcal{B}_3 \subset B\left(z^n, \frac{d}{2}\right)$ for the ball in $\mathcal{B}$.

In the case $d_4 = b(\mathcal{B}_4) - 1$, we use Lemma 10 in the same way as before.

Again it is possible that on some step $\mathcal{B}_j$ becomes binary. In that case we use (5) of Lecture 2 and a consistent choice of the ball center z^n. Check that such a choice always exists. ∎

Theorems 6, 7, and 8 completely solve the problem of determining maximal sets of a given diameter in the Taxi metric, when all components $\mathcal{X}_i$ have even or odd lengths.

Lecture 4 The Diametric Problem for Edges in Hamming Metric

Theorem 5 deals with the vertex-diametric problem: we find the maximal cardinality of a set with given diameter. It was started in [AK00b] to consider the situation where one wants to find a set with given diameter that has maximal number of edges. For a given set $\mathcal{A} \subset \{0,1\}^n$ (in this problem we consider only the binary case, and the general case of an arbitrary alphabet is not solved) the edge set is defined as

$$\mathcal{E}(\mathcal{A}) = \{(a^n, b^n) : a^n, b^n \in \mathcal{A},\ d_H(a^n, b^n) = 1\}.$$

We denote $D(n,d) = I_2(n, n-d)$. The edge-diametric problem is to find the value

$$E(n,d) = \max_{\mathcal{A} \in D(n,d)} |\mathcal{E}(\mathcal{A})|.$$

Theorem 9 gives the complete solution of this problem. As in the case of the vertex diametric problem, while following the proof of this theorem the reader will see that some parts of it use technique from the proof of the Complete Intersection Theorem.
Let

$$\mathcal{W}(n) = \{(a_1, \ldots, a_n) \in \{0,1\}^n : a_1 = 1\},$$
$$\mathcal{G}(r) = \left\{ A \in 2^{[n]} : \left| A \cap [1, t+2r] \right| \geq t+r \right\},\ t = n-d.$$

Note that $\mathcal{G}(r)$ is the Cartesian product of the Hamming ball on the length $2t+r$ with radius r and center in $[1, 2t+r]$ and the whole space $2^{[n-t-2r]}$ on the rest length $n - t - 2r$.

Theorem 9 (Ahlswede and Khachatrian 2000) *Let $t = n - d$. The following relation is valid:*

$$E(n,d) = \begin{cases} |\mathcal{E}(\mathcal{W}(n))|, & \text{if } d = n-1, \\ \left| \mathcal{E}\left(\mathcal{G}\left(\frac{d}{2}\right)\right) \right|, & \text{if } d \leq n-2,\ 2|d, \\ \left| \mathcal{E}\left(\mathcal{G}\left(\frac{d-1}{2}\right)\right) \right|, & \text{if } d \leq n-2,\ 2 \nmid d. \end{cases}$$

For $A \in \mathcal{A}$ we denote $T_j(A) = T_{j0}(A)$. It is easy to see that in addition to the mentioned properties, T_j satisfies the relation

$$|\mathcal{E}(T_j(\mathcal{A}))| \geq |\mathcal{E}(\mathcal{A})|.$$

Let $UD(n,d)$ be the set of all upsets in $D(n,d)$. We have

$$E(n,d) = \max_{\mathcal{A} \in UD(n,d)} |\mathcal{E}(\mathcal{A})|. \tag{1}$$

On the other hand, if $\mathcal{A} \subset 2^{[n]}$ is an upset and has diameter d, then any $A_1, A_2 \in \mathcal{A}$ have at least $(n-d)$ componentwise common 1's.

Hence

$$E(n,d) = \max_{\mathcal{A} \in UI(n,n-d)} |\mathcal{E}(\mathcal{A})|, \tag{2}$$

where $UI(n, n-d)$ denotes the set of all $(n-d)$-intersecting systems, which are also upsets.

Also, it is easy to see that

$$E(n,d) = \max_{\mathcal{A} \in LUI(n,n-d)} |\mathcal{E}(\mathcal{A})|, \tag{3}$$

where $LUI(n, n-d)$ is the set of all left-compressed sets from $UI(n, n-d)$. We define the sets $\mathcal{A}_{i,j}$, $\mathcal{A}'$, etc. in the same way as in Lemma 4. Then all statements of the lemma are still valid in our case. In addition, it is easy to see that the following items (vi) and (vii) are also true.

(vi) Let $A \in \mathcal{A}'$. Then for any $B' \subset [1, \ell]$ with $|B'| < |B|$ and $C' \subseteq C$ we have

$$B' \cup C' \notin \mathcal{A}.$$

(vii) Let $A \in \mathcal{A}'$. It can be shown that for any $C' \subset C$, $B \cup C' \in \mathcal{A}$ implies $B \cup C' \in \mathcal{A}'$.

We will need two more results. The verification of Proposition 8 is left to the reader.

Proposition 8 *Let $\mathcal{A} \subset 2^{[n]}$ be an upset. Then*

$$|\mathcal{E}(\mathcal{A})| = \sum_{A \in \mathcal{A}} (n - |A|).$$

Proposition 9 *The following relation holds:*

$$\max_{\mathcal{A} \subset 2^{[n]}, \, |\mathcal{A}|=2^{n-1}} |\mathcal{E}(\mathcal{A})| = |\mathcal{E}(\mathcal{W}(n))|.$$

The proof of Proposition 9 is given at the end of the lecture. We start with the following lemma.

Lemma 11 *Let $S \subset 2^{[m]}$ have the following properties:*

(i) S is complement closed, that is, from $A \in S$ follows that $\bar{A} \in S$,
(ii) S is convex, that is, from $A, C \in S$ and $A \subset B \subset C$ follows that $B \in S$.

Then there exists an $S' \subset S$ such that $S' \in I(m)$ and

$$\sum_{A \in S'} (m - |A|) \geq \frac{m-1}{2m} \sum_{A \in S} (m - |A|) = \frac{m-1}{4} |S|. \tag{4}$$

Moreover, if $S \neq 2^{[m]}$, then there exists an $S' \subset S$, $S' \in I(m)$ for which strict inequality in (4) holds.

Proof. First we notice that the identity in (4) follows from property (i). In the case $S = 2^{[m]}$, by taking $S' = \{A \in 2^{[m]} : 1 \in A\}$, we have $S' \in I(m)$, $|S'| = \frac{|S|}{2} = 2^{m-1}$, and easily get (4) in this case.

Let now $S \neq 2^{[m]}$, let $B \in S$ be any element with minimal cardinality, and let $i \in B$. We consider the following partition of $S = S_1 \cup S_2 \cup S_3 \cup S_4$, where

$$S_1 = \{A \in S : i \in A, (A \setminus \{i\}) \in S\}, \ S_2 = \{A \in S : i \notin A, A \cup \{i\} \in S\},$$
$$S_3 = \{A \in S : i \in A, (A \setminus \{i\}) \notin S\}, \ S_4 = \{A \in S : i \notin A, A \cup \{i\} \notin S\}.$$

It is easily seen that

$$\bar{S}_1 = S_2, \ \bar{S}_3 = S_4.$$

Hence $|S_1| = |S_2|$ and $|S_3| = |S_4|$. Also $S_3 \neq \emptyset$, since $i \in B \in S$ and B has minimal cardinality. Also, for every $A \in S_4$ and $A' \in S \setminus S_3$, $A \cap A' \neq \emptyset$ holds. Hence $(S_1 \cup S_4), (S_1 \cup S_3) \in I(m)$.

We have

$$\sum_{A \in S_3 \cup S_4} (m - |A|) = m \frac{|S_3| + |S_4|}{2}.$$

Consequently,

$$\max \left\{ \sum_{A \in S_3} (m - |A|), \sum_{A \in S_4} (m - |A|) \right\} \geq m \frac{|S_3| + |S_4|}{4}. \tag{5}$$

On the other hand, by construction of S_1, S_2 and the property $\bar{S}_1 = S_2$, we have

$$m \frac{|S_1| + |S_2|}{2} = \sum_{A \in S_1} (m - |A|) + \sum_{A \in S_2} (m - |A|) = 2 \sum_{A \in S_1} (m - |A|) + \frac{|S_1| + |S_2|}{2}.$$

Hence

$$\sum_{A \in S_1} (m - |A|) = \frac{m-1}{4} (|S_1| + |S_2|). \tag{6}$$

Therefore, from (5) and (6) we get

$$\max\left\{\sum_{A\in\mathcal{S}_1\cup\mathcal{S}_3}(m-|A|),\ \sum_{A\in\mathcal{S}_1\cup\mathcal{S}_4}(m-|A|)\right\}\geq\frac{m}{4}(|\mathcal{S}_3|+|\mathcal{S}_4|)$$

$$+\frac{m-1}{4}(|\mathcal{S}_1|+|\mathcal{S}_2|)\geq\frac{m-1}{4}(|\mathcal{S}_1|+|\mathcal{S}_2|+|\mathcal{S}_3|+|\mathcal{S}_4|)=\frac{m-1}{4}|\mathcal{S}|.$$

□

Corollary 1 *Let* $\mathcal{S}\subset 2^{[m]}$ *be defined as in the previous lemma and let* (4) *hold for* $\mathcal{S}'\subset\mathcal{S}$, $\mathcal{S}'\in I(m)$, $|\mathcal{S}'|=\frac{|\mathcal{S}|}{2}$. *Then for any* $c\in\mathbb{R}$

$$\sum_{A\in\mathcal{S}'}(m-|A|+c)\geq\frac{m+2c-1}{2(m+2c)}\sum_{A\in\mathcal{S}}(m-|A|+c).\tag{7}$$

Proof. We just notice that (7) follows from (4) and the identities

$$\frac{m+2c-1}{2(m+2c)}\sum_{A\in\mathcal{S}}(m-|A|+c)=\frac{m+2c-1}{2(m+2c)}\left(\frac{m}{2}|\mathcal{S}|+c|\mathcal{S}|\right)$$

$$=\frac{m-1}{4}|\mathcal{S}|+\frac{c}{2}|\mathcal{S}|,$$

$$\sum_{A\in\mathcal{S}'}(m-|A|+c)=\sum_{A\in\mathcal{S}'}(m-|A|)+\frac{c}{2}|\mathcal{S}|.$$

□

Now let $\mathcal{A}\in D(n,d)$ and $|\mathcal{E}(\mathcal{A})|=E(n,d)$. We can assume that $\mathcal{A}\in LUI$ $(n,n-d)$. The next lemma plays the central role in the proof of Theorem 9.

Lemma 12 *Let* $\mathcal{A}$ *be the set that was described just above. Then, necessarily,* $\mathcal{A}$ *is invariant under exchange operations in*

   (i) $[1,n]$, *if* $2|d$ *and* $d\leq n-3$
  (ii) $[1,n-2]$, *if* $2|d$ *and* $d=n-2$
 (iii) $[1,n-1]$, *if* $2\nmid d$ *and* $d\leq n-2$.

Proof. The proof of this lemma is quite similar to the proof of Lemma 5. Let ℓ be the largest integer such that $\mathcal{A}_{i,j}=\mathcal{A}$ for all $1\leq i,j\leq\ell$. Assume the opposite to the statement of the lemma is true:

$$\ell<n_1,\tag{8}$$

where $n_1\in\{n-2,n-1,n\}$ depends on the case. We are going to show that, under the assumption (8), there exists a $\mathcal{B}\in I(n,n-d)$ with $|\mathcal{E}(\mathcal{B})|>|\mathcal{E}(\mathcal{A})|$, which is a contradiction. As in the proof of Lemma 5, we start with the partition $\mathcal{A}'=\bigcup_{i=1}^{\ell}\mathcal{A}(i)$. From Lemma 4 it follows that $\mathcal{A}(i)=\emptyset$ for all $1\leq i<n-d=t$. We will show that all $\mathcal{A}(i)$s are empty. Suppose that $\mathcal{A}(i)\neq\emptyset$ for some i, $t\leq i\leq\ell$. From Lemma 4 we know that

$$|\mathcal{A}(i)|=\binom{\ell}{i}|\mathcal{A}^*(i)|.\tag{9}$$

Note that in the case $n = \ell + 1$ we have $\mathcal{A}^*(i) = \emptyset$ and $|\mathcal{A}^*(i)| = 1$. Now as before we consider the sets $\mathcal{B}(i)$. From Lemma 4 it follows that for $B \in \mathcal{B}(i), A \in \mathcal{A}(j)$ with $i + j \neq \ell + t$, $|A \cap B| \geq t$ holds. Hence, using this and (iv) of Lemma 4, we have

$$\mathcal{H}_1 = ((\mathcal{A} \setminus \mathcal{A}(\ell + t - i)) \cup \mathcal{B}(i)) \in I(n, n - d),$$

and

$$\mathcal{H}_2 = ((\mathcal{A} \setminus \mathcal{A}(i)) \cup \mathcal{B}(\ell + t - i)) \in I(n, n - d).$$

Let us show that

$$\max \{|\mathcal{E}(\mathcal{H}_1)|, |\mathcal{E}(\mathcal{H}_2)|\} > |\mathcal{E}(\mathcal{A})| = E(n, d), \tag{10}$$

which will be a contradiction.

From the additional (vi) and (vii) of Lemma 4 one can easily show that the sets $\mathcal{H}_1, \mathcal{H}_2, (\mathcal{A} \setminus \mathcal{A}(j))$ are upsets. Therefore, using Proposition 8, we have

$$|\mathcal{E}(\mathcal{A})| = |\mathcal{E}(\mathcal{A} \setminus \mathcal{A}(\ell + t - i)| + \sum_{A \in \mathcal{A}(\ell + t - i)} (n - |A|)$$

$$= |\mathcal{E}(\mathcal{A} \setminus \mathcal{A}(i))| + \sum_{A \in \mathcal{A}(i)} (n - |A|),$$

$$|\mathcal{E}(\mathcal{H}_1)| = |\mathcal{E}(\mathcal{A} \setminus \mathcal{A}(\ell + t - i))| + \sum_{A \in \mathcal{B}(i)} (n - |A|) \tag{11}$$

$$|\mathcal{E}(\mathcal{H}_2)| = |\mathcal{E}(\mathcal{A} \setminus \mathcal{A}(i))| + \sum_{A \in \mathcal{B}(\ell + t - i)} (n - |A|).$$

Hence the negation of (10) is

$$\sum_{A \in \mathcal{A}(\ell + t - i)} (n - |A|) \geq \sum_{A \in \mathcal{B}(i)} (n - |A|), \tag{12}$$

$$\sum_{A \in \mathcal{A}(i)} (n - |A|) \geq \sum_{A \in \mathcal{B}(\ell + t - i)} (n - |A|).$$

Since we have assumed $\mathcal{A}(i) \neq \emptyset$, then clearly $\mathcal{A}(\ell + t - i) \neq \emptyset$ as well, because otherwise the first inequality of (12) would be false.

Using properties of the sets $\mathcal{A}(i), \mathcal{B}(i)$ we can write (12) in the form

$$\binom{\ell}{\ell + t - i} \sum_{C \in \mathcal{A}^*(\ell + t - i)} (n - \ell - t + i - |C|) \tag{13}$$

$$\geq \binom{\ell}{i - 1} \sum_{D \in \mathcal{A}^*(i)} (n - i - |D|),$$

$$\binom{\ell}{i} \sum_{D \in \mathcal{A}^*(i)} (n - i - |D|)$$

$$\geq \binom{\ell}{\ell + t - i - 1} \sum_{C \in \mathcal{A}^*(\ell + t - i)} (n - \ell - t + i - |C|).$$

However, (13) implies

$$(\ell - i + 1)(i + 1 - t) \geq (\ell + t - i)i,$$

which is false, because $t \geq 2$ and, consequently, $i > i + 1 - t$, $\ell + t - i > \ell - i + 1$.

Hence $\mathcal{A}(i) = \emptyset$ for all $i \neq \ell + t - i$.

Let now $i = \frac{\ell + t}{2}$. Here necessarily $2 | (\ell + t)$ and therefore by assumption (8) we have in Lemma 12 $\ell \leq n - 2$ in the case (i), $\ell \leq n - 4$ in the case (ii), and $\ell \leq n - 3$ in the case (iii). Let us call these conditions "conditions C."

Now we consider any element $A' = B' \cup C'$, where $B' \in \binom{[\ell]}{\frac{\ell + t}{2}}$, $C \subset C' \subset [\ell + 2, n]$, and $C \in \mathcal{A}^* \left(\frac{\ell + t}{2} \right)$. Of course $A' \in \mathcal{A}$, since $\mathcal{A}$ is an upset and $(B' \cup C) \in A' \subset \mathcal{A}$, $(B' \cup C) \subset (B' \cup C')$. It is also clear by the definition that, if $A' \in \mathcal{A}'$, then $A' \in \mathcal{A} \left(\frac{\ell + t}{2} \right)$. Using Lemma 4 we can say more: $A' = B' \cup C' \in \mathcal{A} \left(\frac{\ell + t}{2} \right)$ iff there is a $C'' \in \mathcal{A}^* \left(\frac{\ell + t}{2} \right)$ with $C'' \cap C' = \emptyset$, and hence with every $C \in \mathcal{A}^* \left(\frac{\ell + t}{2} \right)$ we have also $\bar{C} = ([\ell + 2, n] \setminus C) \in \mathcal{A}^* \left(\frac{\ell + t}{2} \right)$. Moreover, it is easily seen that $\mathcal{A}^* \left(\frac{\ell + t}{2} \right)$ is a convex set. Therefore, $\mathcal{A}^* \left(\frac{\ell + t}{2} \right)$ has the properties described in Lemma 11 and we can apply this lemma and the corollary to get an intersecting set $\mathcal{A}_1^* \left(\frac{\ell + t}{2} \right) \subset \mathcal{A}^* \left(\frac{\ell + t}{2} \right)$ for which (7) holds:

$$\sum_{D \in \mathcal{A}_1^* \left(\frac{\ell + t}{2} \right)} (m - |D| + c) \geq \frac{m + 2c - 1}{2(m + 2c)} \sum_{D \in \mathcal{A}^* \left(\frac{\ell + t}{2} \right)} (m - |D| + c) \tag{14}$$

for $m = n - \ell - 1$ and any constant c.

Now denote

$$\mathcal{B}_1 = \left\{ B : |B \cap [1, \ell]| = \frac{\ell + t}{2} - 1, \ell + 1 \in B, (B \cap [\ell + 2, n]) \in \mathcal{A}_1^* \left(\frac{\ell + t}{2} \right) \right\}$$

$$\mathcal{A}_1 \left(\frac{\ell + t}{2} \right) = \left\{ A \in \mathcal{A} \left(\frac{\ell + t}{2} \right) : (A \cap [\ell + 2, n]) \in \mathcal{A}_1^* \left(\frac{\ell + t}{2} \right) \right\} \tag{15}$$

and consider the following competitor of the set $\mathcal{A}$:

$$\mathcal{H}_3 = \left(\left(\mathcal{A} \setminus \mathcal{A} \left(\frac{\ell + t}{2} \right) \right) \cup \mathcal{A}_1 \left(\frac{\ell + t}{2} \right) \cup \mathcal{B}_1 \right).$$

It is easily seen that $\mathcal{H}_3 \in I(n, n - d)$.

We are going to show that

$$|\mathcal{E}(\mathcal{H}_3)| > |\mathcal{E}(\mathcal{A})|, \tag{16}$$

which will be a contradiction.

It is easily verified that both $\mathcal{H}_3$ and $\mathcal{A} \setminus \mathcal{A} \left(\frac{\ell + t}{2} \right)$ are upsets. Therefore, by Proposition 8 we can write

$$|\mathcal{E}(A)| = \left|\mathcal{E}\left(A \setminus A\left(\frac{\ell+t}{2}\right)\right)\right| + \sum_{A \in \mathcal{A}\left(\frac{\ell+t}{2}\right)} (n - |A|),$$

$$|\mathcal{E}(\mathcal{H}_3)| = \left|\mathcal{E}\left(A \setminus A\left(\frac{\ell+t}{2}\right)\right)\right| + \sum_{A \in \mathcal{A}_1\left(\frac{\ell+t}{2}\right) \cup \mathcal{B}_1} (n - |A|).$$

Hence the negation of (16) is

$$\sum_{A \in \mathcal{A}\left(\frac{\ell+t}{2}\right)} (n - |A|) \geq \sum_{A \in \mathcal{A}_1\left(\frac{\ell+t}{2}\right) \cup \mathcal{B}_1} (n - |A|),$$

which can be written in the form

$$\left(\tfrac{\ell}{\frac{\ell+t}{2}}\right) \sum_{D \in \mathcal{A}^*\left(\frac{\ell+t}{2}\right)} (m + c - |D|)$$

$$\geq \left(\left(\tfrac{\ell}{\frac{\ell+t}{2}}\right) + \left(\tfrac{\ell}{\frac{\ell+t}{2}-1}\right)\right) \sum_{D \in \mathcal{A}_1^*\left(\frac{\ell+t}{2}\right)} (m + c - |D|)$$

$$= \left(\tfrac{\ell+1}{\frac{\ell+t}{2}}\right) \sum_{d \in \mathcal{A}_1^*\left(\frac{\ell+t}{2}\right)} (m + c - |D|),$$

$m = n - \ell - 1$, $c = \frac{\ell-t+2}{2}$. This is equivalent to

$$\frac{\ell - t + 2}{2(\ell + 1)} \sum_{D \in \mathcal{A}^*\left(\frac{\ell+t}{2}\right)} (m + c - |D|) \geq \sum_{D \in \mathcal{A}_1^*\left(\frac{\ell+t}{2}\right)} (m + c - |D|). \qquad (17)$$

However, (14) for $m = n - \ell - 1$, $c = \frac{\ell-t+2}{2}$, and (17) imply

$$\frac{n-t}{n-t+1} \leq \frac{\ell-t+2}{\ell+1}, \qquad (18)$$

which is false, since $t \geq 2$ and conditions C can be checked to hold. $\qquad \square$

Now we are ready to make the final step in the proof of the theorem. Let $\mathcal{A} \in D(n,d)$ be a set with $|\mathcal{E}(\mathcal{A})| = E(n,d)$. We can assume that $\mathcal{A} \in LUI(n, n - d)$.

In the case $d = n - 1$ we just notice that any maximal set $\mathcal{B} \in D(n, n-1)$ has cardinality $|\mathcal{B}| = 2^{n-1}$. Now the equality $E(n, n-1) = |\mathcal{E}(\mathcal{H})|$ immediately follows from Proposition 9.

In the case $2|d$, $d \leq n-3$ we get from Lemma 12 $|A| \geq n - \frac{d}{2}$ for all $A \in \mathcal{A}$, since $\mathcal{A}$ is invariant in $[n]$ and at the same time $\mathcal{A} \in I(n, n-d)$. This implies $\mathcal{A} \subset \mathcal{G}\left(\frac{d}{2}\right) \in D(n,d)$, and by maximality of $\mathcal{A}$ we get

$$\mathcal{A} = \mathcal{G}\left(\frac{d}{2}\right).$$

Now we consider the case $2|d$, $d = n - 2$. Looking at the proof of Lemma 12, (ii), we see that in (18) for $t = n - d = 2$, $\ell = n - 2$ we have equality, which means it can be slightly changed to

(ii*) If $2|d$ and $d = n - 2$, then there exists an optimal set that is invariant in $[1, n]$. Therefore, in this case again, we have

$$E(n,d) = \left| \mathcal{G}\left(\mathcal{K}\left(\frac{d}{2}\right) \right) \right|.$$

We verify (for $2|d$, $d = n - 2$) that

$$\left| \mathcal{E}\left(\mathcal{G}\left(\frac{d}{2} - 1\right) \right) \right| = \left| \mathcal{E}\left(\mathcal{G}\left(\frac{d}{2}\right) \right) \right|$$

and hence $\mathcal{G}\left(\frac{d}{2} - 1\right)$ is the second optimal configuration in this case.

Finally, the case $2 \nmid d$, $d \le n - 2$ follows from Lemma 12, (iii), by similar arguments.   ∎

Proof of Proposition 9. First we will make some definitions. A k-subcube of the n-cube is a set of all vertices, which have the same components in some set of $n - k$ positions.

A shadow of a k-subcube is obtained by changing one of the $n - k$ fixed positions. Each k-subcube has $n - k$ shadows.

The following algorithm will number ℓ vertices of the n-cube so that the configuration of these ℓ vertices gives a maximal number of connections: assign one to an arbitrary vertex; having assigned $1, \ldots, \ell - 1$, assign ℓ to an unnumbered vertex (not necessarily unique), which has the most numbered nearest neighbors. We will prove by induction on ℓ and n.

But first we find out which configurations the algorithm delivers. The answer follows from the fact that whenever $\ell = 2^k$, a k-subcube is numbered. This is trivial for $k = 0$. Assume that we have numbered 2^{k-1} vertices of an n-cube and that by the inductive hypothesis they form a $(k - 1)$-subcube. This cube will have $n - (k - 1)$ disjoint shadows. When the $(2^{k-1} + 1)$th vertex is numbered, it will be in any of the shadows. The next $2^{k-1} - 1$ numbers will also fall in this shadow. Since no shadow of that shadow intersects any other shadow, there will always be unnumbered vertices in the first shadow, which will have two or more numbered nearest neighbors. Thus, it is inductively apparent that, for any ℓ, the construction gives a series of cubes, corresponding to the ones in the binary expansion of ℓ, each shadow of every larger cube.

Now we perform induction. It is obviously true for $\ell = 1$. Suppose it is true for $1, \ldots, \ell - 1$ and suppose that we have a maximally connected configuration of ℓ vertices of an n-cube. The n-cube may be divided into two $(n - 1)$-subcubes in n ways. Choose one of them. Suppose we have a numbered vertices contained in one of the halves and $b \le a$ in the other one. If $b = 0$, induction on n completes the proof. If $b > 0$, then the number of connections is maximized by having a maximally connected configuration in each half and b connections between them. By the hypothesis, the maximal configurations for a and b would be built of cubes, so that the smaller one will fit into the shadow of the larger one and so make b connections. This then is the case. Now suppose that 2^k is the largest power of two equal to or less than ℓ. Then if $2^k \leq a$, we have, by the induction hypothesis, a k-subcube in the larger configuration. If $2^k > a$, then 2^{k-1} is the largest power of two equal to or less than both a and b. In this case both a and b configurations contain $(k-1)$-subcubes, and since they are the largest such, each must lie in the other's shadow. In either case we have a k-subcube in maximal configuration. At last we must show that the remaining $\ell - 2^k$ vertices lie in a single shadow of the k-subcube. If not, c vertices lie in one shadow and d lie in another $(c + d \leq \ell - 2^k)$. Let 2^j be the smallest power of two equal to or greater than $c + d$. There can be no connections between the c and d configurations, so that the inductive hypothesis tells us that they are series of subcubes, each in the shadows of all larger ones. Look at a j-subcube which contains the c configurations and lies entirely within the shadow. Note that the complement of the c configurations in that j-subcube is also of the maximally connected type, so that the d configurations could be placed into it without changing its number of connections. But since $c + d > 2^{j-1}$, placing them both in the same j-cube would produce at least one more connection, contradicting our assumption that the configuration was maximally connected with $c, d > 0$.

At last note that the natural numbering of the n-cube assigns to each vertex the number that the vertex represents when considered as a binary digit, plus one. It can be easily seen that this natural numbering produces the above algorithm and hence the first 2^{n-1} vertices in this natural order give a set of 2^{n-1} vertices with the largest possible number of edges. This proves Proposition 9. $\qquad\square$

Lecture 5 Words with Pairwise Common Letter

In this lecture we present a problem that seems to stay apart from the topics of the other lectures. The problem deals with sets of words with pairwise common letter in different positions. It does, however, fall into the general frame of maximizing cardinalities of sets, whose members are pairwise in a certain relation like incomparable, t-intersecting, $t - \mathcal{H}_q^n$ intersecting, having distance d, independent, etc.

We start with some definitions. For an alphabet $\mathcal{X}_q = [q]$ we consider the set $\mathcal{X}_q^n$ of words of length n and also the subset W_q^n of all words without repetition of letters, that is,

$$W_q^n = \{x^n = (x_1, \ldots, x_n) : x_t \in \mathcal{X}_q, \ x_s \neq x_t \ if \ s \neq t\}. \qquad (1)$$

We say that two words x^n and y^n are in "good relation" if $x_s = y_t$ for some $s \neq t$. For this relation we write $x^n \diagup\diagdown y^n$. A set $G \subset \mathcal{X}_q^n$ is good if $x^n \diagup\diagdown y^n$ for all $x^n, y^n \in G$. We will study $\mathcal{G}_q^n$, the family of good sets in $\mathcal{X}_q^n$, and the quantity

$$g_q^n = \max\{|G| : G \in \mathcal{G}_q^n\}. \qquad (2)$$

Denote by $\mathcal{F}_q^n$ the family of good sets in W_q^n and

$$f_q^n = \max\{|F| : F \in \mathcal{F}_q^n\}.$$

Also, denote the set of entries in x^n by

$$E(x^n) = \{x : \text{ for some } t, \ x_t = x\}.$$

Very little is known about the values of g_q^n and f_q^n. In the following we are going to demonstrate the asymptotical behavior of g_q^n as $q \to \infty$. The theorem we will prove states the most significant known result in this area.

§1 Asymptotical Behavior of g_q^n

Here we will show that $g_q^n \sim q^{n-2}\binom{n}{2}$ as $q \to \infty$. We will prove

Theorem 10 (Ahlswede and Cai 1991) *The following relation is valid:*

$$\lim_{q \to \infty} \frac{g_q^n}{q^{n-2}} = \binom{n}{2}. \tag{3}$$

Proof. We begin with the inequality

$$\liminf_{q \to \infty} \frac{g_q^n}{\binom{q-1}{n-2}} \geq \binom{n}{2}(n-2)! \tag{4}$$

Define

$$G_0 = \{x^n \in \mathcal{X}_q^n : |E(x^n)| = n-1 \text{ and } 1 \text{ occurs exactly twice in } x^n\}.$$

Obviously $G_0 \in \mathcal{G}_q^n$, $|G_0| = \binom{n}{2}(n-2)! \cdot \binom{q-1}{n-2}$ $(q-1 \geq n-2)$, and (4) follows.

Next we show that

$$\limsup_{q \to \infty} \frac{g_q^n}{\binom{q-1}{n-2}} \leq \binom{n}{2}(n-2)!. \tag{5}$$

Recall that a partition of an integer n is a finite nonincreasing sequence of positive integers $\lambda_1 \geq \cdots \geq \lambda_r$ with $\sum_{i=1}^{r} \lambda_i = n$. Denote by $\mathcal{P}(n)$ the set of all partitions of n. We partition now $\mathcal{X}_q^n$ according to $\mathcal{P}(n)$ as follows. For $\Lambda = (\lambda_1, \ldots, \lambda_r) \in \mathcal{P}(n)$, set $T(\Lambda) = \{x^n \in \mathcal{X}_q^n : \exists z_1, \ldots, z_r \in \mathcal{X}_q \text{ such that } z_i \text{ occurs in } x^n = (x_1, \ldots, x_n) \text{ exactly } \lambda_i \text{ times }\}$. We subdivide $\{T(\Lambda) : \Lambda \in \mathcal{P}(n)\}$ into three classes. The class 0 consists of $T(\Lambda_0)$, where $\Lambda_0 = (1, \ldots, 1)$. The class 1 consists of $T(\Lambda)$ for $\Lambda = \Lambda_1 = (2, 1, \ldots, 1)$ and the remaining sets belong to the class 2. For all $G \in \mathcal{G}_q^n$ we have by our definitions $|G \cap T(\Lambda_0)| \leq f_q^n$.

It is easy to see that $f_q^n \leq (n!)^2$, the bound being independent of q. Choose any $(x_1, \ldots, x_n) \in F \in \mathcal{F}_\infty^n$. For all $y^n \in F$ it holds $E(y^n) \cap \{x_1, \ldots, x_n\} \neq \emptyset$ and, on the other hand, for fixed j and i, $|F \cap \{y^n : y_j = x_i\}| \leq f_\infty^{n-1}$. This implies $f_\infty^n \leq n^2 f_\infty^{n-1}$ and clearly $f_q^n \leq f_\infty^n$. Therefore,

$$|G \cap T(\Lambda_0)| \le (n!)^2. \tag{6}$$

Now consider the class 2. For all $x^n, y^n \in G$ we have $E(x^n) \cap E(y^n) \ne \emptyset$. So $\{E(x^n) : x^n \in G \cap T(\Lambda)\}$ is an intersecting family of r-element sets (if $\Lambda = (\lambda_1, \ldots, \lambda_r)$). For a $T(\Lambda)$ in the class 2 of partitions n into $r \le n-2$ parts and for all $x^n \in T(\Lambda)$ it holds

$$|\{y^n : E(y^n) = E(x^n)\}| \le r^n \le (n-2)^n. \tag{7}$$

This leads for large q to the estimate

$$\left| G \cap \left(\bigcup_{\Lambda \notin \{\Lambda_0, \Lambda_1\}} T(\Lambda) \right) \right| \le |\mathcal{P}(n)| \binom{q-1}{n-3} (n-2)^n. \tag{8}$$

This inequality uses also equality (3) (Lecture 1). Taking into account relations (6) and (8) ($\ln |\mathcal{P}(n)| = O(\sqrt{n})$), for verification of (5) it suffices to show that for $G \in \mathcal{G}_q^n$

$$\limsup_{q \to \infty} \frac{|G \cap T(\Lambda_1)|}{\binom{q-1}{n-2}} \le \binom{n}{2} (n-2)!. \tag{9}$$

To do this, we have to consider a partition of $T(\Lambda_1) \cap G$ into a few subparts. First of all, we can assume that the intersecting system $\{E(x^n) : x^n \in G \cap T(\Lambda_1)\}$ is not a 2-intersecting family, because otherwise for large q, $|G \cap T(\Lambda_1)| \le \binom{q-2}{n-3} \binom{n}{2}(n-1)! \sim q^{n-3}$, which follows from the equality

$$|\{y^n : E(y^n) = E(x^n)\}| = \binom{n}{2}(n-1)!, \quad x^n \in T(\Lambda_1) \tag{10}$$

and (3) (Lecture 1). From this (9) follows.

Using (6), (8), and (9) and taking into account that the set on the LHS of (10) is intersecting (which gives the factor $\binom{q-1}{n-2}$), we obtain the relation

$$\limsup_{q \to \infty} \frac{g_q^n}{\binom{q-1}{n-2}} \le \binom{n}{2}(n-1)!. \tag{11}$$

Now suppose that $|E(a^n) \cap E(b^n)| = 1$ for some $a^n, b^n \in G \cap T(\Lambda_1)$. W.l.o.g. let $E(a^n) = \{1, 2, \ldots, n-1\}$ and $E(b^n) = \{1, n, n+1, \ldots, 2n-3\}$. Denote $\mathcal{Z} = \{x^n \in G \cap T(\Lambda_1) : 1 \notin E(x^n)\}$. Since $E(x^n) \cap E(a^n) \ne \emptyset$ and $E(x^n) \cap E(b^n) \ne \emptyset$ for all $x^n \in \mathcal{Z}$, we have $|\{E(x^n) : x^n \in \mathcal{Z}\}| < 2^{2(n-2)} \binom{q-2n+2}{n-3}$. Consequently, by (10),

$$|\mathcal{Z}| < 2^{2(n-1)} \binom{n}{2}(n-1)! \binom{q-2n+2}{n-3}. \tag{12}$$

Let now $C_i = \{(c_1, \ldots, c_n) \in T(\Lambda_1) : c_i = 1, c_j \ne 1, j \ne i\}$ for $i = 1, \ldots, n$. Then

$$T(\Lambda_1) \cap G = (G \cap G_0) \cup \mathcal{Z} \cup (C_1 \cap G) \cup \ldots \cup (C_n \cap G). \tag{13}$$

As $\{(c_1, \ldots, c_{i-1}, c_{i+1}, \ldots, c_n) : (c_1, \ldots, c_{i-1}, 1, c_{i+1}, \ldots, c_n) \in C_i \cap G\} \in \mathcal{G}_q^{n-1}$, we obtain

$$|C_i \cap G| = O(q^{n-3}), \quad q \to \infty \tag{14}$$

by inequality (11).

Finally,

$$|G_0 \cap G| \leq |G_0| = \binom{n}{2}(n-2)!\binom{q-1}{n-2} \tag{15}$$

and (12)–(15) imply (9). This completes the proof of (5) and the theorem.   ■

Lecture 6 Constant Distance Code Pairs

For an alphabet $\mathcal{X}_q = [q]$ consider the Hamming metric d_H on $\mathcal{X}_q^n$: $d_H(x^n, y^n) = |\{i : x_i \neq y_i\}|$.

A pair (A, B) of sets $A, B \subset \mathcal{H}_q^n$ is an (n, δ) constant distance code pair if

$$d_H(a^n, b^n) = \delta, \quad \text{for all } a^n \in A, \ b^n \in B.$$

The set of all such pairs we denote by $S_q(n, \delta)$. In this lecture we give a partial solution to the problem of determining the value

$$M_q(n, \delta) = \max\{|A||B| : (A, B) \in S_q(n, \delta)\}.$$

We will find an explicit formula for $M_q(n, \delta)$ only in the cases $q = 2, 4, 5$ and will formulate a conjecture for the values of $M_q(n, \delta)$, when $q = 3$ and $q \geq 6$. The explicit formula for $M_q(n, \delta)$ will be expressed in terms of the following functions

$$F_2(n, \delta) = \max_{d_1 + d_2 = \delta} 4^{d_1}\binom{n - 2d_1}{d_2}, \tag{1}$$

$$F_3(n, \delta) = \max_{2\ell + d = \delta} 18^{\ell}\binom{n - 3\ell}{d}2^d, \tag{2}$$

$$F_q(n, \delta) = \max_{d_1 + d_2 = \delta} \bar{q}^{d_1}\binom{n - d_1}{d_2}(q - 1)^{d_2}, \quad q \geq 4, \tag{3}$$

$$\bar{q} = \left\lfloor \frac{q}{2} \right\rfloor\left\lceil \frac{q}{2} \right\rceil.$$

§1 The Exact Value of $M_q(n, \delta)$

The main result we are going to prove here is contained in the following:

Theorem 11 (Ahlswede 1987) *For $q = 2, 4, 5$ the following equality holds:*

$$M_q(n, \delta) = F_q(n, \delta). \tag{4}$$

Proof. First we show for arbitrary q the validity of the inequality

$$M_q(n,\delta) \geq F_q(n,\delta). \tag{5}$$

To do this we present explicit constructions of the sets A and B such that $(A,B) = (A,B)_{q,n,\delta}$ and $|A||B| = F_q(n,\delta)$.

First of all we define the following sets:

$$E_1(q,m) = \{(1,\ldots,1),\ldots,(q,\ldots,q)\} \subset [q]^m,$$
$$E_2(q) = \{\pi(1),\ldots,\pi(q) : \pi \in S_q\} \subset [q]^q,$$
$$E_3(q,m,d) = \{x^m \in [q]^m : d_H(x^m,(1,\ldots,1)) = d\},$$
$$E_4 = \{1,2,\ldots,\beta\},\ \bar{E}_4 = \{\beta+1,\ldots,q\},\ \beta = \left\lfloor \frac{q}{2} \right\rfloor,$$

where S_q is the set of all permutations on $[q]$.

We treat first the case $q = 2$ and consider the sets

$$A = (E_1(2,2))^{d_1} \times E_1(1,n-2d_1),$$
$$B = (E_2(2))^{d_1} \times E_3(2,n-2d_1,d_2).$$

We have $d_H(a,b) = d_1 + d_2$, when $a \in A$, $b \in B$ and $|A| = 2^{d_1}$, $|B| = 2^{d_1}\binom{n-2d_1}{d_2}$. Thus an optimal choice of d_i gives

$$|A||B| = F_2(n,\delta).$$

Next, suppose $q = 3$. This time we define the sets A and B as follows:

$$A = (E_1(3,3))^{\ell} \times E_1(1,n-3\ell),$$
$$B = (E_2(3))^{\ell} \times E_3(3,n-3\ell,d).$$

We have $d_H(a,b) = 2\ell + d$, $a \in A$, $b \in B$, and

$$|A| = 3^{\ell},\ |B| = 6^{\ell}\binom{n-3\ell}{d}2^d.$$

An optimal choice of d,ℓ with $2\ell + d = \delta$ gives

$$|A||B| = F_3(n,\delta).$$

In the case $q \geq 4$ define

$$A = (E_4(q))^{d_1} \times E_1(1,n-d_1),$$
$$B = (\bar{E}_4(q))^{d_1} \times E_3(d,n-d_1,d_2).$$

Again, $d_H(a,b) = d_1 + d_2$, $a \in A$, $b \in B$, and

$$|A| = \left\lfloor \frac{q}{2} \right\rfloor^{d_1}, \ |B| = \left\lceil \frac{q}{2} \right\rceil^{d_1} \binom{n-d_1}{d_2} (q-1)^{d_2}.$$

An optimal choice of d_i with $d_1 + d_2 = \delta$ yields

$$|A||B| = F_q(n,\delta).$$

Now we start to prove for $q = 2,4,5$ the inequality

$$M_q(n,\delta) \leq F_q(n,\delta),$$

which together with (5) gives the proof of Theorem 11. We need the following lemma, which we then use in the inductive proof of the theorem.

Lemma 13 *The following relations are valid:*

$$F_2(n,\delta) = F_2(n-2,\delta-1) \max\left(4, \frac{n(n-1)}{\delta(n-\delta)}\right), \tag{6}$$

$$n \geq 3, \ 1 \leq \delta \leq n-1,$$

$$F_q(n,\delta) = F_q(n-1,\delta-1) \max\left(\bar{q}, \frac{n(q-1)}{\delta}\right), \tag{7}$$

$$q \geq 4, \ n \geq 2, \ \delta \geq 1.$$

Proof. First we show that the LHS of equalities (6), (8) do not exceed their RHS. Choose d_1, d_2 such that $d_1 + d_2 = \delta$ and

$$F_2(n,\delta) = 2^{2d_1} \binom{n-2d_1}{d_2}.$$

If $d_1 = 0$, then

$$F_2(n,\delta) = \binom{n}{\delta} = \binom{n-2}{\delta-1} \frac{n(n-1)}{\delta(n-\delta)} \leq F_2(n-2,\delta-1) \frac{n(n-1)}{\delta(n-\delta)},$$

and if $d_1 \geq 1$, then

$$F_2(n,\delta) = 2^2 2^{2(d_1-1)} \binom{n-2-2(d_1-1)}{d_2} \leq 4F_2(n-2,\delta-1).$$

For $q \geq 4$ we have

$$F_q(n,\delta) = \bar{q}^{d_1} \binom{n-d_1}{d_2} (q-1)^{d_2}$$

and in the case $d_1 = 0$

$$F_q(n, \delta) = \binom{n}{\delta}(q-1)^\delta = \frac{n(q-1)}{\delta}\binom{n-1}{\delta-1}(q-1)^{\delta-1}$$

$$\leq F_q(n-1, \delta-1)\frac{n(q-1)}{\delta}.$$

If $d_1 \geq 1$, then

$$F_q(n, \delta) = \bar{q}^{d_1}\binom{n-d_1}{d_2}(q-1)^{d_2}$$

$$= \bar{q}\bar{q}^{d_1-1}\binom{(n-1)-(d-1)}{d_2}(q-1)^{d_2} \leq \bar{q}F_q(n-1, \delta-1).$$

Next we prove that the RHS of (8) does no exceed its LHS. Let d_1, d_2 satisfy $d_1 + d_2 = \delta - 1$ and

$$F_q(n-1, \delta-1) = \bar{q}^{d_1}\binom{n-1-d_1}{d_2}(q-1)^{d_2}.$$

Then $d_1 + 1 + d_2 = \delta$ and we have

$$\bar{q}F_q(n-1, \delta-1) = \bar{q}^{d_1+1}\binom{n-(d_1+1)}{d_2}(q-1)^{d_2} \leq F_q(n, \delta).$$

Furthermore, since

$$\frac{n(q-1)}{\delta}F_q(n-1, \delta-1) = \bar{q}^{d_1}\binom{n-1-d_1}{d_2}\frac{n}{\delta}(q-1)^{d_2+1},$$

it suffices to show that

$$\binom{n-1-d_1}{d_2}\frac{n}{\delta} \leq \binom{n-d_1}{d_2+1}.$$

But

$$\binom{n-1-d_1}{d_2}\frac{n}{\delta} = \binom{n-d_1}{d_2+1}\frac{d_2+1}{n-d_1}\frac{n}{\delta}.$$

Therefore, it suffices to show that

$$\frac{n}{n-d_1} \leq \frac{\delta}{d_2+1} = \frac{\delta}{\delta-d_1},$$

which is true, because for $x \geq y \geq 0, z \geq 0$ with $xyz \neq 0$ it holds $\frac{x+z}{y+z} \leq \frac{x}{y}$.

Now we prove that the RHS of (6) does not exceed its LHS. Suppose that

$$F_2(n-2, \delta-1) = 2^{2d}\binom{n-2-2d}{\delta-1-d}, \tag{8}$$

then

$$4F_2(n-2,\delta-1) = 2^{2(d+1)}\binom{n-2(d+1)}{\delta-(d+1)} \leq F_2(n,\delta)$$

and, to finish the proof, we have to consider the case

$$4 < \frac{n(n-1)}{\delta(n-\delta)}. \tag{9}$$

From (8) it follows that

$$F_2(n-2,\delta-1)\frac{(n-2d)(n-2d-1)}{(\delta-d)(n-d-\delta)} = 2^{2d}\binom{n-2d}{\delta-d} < F_2(n,\delta).$$

It remains to prove that under condition (9) either

$$\frac{n(n-1)}{\delta(n-\delta)} \leq \frac{(n-2d)(n-2d-1)}{(\delta-d)(n-d-\delta)}$$

or

$$\delta(n-\delta)(n^2-4nd+4d^2-n+2d) \geq (n^2-n)((n-\delta)\delta-(n-\delta)d-\delta d+d^2)$$

or

$$\frac{n(n-1)}{\delta(n-\delta)} \geq 4 - \frac{2}{n-d} \tag{10}$$

holds, which is true under condition (9). The proof of Lemma 13 is completed.

$\square$

Next we give the following definitions. For a set $C \subset [q]^n$ and $i,j \in [q]$, $J \subset [q]$, define

$$C_i^t = \{(c_1,\ldots,c_{t-1},c_{t+1},\ldots,c_n) : (c_1,\ldots,c_{t-1},i,c_{t+1},\ldots,c_n) \in C\},$$
$$C^t(J) = \{(c_1,\ldots,c_n) \in C : c_i \in J\} \subset C, \ n \geq 2,$$
$$C_{ij}^{st} = \{(c_1,\ldots,c_{s-1},c_{s+1},\ldots,c_{t-1},c_{t+1},\ldots,c_n) :$$
$$(c_1,\ldots,c_{s-1},i,c_{s+1},\ldots,c_{t-1},j,c_{t+1},\ldots,c_n) \in C\}, s \neq t, \ n \geq 3.$$

Denote also $\mathcal{J}_q = \binom{[q]}{\lfloor\frac{q}{2}\rfloor}$. We need two lemmas.

Lemma 14 *For $(A,B) \in S_2(n,\delta)$ there exist $s,t \in [n]$ such that*

$$(|A_{11}^{st}|+|A_{22}^{st}|)(|B_{12}^{st}|+|B_{21}^{st}|) + (|A_{12}^{st}|+|A_{21}^{st}|)(|B_{11}^{st}|+|B_{22}^{st}|) \tag{11}$$
$$\geq \frac{2\delta(n-\delta)}{n(n-1)}|A||B|.$$

Proof. Let

$$C_{ij}(s,t) = \{(c_1,\ldots,c_n) \in C : c_s = i, \ c_t = j\}.$$

Then $|C_{ij}(s,t)| = |C^{st}_{ij}|$ and if $I_A(x^n)$ is the indicator function of the set A, then

$$\sum_{s \neq t} \left[(|A^{st}_{11}| + |A^{st}_{22}|)(|B^{st}_{12}| + |B^{st}_{21}|) + (|A^{st}_{12}| + |A^{st}_{21}|)(|B^{st}_{11}| + |B^{st}_{22}|) \right]$$

$$= \sum_{(x^n,y^n) \in (A,B),\, s \neq t} \left[(I_{A_{11}(s,t)}(x^n) + I_{A_{22}(s,t)}(x^n))(I_{B_{12}(s,t)}(y^n) + I_{B_{21}(s,t)}(y^n)) \right.$$

$$\left. + (I_{A_{12}(s,t)}(x^n) + I_{A_{21}(s,t)}(x^n))(I_{B_{11}(s,t)}(y^n) + I_{B_{22}(s,t)}(y^n)) \right].$$

Since $d_H(x^n,y^n) = \delta$ for $x^n \in A$ and $y^n \in B$, the contribution of (A,B) is $|A||B|\delta(n-\delta)$ and there exists at least one pair (s,t) with contribution at least $|A||B|\delta(n-\delta)/\binom{n}{2}$. The lemma is proved. $\square$

Lemma 15 *For $(A,B) \in S_q(n,\delta)$ there exists a $t \in [n]$, such that*

$$\sum_{J \in \mathcal{J}_q} |A^t(J)||B^t(J^c)| \geq |A||B| \frac{\delta}{n(q-1)} \frac{\bar{q}}{q} \binom{q}{\lfloor \frac{q}{2} \rfloor}, \tag{12}$$

where $J^c = [q] \setminus J$.

Proof. We have

$$\sum_{t=1}^{n} \sum_{J \in \mathcal{J}_q} |A^t(J)||B^t(J^c)| = \sum_{t=1}^{n} \sum_{J \in \mathcal{J}_q} \sum_{x^n \in A,\, y^n \in B} I_{A^t(J)}(x^n) I_{B^t(J^c)}(y^n)$$

$$= \sum_{x^n \in A,\, y^n \in B} \sum_{t=1}^{n} \sum_{J \in \mathcal{J}_q} I_{A^t(J)}(x^n) I_{B^t(J^c)}(y^n)$$

$$= \sum_{x^n \in A,\, y^n \in B} \delta \binom{q-2}{\lfloor \frac{q}{2} \rfloor - 1} = |A||B|\delta \binom{q-2}{\lfloor \frac{q}{2} \rfloor - 1}.$$

Therefore, there exists a t with

$$\sum_{J \in \mathcal{J}_q} |A^t(J)||B^t(J^c)| \geq |A||B| \frac{\delta}{n} \binom{q-2}{\lfloor \frac{q}{2} \rfloor - 1}$$

and (12) follows due to the identity

$$\binom{q}{\lfloor \frac{q}{2} \rfloor} = \frac{q(q-1)}{\bar{q}} \binom{q-2}{\lfloor \frac{q}{2} \rfloor - 1}.$$

By symmetry also

$$\sum_{J \in \mathcal{J}_q} |A^t(J^c)||B^t(J)| \geq |A||B| \frac{\delta}{n(q-1)} \frac{\bar{q}}{q} \binom{q}{\lfloor \frac{q}{2} \rfloor}.$$

Thus there exists a t for which we have

$$\sum_{J \in \mathcal{J}_q} (|A^t(J)||B^t(J^c)| + |A^t(J^c)||B^t(J)|) \geq |A||B| \frac{2\delta}{n(q-1)} \frac{\bar{q}}{q} \binom{q}{\lfloor \frac{q}{2} \rfloor}. \tag{13}$$

$\square$

Now we continue to prove the theorem. First we prove (4) for $q = 2$. In the cases $\delta = 0$ and $\delta = n$, it can be easily verified that

$$M_2(n,0) = F_2(n,0) = M_2(n,n) = F_2(n,n) = 1.$$

In the other cases we proceed by induction on n and we assume that $\delta \neq 0, n$. For $n = 1, 2$ only the case

$$M_2(2,1) = F_2(2,1) = 4$$

is relevant. An optimal configuration here is $(A,B) = (\{11,22\}, \{21,12\})$.

Let (4) be valid for $n - 2$. We show that it holds also for n. We use the sets $A_{\alpha\beta}^{st}, B_{\alpha\beta}^{st}$ with property (11). For simplicity we omit the indices s, t and make the following conventions:

$$I = (|A_{11}| + |A_{22}|)(|B_{11}| + |B_{22}|),$$
$$II = (|A_{12}| + |A_{21}|)(|B_{11}| + |B_{22}|),$$
$$III = (|A_{11}| + |A_{22}|)(|B_{12}| + |B_{21}|),$$
$$IV = (|A_{12}| + |A_{21}|)(|B_{12}| + |B_{21}|).$$

Lemma 14 says that

$$|A||B| \leq \frac{n(n-1)}{2\delta(n-\delta)}(II + III). \tag{14}$$

W.l.o.g. we can assume that

$$II \leq III. \tag{15}$$

First we consider the case $A_{11} \cap A_{22} \neq \emptyset$. Then

$$d_H(a_{11}^n, b_{\beta\beta}^n) \neq d_H a_{22}^n, b_{\beta\beta}^n), \ a_{\alpha\alpha}^n \in A_{\alpha\alpha}(s,t), \ b_{\beta\beta}^n \in B_{\beta\beta}(s,t)$$

and we have $B_{11} = B_{22} = \emptyset$ and therefore $I = II = 0$. If now $B_{12} \cap B_{21} \neq \emptyset$, then by the same argument $A_{12} = A_{21} = \emptyset$ and thus also $IV = 0$. Therefore,

$$|A||B| = III \leq 4M_2(n-2, \delta - 1) = 4F_2(n-2, \delta - 1) \leq F_2(n, \delta).$$

Here, the last equality follows from the induction hypothesis and the last inequality follows from (6).

On the other hand, if $B_{12} \cap B_{21} = \emptyset$, then $(A_{\alpha\alpha}, B_{12} \cup B_{21}) \in S_2(n-2, \delta - 1)$, $\alpha = 1, 2$ and therefore $III \leq 2M_2(n-2, \delta - 1)$. Since $II = 0$, we conclude that

$$II + III \leq 2M_2(n-2, \delta - 1) = 2F_2(n-2, \delta - 1)$$

and by (14) we have

$$|A||B| \leq \frac{n(n-1)}{\delta(n-\delta)} F_2(n-2,\delta-1).$$

Then (6) implies $|A||B| \leq F_2(n,\delta)$.

Suppose now $A_{11} \cap A_{22} = \emptyset$. If $B_{12} \cap B_{21} \neq \emptyset$, then, as previously, $A_{12} = A_{21} = \emptyset$ and $II = 0$, $II + III = III \leq 2M_2(n-2,\delta-1)$, and $|A||B| \leq F_2(n,\delta)$. Finally, if $B_{12} \cap B_{21} = \emptyset$, then

$$(A_{11} \cup A_{22}, B_{12} \cup B_{21}) \in S_2(n-2,\delta-1)$$

and thus $III \leq M_2(n-2,\delta-1)$. From the assumption (15) it follows that $II + III \leq 2M_2(n-2,\delta-1)$ and the proof can be completed as in the previous case.

Now we prove (4) for $q = 4$. The case $n = 1$ is settled by inspection. We assume that $J = \{0,1\}$, $J^c = \{2,3\}$ and consider the following scheme (we omit index t in the notations A_i^t, B_i^t):

For $q = 4$ inequality (13) can be written in the form

$$|A||B| \leq \frac{3n}{2\delta}((|A_1|+|A_2|)(|B_3|+|B_4|) \tag{16}$$
$$+ (|A_3|+|A_4|)(|B_1|+|B_2|)) = \frac{3n}{2\delta}(II+III),$$

where

$$I = (|A_1|+|A_2|)(|B_1|+|B_2|),$$
$$II = (|A_3|+|A_4|)(|B_1|+|B_2|),$$
$$III = (|A_1|+|A_2|)(|B_3|+|B_4|),$$
$$IV = (|A_3|+|A_4|)(|B_3|+|B_4|).$$

Now we proceed as in the proof of the case $q = 2$ with substitutions $11 \to 1$, $22 \to 2$, $12 \to 3$, $21 \to 4$, $A_{11} \to A_1$, $B_{11} \to B_1$, etc.; $F_2(n,\delta) \to F_4(n,\delta)$, (14)$\to$ (16). Repeating all arguments from the previous proof and taking into account that $\bar{q} = 4$, we are done with the case $q = 4$.

Now let $q = 5$. We need one simple preliminary result, which we state in the forthcoming Lemma 16, whose proof we leave to the reader (Exercise 12). For simplicity we again omit the index t in the notations A_i^t, B_i^t. Define the numbers r, s, p by

$$r = |\{1 \leq i \leq q : |A_i||B_i| > 0\}|,$$
$$s = |\{1 \leq i \leq q : |A_i| > 0\}| - r,$$
$$p = |\{1 \leq i \leq q : |B_i| > 0\}| - r.$$

After relabeling we have $|A_i||B_i| > 0$ for $1 \leq i \leq r$, $|A_i| > 0$ for $1 \leq i \leq r+s$, $|B_i| > 0$ for $1 \leq i \leq r$, and $r+s+1 \leq i \leq r+s+p$.

Lemma 16 *Let $n \geq 2$. If $(A,B) \in S_q(n,\delta)$ and $r+s, r+p \geq 2$, then for $1 \leq i \leq r$, $1 \leq j \leq q$, $i \neq j$ we have $A_i \cap A_j = \emptyset$ and $B_i \cap B_j = \emptyset$.*

Denote

$$X = \{1,\ldots,r\}, \ Y = \{r+1,\ldots,r+s\},$$
$$Z = \{r+s+1,\ldots,r+s+p\}.$$

It is easy to see that if we replace A_i, $i \in Y$ by $E = \bigcup_{i \in Y} A_i$ and B_i, $i \in Z$ by $F = \bigcup_{i \in Z} B_i$, then we again obtain a pair in $S_q(n,\delta)$. Note also that if $s + p \neq 0$ we can enlarge Y or Z so that $r+s+p = q$.

Denote

$$e = \sum_{J \in \mathcal{J}_q} |A(J)||B(J^c)|.$$

For $J \in \mathcal{J}_q$ we define

$$U = J \cap X, \ V = J \cap Y, \ W = J \cap Z, \ E = \bigcup_{i \in Y} A_i, \ F = \bigcup_{i \in Z} B_i,$$

$$a(J) = \sum_{i \in J} |A_i|, \ b(J) = \sum_{i \in J} |B_i|, \ J \subset [q].$$

If $s + p = 0$, then

$$e = \sum_{U \subset X, \ 1 \leq |U| \leq \min\{\beta k - 1\}} a(U) b(X \setminus U) \binom{q-r}{\beta - |U|}. \tag{17}$$

If $r + s + p = q$, then

$$e = \sum_{U \subset X, \ V \subset Y, \ W \subset Z, \ |U|+|V|+|W|=\beta} (a(U) + |V||E|)(b(X \setminus U) + |Z \setminus W||F|).$$

Opening the brackets on the RHS of the expression for e we obtain four sums

$$e_1 = \sum_{U \subset X, \ V \subset Y, \ W \subset Z, \ |U|+|V|+|W|=\beta} a(U) b(X \setminus U)$$

$$= \sum_{U \subset X, \ell, |U|+\ell=2} \binom{s+p}{\ell} a(U) b(x \setminus U)$$

$$= \sum_{U \subset X, \ 1 \leq |U| \leq \min\{\beta, r-1\}} \binom{q-r}{\beta - |U|} a(U) b(X \setminus U),$$

$$e_2 = \sum_{U \subset X, \ V \subset Y, \ W \subset Z, \ |U|+|V|+|W|=\beta} a(U)|Z \setminus W||F|$$

$$= \sum_{U \subset X, \ |U|+|V|+|W|=\beta} \binom{s}{|V|}\binom{p}{|W|}(p - |W|)a(U)|F|$$

$$= \sum_{U \subset X,\, 1 \le |U| \le \beta} \binom{q-k-1}{\beta - |U|} pa(U)|F|,$$

$$e_3 = \sum_{U \subset X,\, V \subset Y,\, W \subset Z,\, |U|+|V|+|W|=\beta} b(X \setminus U)|V||E|$$

$$= \sum_{U \subset X,\, |U|+|V|+|W|=\beta} \binom{s}{|V|}\binom{p}{|W|}|V|b(X \setminus U)|E|$$

$$= \sum_{U \subset X,\, 1 \le |U| \le \min\{\beta, r-1\}} \binom{q-r-1}{\beta - |U| - 1} b(X \setminus U)s|E|,$$

$$e_4 = \sum_{U \subset X,\, V \subset Y,\, W \subset Z,\, |U|+|V|+|W|=\beta} |V||E||Z \setminus W||F|$$

$$= \sum_{U \subset X,\, |U|+|V|+|W|=\beta} \binom{s}{|V|}\binom{p}{|W|}|V||Z \setminus W||E||F|$$

$$= \sum_{U \subset X,\, |U| \le \beta - 1} \binom{q-r-2}{\beta - |U| - 1} sp|E||F|.$$

Note that in the case $s + p = 0$ we obtain the same final relations for e_i and e. Now by (8) and Lemma 15, in the case $q = 5$ the relation

$$e \le F_5(n-1, \delta - 1)12 \tag{18}$$

is sufficient for induction to work. To prove this inequality we go through the cases defined by the value of r.

$\mathbf{r = 5}$. Since $s = p = 0$, we have $e_2 = e_3 = e_4 = 0$. Therefore,

$$e = e_1 = \sum_{U \subset [5],\, 1 \le |U| \le 2} \binom{5-5}{2-|U|} a(U)b(X \setminus U).$$

As $\left(\bigcup_{i \in U} A_i, \bigcup_{i \in X \setminus U} B_i \right) = S_5(n-1, \delta - 1)$ and by Lemma 16

$$\left| \bigcup_{i \in U} A_i \right| = a(U), \quad \left| \bigcup_{i \in X \setminus U} B_i \right| = b(X \setminus U),$$

we conclude by using the induction hypothesis that

$$e \le \binom{5}{2} M_2(n-1, \delta - 1) \le 10 F_5(n-2, \delta - 1).$$

r = 4. In this case either $s = 1, p = 0$ or $s = 0, p = 1$ holds. By symmetry, it suffices to consider only the first case. Then $e_2 = e_4 = 0$ and

$$
\begin{aligned}
e = e_1 + e_2 = {} & \sum_{U \subset [4], \, 1 \leq |U| \leq 2} \binom{2}{2 - |U|} a(U) b(X \setminus U) \\
& + \sum_{U \subset [4], \, |U| \leq 2} \binom{0}{2 - |U| - 1} b(X \setminus U) s |E| \\
\leq {} & \binom{4}{2} F_5(n - 1, \delta - 1) + \sum_{U \subset [4], |U| = 1} (a(U) + |E|) b(X \setminus U).
\end{aligned}
$$

By Lemma 16 and the induction hypothesis the second summand is smaller than $4F_5(n - 1, \delta - 1)$ and therefore $e \leq 10F_5(n - 1, \delta - 1)$.

r = 3. Here we have

$$
\begin{aligned}
e_1 = {} & \sum_{U \subset [3], \, 1 \leq |U| \leq 2} \binom{2}{2 - |U|} a(U) b(X \setminus U) \\
= {} & 2(|A_1|(|B_2| + |B_3|)|A_2|(|B_1| + |B_3|) + |A_3|(|B_1| + |B_2|)) \\
& + ((|A_1| + |A_2|)|B_3| + (|A_1| + |A_3|)|B_2| + (|A_2| + |A_3|)|B_1|), \\
e_2 = {} & 3(|A_1| + |A_2| + |A_3|) p |F|, \\
e_3 = {} & 3(|B_1| + |B_2| + |B_3|) s |E|, \\
e_4 = {} & 3 s p |E||F|.
\end{aligned}
$$

Now we consider a few subcases.

s = 2, p = 0. Then

$$
\begin{aligned}
e = e_1 + e_2 = {} & 3(|B_1| + |B_2|)(|A_3| + |E|) \\
& + 3(|B_1| + |B_3|)(|A_2| + |E|) + 3(|B_2| + |B_3|)(|A_1| + |E|).
\end{aligned}
$$

Since $(B_1 \cup B_2, A_3 \cup E) \in S_5(n - 1, \delta - 1)$, we have $3(|B_1| + |B_2|)(|A_3| + |E|) \leq F_5(n - 1, \delta - 1)$. The remaining terms in the expression for e are estimated in the same manner. Thus we have $e \leq 9F_5(n - 1, \delta - 1)$.

s = 1, p = 1. Then

$$
\begin{aligned}
e = {} & e_1 + e_2 + e_3 + e_4, \\
e_2 = {} & 3(|A_1| + |A_2| + |A_3|)|F|, \; e_3 = 3(|B_1| + |B_2| + |B_3|)|E|, \; e_4 = 3|E||F|
\end{aligned}
$$

and e_1 has the same expression as in the previous subcase. We can assume that $e_2 \leq e_3$, because otherwise we can exchange the roles of A and B. Thus, by the previous subcase, $e_1 + e_2 + e_3 \leq 9F_5(n - 1, \delta - 1)$, and since $e_4 \leq 3F_5(n - 1, \delta - 1)$, we conclude $e \leq 12F_5(n - 1, \delta - 1)$.

s = 0, p = 2. Since e_1 and e_2 are symmetric in A and B, replacement of e_3 by e_2 in the case $s = 2, t = 0$ gives again the bound $e \leq 9F_5(n - 1, \delta - 1)$.

r = 2. In this case

$$e_1 = \sum_{U \subset [2], |U|=1} \binom{3}{|U|} a(U)b(X \setminus U) = 3(|A_1||B_2| + |A_2||B_1|),$$

$$e_2 = \sum_{U \subset [2], 1 \le |U| \le 2} \binom{2}{2-|U|} a(U)p|F| = 3(|A_1| + |A_2|)p|F|,$$

$$e_3 = \sum_{U \subset [2], |U| \le 1} \binom{2}{1-|U|} b(X \setminus U)s|E| = 3(|B_1| + |B_2|)s|E|,$$

$$e_4 = \sum_{U \subset [2], |U| \le 1} \binom{1}{1-|U|} sp|E||F| = 3sp|E||F|.$$

Here also we have some subcases.
s = 3, p = 0. Then

$$e = e_1 + e_2 = 3(|B_1|(|A_2| + |E|)$$
$$+ 3|B_2|(|A_1| + |E|) + 6(|B_1| + |B_2|)|E|) \le 12F_5(n-1, \delta - 1).$$

s = 2, p = 1. Then

$$e = e_1 + e_2 + e_3 + e_4 = \left(e_1 + e_2 + \frac{1}{2}e_3 + e_4\right) + \frac{1}{2}e_3$$
$$= 3(|B_1| + |F|)(|A_2| + |E|) + 3(|B_2| + |F|)(|A_1| + |E|)$$
$$+ 3(|B_1| + |B_2|)|E| \le 9F_5(n-1, \delta - 1).$$

The other subcases are symmetrically the same.
r = 1. In this case we can write

$$e_1 = 0, \; e_2 = \binom{5-1}{2-1}|A_1||F|,$$

$$e_3 = \binom{5-1}{2-1}|B_1|s|E|,$$

$$e_4 = \binom{5-3}{2-1}sp|E||F| \binom{5-3}{2-2}sp|E||F| = 5q - 2\beta - 1sp|E||F|$$

and

$$e = 3(|A_1|p|F| + |B_1|s|E| + s|E|p|F|).$$

But

$$\lambda = |A_1|p|F| + |B_1|s|E| + s|E|p|F| = s|E|(|B_1| + |F|) + (|A_1| + |E|)p|F|$$
$$+ (s-1)p|E||F| - s|E||F|$$
$$= s|E|(|B_1| + |F|) + (|A_1| + |E|)p|F| + (sp - s - p)|E||F|.$$

As $(E,(B_1 \cup F))$, $(A_1 \cup E, F)$, $(E, F) \in S_5(n-1, \delta-1)$, the induction hypothesis gives

$$\lambda \leq (s + p + (sp - s - p))F_5(n-1, \delta-1) \leq 4F_5(n-1, \delta-1)$$

and therefore $e \leq 12F_5(n-1, \delta-1)$.

r = 0. We have

$$|A||B| = sp|E||F| \leq 6|E||F| \leq 6F_5(n-1, \delta-1).$$

This completes the proof of the theorem.   ∎

§2 Four-Words Property

We formulate a generalization of the property of the pair (A, B) to be a constant distance pair. We say that the pair of sets (A, B), $A, B \subset \mathcal{X}_q^n$, $\mathcal{X}_q = [q]$ satisfies the four-words property (4-WP) if

$$d_H(a^n, b^n) - d_H(a^n, b'^n) + d_H(a'^n, b'^n) - d_H(a'^n, b^n) \neq 1, 2$$

for all $a^n, a'^n \in A$, $b^n, b'^n \in B$.

Proposition 10 *If a pair (A, B) satisfies the 4-WP, then*

$$|A||B| \leq q^{*n}, \quad q^* = \begin{cases} q, & q = 2, 3, 4, \\ \bar{q} = \left\lfloor \frac{q}{2} \right\rfloor \cdot \left\lceil \frac{q}{2} \right\rceil, & q \geq 4 \end{cases}$$

and this bound is best.

Next we consider a further generalization of the 4-WP and prove Theorem 12 below, from which also follows the statement of Proposition 10.

Let $\mathcal{X}$ and $\mathcal{Y}$ be two finite sets. We consider the function

$$f : \mathcal{X} \times \mathcal{Y} \to \mathbb{Z}.$$

With f we associate the sum-type function $f_n : \mathcal{X}^n \times \mathcal{Y}^n \to \mathbb{Z}$:

$$f_n(x^n, y^n) = \sum_{i=1}^{n} f(x_i, y_i),$$

$x^n = (x_1, \ldots, x_n) \in \mathcal{X}^n$, $y^n = (y_1, \ldots, y_n) \in \mathcal{Y}^n$.

We say that the pair (A, B) with $A \subset \mathcal{X}^n$, $B \subset \mathcal{Y}^n$ satisfies the $\mathcal{R}$-four-word property ($\mathcal{R} - 4$-WP), if

$$f_n(a^n, b^n) - f_n(a^n, b'^n) + f_n(a'^n, b'^n) - f_n(a'^n, b^n) \in \mathcal{R}, \tag{19}$$

for all $a^n, a'^n \in A$, $b^n, b'^n \in B$. Let $\mathcal{P}(f, \mathcal{R}, n)$ be the set of all those pairs. We are interested in

$$M(f, \mathcal{R}, n) = \max\{|A||B| : (A, B) \in \mathcal{P}(f, \mathcal{R}, n)\}.$$

Let $\mathcal{P}^*(f, \mathcal{R}, n)$ be the set of those pairs in $\mathcal{P}(f, \mathcal{R}, n)$ on which the maximum $M(f, \mathcal{R}, n)$ is achieved. The following theorem is the basis in the study of the $\mathcal{R} - 4$-WP [ACZ89].

Theorem 12 (Ahlswede, Cai, and Zhang 1989) *For any $\mathcal{R} \subset \mathbb{Z}$*

$$M(f, \mathcal{R}, n) \leq M^n(f, \mathcal{R}, 1). \tag{20}$$

Furthermore, if $0 \in \mathcal{R}$ and $M(f, \{0\}, 1) = M(f, \mathcal{R}, 1)$, then equality holds in (20).

The proof of this theorem proceeds by induction on n and is based on two simple lemmas, which we first state and prove.

For the set C of sequences of length n from some finite alphabet denote

$$C_c = \{(c_1, \ldots, c_{n-1}) : (c_1, \ldots, c_{n-1}, c) \in C\},$$
$$J(C) = \{c : C_c \neq \emptyset\},$$
$$L(C) = \max\left\{|D| : D \in J(C), \bigcap_{c \in D} C_c \neq \emptyset\right\}.$$

Lemma 17 *For $(A, B) \in \mathcal{P}(f, \mathcal{R}, n)$ we have $L(A)|J(B)| \leq M(f, \mathcal{R}, 1)$.*

Proof. It suffices to show that for every $D \subset J(A)$ with $\bigcap_{a \in D} A_a \neq \emptyset$ necessarily $(D, J(B)) \in \mathcal{P}(f, \mathcal{R}, 1)$.

To see this choose $a, a' \in D$, $b, b' \in J(B)$ and note that by assumptions there are a^{n-1}, b^{n-1}, b'^{n-1} such that $a^{n-1}a, a^{n-1}a' \in A$, and $b^{n-1}b, b'^{n-1}b' \in B$. Now obviously

$$\begin{aligned} \mathcal{R} \ni\ & f_n(a^{n-1}a, b^{n-1}b) - f_n(a^{n-1}a, b'^{n-1}b') \\ & + f_n(a^{n-1}a', b'^{n-1}b') - f_n(a^{n-1}a', b^{n-1}b) \\ =\ & f(a, b) - f(a, b') + f(a', b') - f(a', b). \end{aligned}$$

$\square$

Lemma 18 *If $(A, B) \in \mathcal{P}(f, \mathcal{R}, n)$, then $\left(\bigcup_{d \in J(A)} A_d, B_b\right) \in \mathcal{P}(f, \mathcal{R}, n-1)$ for all $b \in J(B)$.*

Proof. For $a^{n-1}, a'^{n-1} \in \bigcup_{d \in J(A)} A_d$ choose $a, a' \in J(A)$ such that $a^{n-1} \in A_a$, $a'^{n-1} A_{a'}$. Now for any $b^{n-1}, b'^{n-1} \in B_b$,

$$\begin{aligned} \mathcal{R} \ni\ & f_n(a^{n-1}a, b^{n-1}b) - f_n(a^{n-1}a, b'^{n-1}b) \\ & + f_n(a'^{n-1}a', b'^{n-1}b) - f_n(a'^{n-1}a', b^{n-1}b) \end{aligned}$$

$$= f_{n-1}(a^{n-1}, b^{n-1}) - f_{n-1}(a^{n-1}, b'^{n-1})$$
$$+ f_{n-1}(a'^{n-1}, b'^{n-1}) - f_{n-1}(a'^{n-1}, b^{n-1}).$$

$\square$

Proof of Theorem 12. Obviously, if for $(A, B) \in \mathcal{P}^*(f, \{0\}, 1)$ we have $|A||B| = M(f, \mathcal{R}, 1)$, then $(\prod_{i=1}^{n} A, \prod_{i=1}^{n} B) \in \mathcal{P}(f, \{0\}, n)$, where $\prod_{i=1}^{n} C$ is the set of n-tuples of elements from C. Therefore, if $0 \in \mathcal{R}$, then $M(f, \mathcal{R}, n) \geq (|A||B|)^n = M^n(f, \mathcal{R}, 1)$.

To prove (20) we use induction. For $n = 1$ nothing needs to be proved. For $(A, B) \in \mathcal{P}(f, \mathcal{R}, n)$ we have

$$|A||B| = \sum_{a \in J(A)} |A_a| \sum_{b \in J(B)} |B_b|$$

$$\leq L(A) \left| \bigcup_{a \in J(A)} A_a \right| |J(B)| \max_{b \in J(B)} |B_b|$$

$$\leq M(f, \mathcal{R}, 1) \left| \bigcup_{a \in J(A)} A_a \right| \max_{b \in J(B)} |B_b|.$$

The last inequality here follows from Lemma 17. The result $|A||B| \leq M^n(f, \mathcal{R}, 1)$ now follows from Lemma 18 and the induction hypothesis. ∎

From this Theorem follows Proposition 10. Indeed the 4-WP means that $(A, B) \in \mathcal{P}(d_H, \mathbb{Z} - \{1, 2\}, n)$. We have $\mathcal{P}(d_H, \mathbb{Z} - \{1, 2\}, 1) = \mathcal{P}(d_H, \{0\}, 1)$ and therefore $M(d_H, \mathbb{Z} - \{1, 2\}, n) = M^n(d_H, \{0\}, 1)$. Finally, equality $M(d_H, \{0\}, 1) = q^*$ is easily verified.

The following fact easily follows from Theorem 12 (Exercise 13). If $A, B \subset [0, q-1]^n$ and the set $[0, q-1]^n$ is equipped with a Lee metric d_L, which is defined as follows:

$$d_L(x^n, y^n) = \sum_{i=1}^{n} \min\{|x_i - y_i|, q - |x_i - y_i|\},$$

and for all $a^n, a'^n \in A$, $b^n, b'^n \in B$

$$d_L(a^n, b^n) - d_L(a^n, b'^n) + d_L(a'^n, b'^n) - d_L(a'^n, b^n) \neq 1, 2, \ldots q, \qquad (21)$$

then

$$|A||B| \leq \left(\max\left\{ q, \left(\left\lfloor \frac{q}{4} \right\rfloor + 1 \right) \left(\left\lceil \frac{\lfloor q/2 \rfloor}{2} \right\rceil + 1 \right) \right\} \right)^n . \, . \qquad (22)$$

The next fact is also the consequence of Theorem 12 (Exercise 13). Let d_T be a Taxi metric on $[0, q-1]^n$. If $A, B \subset [0, q-1]^n$ and for $a^n, a'^n \in A$, $b^n, b'^n \in B$,

$$d_T(a^n, b^n) - d_T(a^n, b'^n) + d_T(a'^n, b'^n) - d_T(a'^n, b^n) \neq 1, 2, \ldots 2q,$$

then

$$|A||B| \leq \left(\max\left\{ q, \left(\left\lfloor \frac{q}{2} \right\rfloor + 1 \right) \left\lceil \frac{q}{2} \right\rceil \right\} \right)^n \qquad (23)$$

and this bound is best possible.

Notes to Chapter II

As already said, Theorems 6, 7, and 8 completely solve the problem of determining the maximal sets of given diameter in the Taxi metric when all components $\mathcal{X}_j$ have even or odd lengths. In the mixed case, when some of the components have even length and some odd length, in general some partial results are known when $d < b(\mathcal{B})$ and $d \geq b(\mathcal{B}) + e(\mathcal{B}) - 1$, where $e(\mathcal{B})$ is the number of components with odd q_i's. In this case it is known that the ball of radius $d/2$ with some center in L^1 is a maximal set of diameter d. The proof here is the same as for all-even q_i's in Theorem 8, but the splitting of $\mathcal{B}$ is different. For details about such splittings we refer to [ACZ92a] and [DK90], see also [KF88]. In [BL93] a direct approach was used to the diametric problem in Taxi metric. A complete solution was presented for the problem in the space $\mathcal{B}$, where all q_i are equal. There the diametric problem on the torus also has been considered. Relation (5) (Lecture 2) was first proved in [K66a].

Theorem 3 was proved in [AK97b] and Theorem 5 was proved in [AK98]. We reproduced here their proofs. The Intersection Theorem 4 was first proved by Katona by using another method in [K64]. Relation (3) (Lecture 1) for $t \geq 15$ was first established by Frankl [F78] and subsequently by Wilson [W84] for all t. We took the proof of Proposition 9 from [H64]. Theorems 6 and 7 were proved in [ACZ92a]. Theorem 9 is taken from [AK00b].

An $\mathcal{A} \in \mathcal{I}(n,k,t)$ is called *nontrivial* if $\left| \bigcap_{A \in \mathcal{A}} A \right| < t$ and $\tilde{\mathcal{I}}(n,k,t)$ denotes all nontrivial families from $\mathcal{I}(n,k,t)$. Let

$$\tilde{M}(n,k,t) = \max_{\mathcal{A} \in \tilde{\mathcal{I}}(n,k,t)} |\mathcal{A}|, \ 1 \leq t \leq k \leq n.$$

Let also

$$\mathcal{V}_1(n,k,t) = \left\{ V \in \binom{[n]}{k} : [1,t] \subset V, \right.$$

$$\left. V \cap [1+t, k+1] \neq \emptyset \right\} \cup \{[1, k+1] \setminus \{i\} : i \in [1,t]\}.$$

In [AK96d] the following equalities are proved, which give the complete solution of the determination of the maximal cardinality of a nontrivial family. This settles the Hilton-Milner problem, whose investigation was initiated in [HM67b].

(i) If $2k - t < n \leq (t+1)(k-t+1)$, then

$$\tilde{M}(n,k,t) = M(n,k,t).$$

(ii) If $(t+1)(k-t+1) < n$ and $k \leq 2t+1$, then

$$\tilde{M}(n,k,t) = |\mathcal{F}(1)|$$

and $\mathcal{F}(1)$ is – up to permutations – the unique optimum.

(iii) If $(t+1)(k-t+1) < n$ and $k > 2t+1$, then

$$\tilde{M}(n,k,t) = \max\{|\mathcal{F}(1)|, |\mathcal{V}_1|\},$$

and – up to permutations – $\mathcal{F}(1)$ or $\mathcal{V}_1$ are the only solutions.

Consider the following sets:

$$\binom{[n]}{\geq k} = \bigcup_{i=k}^{n} \binom{[n]}{i},$$

$$\binom{[n]}{\leq k} = \bigcup_{i=0}^{k} \binom{[n]}{i},$$

$$\mathcal{I}(n, \geq k, t) = \mathcal{I}(n,t) \cap 2^{\binom{[n]}{\geq k}},$$

$$\mathcal{I}(n, \leq k, t) = \mathcal{I}(n,t) \cap 2^{\binom{[n]}{\leq k}},$$

$$\mathcal{F}(i, \geq k) = \mathcal{G}(i) \cap \binom{[n]}{\geq k},$$

$$\mathcal{F}(i, \leq k) = \mathcal{G}(i) \cap \binom{[n]}{\leq k}, i = 0, \ldots \left\lfloor \frac{n-t}{2} \right\rfloor.$$

The description of the following results can be found in [ABEK02]. Using Katona's Theorem 4 it is not difficult to prove that

$$\max_{\mathcal{A} \in \mathcal{I}(n,t,\geq k)} |\mathcal{A}| = \left| \mathcal{F}\left(\left\lfloor \frac{n-t}{2} \right\rfloor, \geq k \right) \right|.$$

The problem of determination of the value $\max_{\mathcal{A} \in \mathcal{I}(n,t,\leq k)} |\mathcal{A}|$ is still open. In Research Problem 3 at the end of the chapter the corresponding conjecture is formulated.

In [AAK98], Ahlswede et al. consider the problem of maximal intersecting systems for direct products. This problem was initiated by Frankl and arose in connection with a result of Sali. Let $n = n_1 + \cdots + n_m$, $k = k_1 + \cdots + k_m$, $[n] = [n_1] \cup [n_2] \cdots \cup [n_m]$, $\mathcal{H} = \left\{ F \in \binom{[n]}{k} : |F \cap [n_i]| = k_i \text{ for } i = 1, \ldots, m \right\}$. For given integers t_i, $1 \leq t \leq t_i \leq k_i$, $1 \leq i \leq m$, we may say that $\mathcal{A} \subset \mathcal{H}$ is $(t_1, \ldots, t_m)$-intersecting, if for every $A, B \in \mathcal{A}$ there exists an i, $1 \leq i \leq m$, such that $|A \cap B \cap \Omega_i| \geq t_i$ holds.

Denote the set of such systems by $I(\mathcal{H}, t_1, \ldots, t_m)$. The problem is to determine $\max_{\mathcal{A} \in I(\mathcal{H}, t_1, \ldots, t_m)} |\mathcal{A}|$.

The case $t_1 = t_2 = \cdots = t_m = 1$ has been solved by Frankl. Here is the complete solution.

Theorem 13 (Ahlswede, Aydinian, and Khachatrian 1998) *Let $n_i \geq k_i \geq t_i \geq 1$ for $i = 1, \ldots, m$, then*

$$\max_{\mathcal{A} \in I(\mathcal{H}, t_1, \ldots, t_m)} |\mathcal{A}| = \max_i \frac{M(n_i, k_i, t_i)}{\binom{n_i}{k_i}} |\mathcal{H}|.$$

We emphasize that the combination of this Theorem and Theorem 3 gives an explicit value. The proof is heavily (but not only!) based on ideas and methods from [A96], in particular the method of "generated sets" (c.f. [N] Bey/Engel, "Old and New Results for the Weighted t-Intersection Problem via AK-Methods", 45-74;) takes a central role in the book [E97b].

We took Theorem 10 from [AC91]. Also the following relations were proved there:

$$f_q^{q-1} = \frac{1}{2} |W_q^{q-1}| = \frac{1}{2} q!,$$
$$f_q^3 = 12, \; q \geq 4,$$
$$g_q^3 = 3q + 7, \; 3 \leq q < \infty.$$

We took Theorem 11 from [A87].

For the matrix with entries $a_{ij} = d(x_i^n, y_j^n)$, where x_i^n are n-tuples with elements from some finite set $\mathcal{X}$ and $d(\cdot, \cdot)$ is a metric on $\mathcal{X}^n$, consider the area $i \cdot j$ of an $i \times j$ minors with constant entries. This concept was introduced in [Y79] for estimating communication complexity. It inspired the work reported in Lecture 6.

Proposition 10 was first proved in [AM88]. Inequality (22) was first proved in [C86]. Theorem 12 was proved in [ACZ89].

A pair (A, B), $A, B \subset \{0, 1\}^n$ is said to be ℓ-cross-intersecting iff $\lambda(a^n, b^n) = \sum_{i=1}^n \min\{a_i, b_i\} = \ell$ for all $a^n \in A$, $b^n \in B$. If one considers a^n, b^n as subsets of $[n]$, then $\lambda(a^n, b^n)$ is their intersection. How large can $|A||B|$ be? A simple construction in [ACZ89] gives a lower bound stated in Exercise 14. Moreover, it is conjectured that the construction is best possible. In [AL06] this conjecture is proved for sufficiently large $\ell > \ell_0$.

Exercises

1. Erdös, Ko, and Rado [EKR61] proved the Theorem 3 for the case $t = 1$. Give a proof using the Kruskal/Katona Theorem ([K63],[K68],[D74]). A formulation of the result of [K63] and [K68] can be found without proof in [S59].
2. Give another proof with Katona's cycle method ([K72], see also the book [N] with the survey [K00]).

3. Prove relation (4) (Lecture 1).
4. Prove Lemma 4 and Proposition 1.
5. Determine all optimal unrestricted t-intersecting families for $t = 1$. *Hint:* Among them is always $\mathcal{A} = \{A \in 2^{[n]} : \{1\} \in A\}$, $\mathcal{A} = \bigcup\limits_{i \geq \frac{n+1}{2}} \binom{[n]}{i}$ for n odd,

   and $\mathcal{A} = \bigcup\limits_{i \geq \frac{n}{2}} \binom{[n-1]}{i}$ for n even.
6. For $t = 1$ and $q = 2$ find all optimal configurations for the Hamming distance problem. *Hint:* see considerations before Lemma 6.
7. One can see that $M(n,t) = N_2(n,t)$. Using operations T_{ji}, relation (8), and the method of the proof of Theorem 4, prove the validity of (5) (Lecture 2).
   For $t > 1$ prove that the set on which $N_2(n,t)$ is achieved is unique up to changing $0 \leftrightarrow 1$ symbols in components and permutations of components.
8. Prove equality (4) (Lecture 2) directly. Consider *mod q* componentwise summation in $\mathcal{H}_q^n$ and prove that if a^n is in intersection family $\mathcal{A}$, then $a^n + b^n \notin \mathcal{A}$ for all $b^n = (b, \ldots, b)$, $b \in \{1, \ldots, q-1\}$.
9. Prove that for $t > 1$ or $t = 1$, $q > 2$ up to permutations of the components and elements of the alphabet in the components there is only one optimal configuration in Theorem 5, unless $t > 1$, $t + 2(t-1)/(q-1) \leq n$, and $(t-1)/(q-2)$ is an integer in which case we have two optimal configurations $\mathcal{K}\left(\frac{t-1}{q-2}\right)$ and

   $\mathcal{K}\left(\frac{t-1}{q-2} - 1\right)$.
   In addition to the optimal configuration in Theorem 9 we have in the case $d = n - 2$, $2|d$ also the optimal configuration $\mathcal{G}_{d/2-1}(n, n-d)$. Prove that up to permutations of the components and elements of the alphabet in the components these configurations are unique.
10. Prove Propositions 2, 3, 4, and 5.
11. Prove relation (18) (Lecture 2): if $E_1, E_2 \in \mathcal{M}_0(\mathcal{A})$ and $|E_1 \cap E_2| = t$, then $|E_1| + |E_2| = \ell + t$.
12. Prove Lemma 16.
13. Using Theorem 12 prove inequalities (22) and (23).
14. Give a construction of an l-cross-intersecting pair (A,B), $A, B \subset \{0,1\}^n$ with

$$|A||B| \geq \begin{cases} \binom{2l}{l} 2^{n-2l} & \text{if } n \geq 2l, \\ \binom{n}{l} & \text{if } n < 2l. \end{cases}$$

15. Actually, it was originally conjectured in [A87] that

$$\max_{\substack{A,B \subset \{0,1\}^n \\ l\text{-cross-inters.}}} |A||B| = \max_{l \leq x \leq n} 2^{n-x} \binom{x}{l}.$$

Show that this bound equals the bound in exercise 14, which was conjectured in [ACZ89].

16. For $B \subset \{0,1\}^n$, $\mathcal{X}^t = \{0,1\}$, and $X = (x_1 < x_2 < \ldots, < x_{|X|}) \subset [n]$, we say that B has parity on X if for all $b^n \in B$, $\|X\|$-tuples $b^{|X|} = (b_{x_1}, \ldots, b_{x_{|X|}})$ have number of units of the same parity. Prove that [A87]

$$\sum_{X \subset [n], \, B \text{ has parity on } X} 2^{|X|} |B| \leq (2^n + 1) 2^{n-1}.$$

This bound achieves equality, for instance, on the set B of all n-tuples with even number of ones.

Research Problems

1. **Conjecture** The lower bound in Exercise 14 is the maximum value for $|A||B|$. This was proved in [AL06] for large n.
2. **Conjecture** Theorem 11 holds also for values of q different from $2, 4, 5$.
3. **Conjecture** If $k \leq \frac{n+t}{2}$, then the following relation is valid

$$\max_{A \in \mathcal{I}(n,t,\leq k)} |A| = \max \left\{ |\mathcal{F}(i, \leq k)| : i = 0, \ldots, \left\lfloor \frac{n-t}{2} \right\rfloor \right\}.$$

Chapter III
Covering, Packing, and List Codes

In this chapter we investigate two basic notions from Coding Theory: covering and packing. Usually if one considers a finite metric space, the main problem in Coding Theory (see, e.g., [L98]) is to find the maximal number of points in this space such that the balls of a given radius with centers in those points do not intersect. This is the packing problem. The dual problem in Coding Theory is the covering problem: find the minimal cardinality of a subset of the metric space such that the union of the balls with centers in the points of that set is the whole space. Usually the covering problem is much simpler than the packing problem (we see this in Lecture 8) and only asymptotic bounds on the cardinality of a packing are known in the general case.

The situation is quite different if one considers the packing and covering problems for k-uniform hypergraphs, when k is small (fixed). We see in the next lecture that in that case it is possible to find the exact asymptotical growth of the cardinalities of optimal coverings and packings (and they coincide).

Lecture 7 Covering and Packing of Hypergraphs

Results of this lecture will find application in Lecture 11, where we consider higher level extremal problems. We start with one useful theorem, which is based on the probabilistic method. For a k-uniform hypergraph $\mathcal{H} = (\mathcal{V}, \mathcal{E})$, $|\mathcal{V}| = n$, define the packing number $p(\mathcal{H})$ as the maximal number of pairwise disjoint edges and the covering number $c(\mathcal{H})$ as the minimal number of edges such that their union is the whole set $\mathcal{V}$. Evidently

$$p(\mathcal{H}) \leq \frac{n}{k} \leq c(\mathcal{H}). \tag{1}$$

For a set of vertices $A \subset \mathcal{V}$, denote by $\mathcal{H}(A)$ the set of edges $E \in \mathcal{E}$ in hypergraph $\mathcal{H} = (\mathcal{V}, \mathcal{E})$ such that $A \subset E$. Denote also

$$\deg_{\mathcal{H}}(A) = |\mathcal{H}(A)|.$$

In other words, $\mathcal{H}(A)$ consists of the edges containing all the elements of A and the number of those edges is the degree $\deg_{\mathcal{H}}(A)$ of A in $\mathcal{H}$. Instead of $\mathcal{H}(\{x\})$, $\mathcal{H}(\{x,y\})$, $\deg_{\mathcal{H}}(\{x\})$, and $\deg_{\mathcal{H}}(\{x,y\})$ we write $\mathcal{H}(x)$, $\mathcal{H}(x,y)$, $\deg_{\mathcal{H}}(x)$, and $\deg_{\mathcal{H}}(x,y)$.

Sometimes we allow ourselves to identify hypergraphs and sets of edges. Namely, any family $\mathcal{F}$ of sets can be viewed as a hypergraph with vertex set $V = \bigcup_{F \in \mathcal{F}} F$ (shortly denoted $\cup F$) and edge set $\mathcal{F}$. For the resulting hypergraph we use the same letter $\mathcal{F}$. Thus, for example, $\mathcal{F}(x)$ is the set of all $F \in \mathcal{F}$ containing x. Sometimes we write a union of edges thinking of it as a set of edges included in this union. We think that these conventions will not make difficulties and will be clear in every particular case. We frequently use inequalities (1), (2) (Chapter I) without directly referring to them. For example, this is the case when we come from the estimation of the average $\mathbb{E}(|\mathcal{R}|)$ in (6) to the estimation (7) of the random variable itself.

Theorem 14 (Frankl and Rödl 1985) *Suppose that $\varepsilon > 0$ is arbitrary, $\mathcal{H}$ is a k-uniform hypergraph, $a > 3$. If there exists $\delta = \delta(\varepsilon) > 0$ such that for some d one has $|\deg_{\mathcal{H}}(x) - d| < \delta d$ for all $x \in V$ and $\deg_{\mathcal{H}}(x,y) < d/(\ln n)^a$ holds for all distinct $x,y \in V$, then for all $n > n_0(\delta)$,*

$$c(\mathcal{H}) \le (1+\varepsilon)\frac{n}{k}. \tag{2}$$

It is easy to see that (1) and (2) imply

$$p(\mathcal{H}) \ge (1-\varepsilon k)\frac{n}{k}. \tag{3}$$

This theorem shows that the asymptotic behavior of $c(\mathcal{H})$ and $p(\mathcal{H})$ when k is fixed is equivalent to n/k. We use the convenient notation $X \overset{\rho}{\sim} Y$ iff $|X - Y| \le \rho Y$.

To prove the theorem we need the following:

Lemma 19 *Suppose that $\varepsilon \in (0,1)$ is a small number and $\mathcal{F}$ is a k-uniform hypergraph on vertices V, $|V| = n$, and there exists a special choice of $\rho = \rho(\varepsilon) \overset{\varepsilon \to 0}{\to} 0$ such that the following two properties are satisfied for all $x,y \in V$:*

(i) $|\mathcal{F}(x)| \overset{\rho}{\sim} d$,
(ii) $|\mathcal{F}(x,y)| < d/(\ln n)^a$, $a > 3$.

Then for $n > n_0(\varepsilon,\rho)$ there exists a subhypergraph $\mathcal{R} \subset \mathcal{F}$ such that

$$|\mathcal{R}| \overset{2\rho}{\sim} \varepsilon\frac{n}{k}, \tag{4}$$

$$\left|\cup\mathcal{R}\right| \overset{4\rho}{\sim} n(1 - e^{-\varepsilon}) \tag{5}$$

and $\tilde{\mathcal{F}} = \{F \in \mathcal{F} : F \cap R = \emptyset \text{ for all } R \in \mathcal{R}\}$ with vertex set $\tilde{V} = V - \cup\mathcal{R}$ satisfying (i), (ii) for $\tilde{\rho} = 6\rho$, $\tilde{d} = e^{-(d_0-1)\varepsilon}d$, $\tilde{a} > a - o(1)$. Here $\cup\mathcal{R}$ is the set of the vertices each contained in some edge from $\mathcal{R}$.

This lemma shows that we can divide the whole hypergraph into two parts: hypergraph $\mathcal{R}$, which has a proper covering of $|\cup\mathcal{R}|$ vertices, and hypergraph $\tilde{\mathcal{F}}$, with noncovered vertices, satisfying conditions (i) and (ii). To prove the theorem we iterate this procedure t times, with t large enough, and come to the noncovered hypergraph $\tilde{\mathcal{F}}$ with a small number of vertices and we cover each vertex from this hypergraph by one edge. After proving the lemma we show more precisely how this technique works.

Proof. Note that $|\mathcal{F}|k \overset{\rho}{\sim} dn$. Let $\mathcal{R}$ be a random hypergraph, which is obtained by choosing each edge of $\mathcal{F}$ independently with probability ε/d. The expected number of edges of $\mathcal{R}$ is

$$\mathbb{E}(|\mathcal{R}|) = \varepsilon|\mathcal{F}|/d \overset{\rho}{\sim} \varepsilon n/k \tag{6}$$

and with exponentially high probability

$$|\mathcal{R}| \overset{2\rho}{\sim} \varepsilon n/k. \tag{7}$$

Next we show that with high probability (5) holds. For a given $x \in V$ we have $P(x \in \cup\mathcal{R}) = 1 - (1-\varepsilon/d)^{|\mathcal{F}(x)|} \overset{\rho}{\sim} 1 - e^{-\varepsilon}$. Hence $\mathbb{E}(|\cup\mathcal{R}|) \overset{\rho}{\sim} n(1-e^{-\varepsilon})$.

Let Z_i be the indicator random variable

$$Z_i = \begin{cases} 1, & \text{if } x_i \in \cup\mathcal{R}, \\ 0, & \text{otherwise,} \end{cases}$$

and $V = \{x_1,\ldots,x_n\}$. We have $|\cup\mathcal{R}| = \sum Z_i$ and we break up this sum into $t \simeq n/(\ln n)^{a/3}$ parts, by partitioning $\{1,2,\ldots,n\}$ into $I_1 \cup \ldots \cup I_t$ with $||I_j| - (\ln n)^{a/3}| \le 1$.

First we estimate $\sum_{i\in I_j} Z_i$. Let I be one of I_j. Note that for distinct $i, i' \in I$,

$$|\mathcal{F}(x_i) \cap \mathcal{F}(x_{i'})| = |\mathcal{F}(x_i, x_{i'})| < d/(\ln n)^a.$$

Thus

$$\sum_{i,i'\in I\ i>i'} |\mathcal{F}(x_i) \cap \mathcal{F}(x_{i'})| < \binom{|I|}{2} \frac{d}{(\ln n)^a} < \frac{d}{2(\ln n)^{a/3}}. \tag{8}$$

Now set $\mathcal{Y} = \bigcup_{i\in I} \mathcal{F}(x_i)$ as the set of edges and consider the partition $\mathcal{Y} = \mathcal{Y}_1 \cup \ldots \cup \mathcal{Y}_k$, where $\mathcal{Y}_m$ consists of those elements that appear m times in the union. From (8) it follows

$$\sum_{\ell=2}^{k} \ell|\mathcal{Y}_\ell| < \frac{d}{(\ln n)^{a/3}} \tag{9}$$

and hence

$$|\mathcal{F}(x_i) \cap \mathcal{Y}_1| \overset{2\rho}{\sim} d \tag{10}$$

for all $i \in I$.

Define the indicator random variable Z_i' by

$$Z_i' = \begin{cases} 1, & \text{if } x_i \in R \in \mathcal{R} \text{ for some } R \in \mathcal{Y}_1, \\ 0, & \text{otherwise.} \end{cases}$$

Next we have $\sum_{i \in I}(Z_i - Z_i') \leq \sum_{\ell=2}^{k} \ell |\mathcal{Y}_\ell \cap \mathcal{R}|$ and $Z_i' \leq Z_i$. The random variables Z_i', $i \in I$ are independent. Therefore, $\sum Z_i' \overset{3\rho}{\sim} |I|(1 - e^{-\varepsilon})$ holds with high probability (exponentially with $|I|$, i.e., with probability greater than $1 - n^{-b}$ for some $b > 1$ and large n). From this it follows that $\sum_{i=1}^{n} Z_i' \overset{3\rho}{\sim} n(1 - e^{-\varepsilon})$ holds with probability greater than $1 - n^{-(b-1)}$. If $\bar{\mathcal{Y}}$ is the union of all $\mathcal{Y}_2 \cup \ldots \cup \mathcal{Y}_k$ when I takes values $I_1, \ldots, I_t$, then using (9) we have

$$|\bar{\mathcal{Y}}| < n \frac{d}{(\ln n)^{2a/3}}.$$

Hence from (2) in Chapter I follows that

$$|\bar{\mathcal{Y}} \cap \mathcal{R}| < \frac{n}{(\ln n)^{a/3}}$$

with probability greater than $1 - n^{-b}$, that is, with this probability $\sum(Z_i - Z_i') < kn/(\ln n)^{a/3}$. Thus we get

$$|\cup \mathcal{R}| \overset{4\rho}{\sim} n(1 - e^{-\varepsilon})$$

with probability greater than $1 - 2n^{-(b-1)}$.

Next we prove the lemma for the hypergraph $\tilde{\mathcal{F}}$ on vertex set $\tilde{\mathcal{V}}$. Suppose that $F \in \mathcal{F}(x)$. Then for given $x \in \tilde{\mathcal{V}}$, $F \in \tilde{\mathcal{F}}(x)$ iff none of the edges from $\mathcal{F}(F - \{x\})$ was chosen.

We have

$$(k-1)d(1-\rho) - \binom{k-1}{2}\frac{d}{(\ln n)^a} \leq |\mathcal{F}(F - \{x\})| < (k-1)d(1+\rho).$$

Thus for n large enough

$$P(F \in \tilde{\mathcal{F}}(x)) = \left(1 - \frac{\varepsilon}{d}\right)^{|\mathcal{F}(F - \{x\})|} \overset{\rho}{\sim} e^{-(k-1)\varepsilon}$$

and

$$\mathbb{E}(|\tilde{\mathcal{F}}(x)|) \overset{2\rho}{\sim} e^{-(k-1)\varepsilon}d.$$

We need to prove that with high probability $|\tilde{\mathcal{F}}(x)| \overset{6\rho}{\sim} e^{-(k-1)\varepsilon}d$. To do this, we consider a partition of $\mathcal{F}(x)$ into stars. A star in $\mathcal{F}(x)$ is a sub-family $\{F_1, \ldots, F_t\}$ satisfying $F_i \cap F_{i'} = \{x\}$ with $i \neq i'$. Let $t = [(\ln n)^{a/3}]$ and suppose that we have already fixed a partition of stars $\mathcal{F}^1, \ldots, \mathcal{F}^s$ in $\mathcal{F}(x)$. Continuing the process we finally want to find a star $\mathcal{F}^{s+1}$ in $\mathcal{J} = \mathcal{F}(x) - \cup_{i=1}^{s}\mathcal{F}^i$ with $|\mathcal{F}^{s+1}| = t$. Suppose that this is impossible, that is, we find a set $\{F_1, \ldots, F_\ell\}$ with $l < t$ forming a star and for all other $F \in \mathcal{J}$ we have that $\{F_1, \dots, F_\ell, F\}$ is not a star. This means that the $(k-1)\ell$-element set $G = (F_1 \cup \dots \cup F_\ell) - \{x\}$ meets every $F \in \mathcal{J}$. Thus $\mathcal{J} \subset \bigcup_{g \in G} \mathcal{F}(x,g)$ and hence

$$|\mathcal{J}| \leq (k-1)\ell \frac{d}{(\ln n)^a} < k \frac{d}{(\ln n)^{2a/3}}.$$

Now we consider the indicator random variable

$$Y_F = \begin{cases} 1, & \text{if } F \in \tilde{\mathcal{F}} \\ 0, & \text{otherwise.} \end{cases}$$

Then

$$\sum_{F \in (\mathcal{F}(x) - \mathcal{J})} Y_F \leq |\tilde{\mathcal{F}}(x)| \leq \sum_{F \in (\mathcal{F}(x) - \mathcal{J})} Y_F + k \frac{d}{(\ln n)^{2a/3}}. \tag{11}$$

The variables Y_F are not independent, but their dependence for the sets from $\mathcal{F}^i$ is caused only by sets $H \in \mathcal{F}$, $x \notin H$ such that there exist $y, y' \in H$ with $y \in F$, $y' \in F'$ for some $F, F' \in \mathcal{F}^i, F \neq F'$.

Hence for a given star $\mathcal{F}^i$ the number of such "bad" H counted with multiplicity is upperbounded by the value

$$\binom{t}{2}(k-1)^2 \frac{d}{(\ln n)^a} \leq (k-1)^2 \frac{d}{2(\ln n)^{a/3}}.$$

Denote by $\mathcal{H}^i$ the union of such sets for given $\mathcal{F}^i$ and $\mathcal{H} = \bigcup_i \mathcal{H}^i$. Let also $\mu(H)$ be the multiplicity of H (showing how many times it is "bad" for different y, y'). Then we have

$$\sum_{H \in \mathcal{H}} \mu(H) \leq (k-1)^2 \frac{d^2}{(\ln n)^{2a/3}}. \tag{12}$$

We state that

$$\mu(H) \leq \binom{k}{2} \frac{d}{(\ln n)^a}. \tag{13}$$

Let $\mathcal{N}(H) = \{F \in \mathcal{F}(x) : (F - \{x\}) \cap H \neq \emptyset\}$. Then $|\mathcal{N}(H)| < kd/(\ln n)^a$. But for $F, F' \in (\mathcal{N}(H) \cap \mathcal{F}^i)$ we have $((F - \{x\}) \cap H) \cap ((F' - \{x\}) \cap H) = \emptyset$, since $F \cap F' = \{x\}$. This means that each $F \in \mathcal{N}(H)$ adds at most $k-1$ to the multiplicity of H. This yields (13).

Next we define new random variables. For $F \in \mathcal{F}^i$ we set

$$Y_F^* = \begin{cases} 1, & \text{if } F \cap R = \emptyset \text{ for all } R \in (\mathcal{R} - \mathcal{H}^i), \\ 0, & \text{otherwise.} \end{cases}$$

We have $Y_F^* \geq Y_F$ and $\{Y_F^* : F \in \mathcal{F}^i\}$ are independent random variables. Also we have

$$\sum_{F \in \mathcal{F}(x)} (Y_F^* - Y_F) \leq \sum_{H \in \mathcal{H} \cap \mathcal{R}} \mu(H). \tag{14}$$

Define $\mathcal{H}_i = \{H \in \mathcal{H} : \mu(H) = i\}$ and $\ell = \max_{H \in \mathcal{H}} \mu(H)$. It holds

$$\ell \le \binom{k}{2} \frac{d}{(\ln n)^a}.$$

Next we estimate the RHS of (14). We get

$$\sum_{H \in \mathcal{H} \cap \mathcal{R}} \mu(H) = \sum_{i=1}^{\ell} i |\mathcal{H}_i \cap \mathcal{R}| = \sum_{i=1}^{\ell} |(\mathcal{H}_i \cup \ldots \cup \mathcal{H}_\ell) \cap \mathcal{R}|. \tag{15}$$

From (12) it follows

$$|\mathcal{H}_i \cup \ldots \cup \mathcal{H}_\ell| \le \frac{(k-1)^2 d^2}{i(\ln n)^{2a/3}}.$$

Since $R \in \mathcal{R}$ are chosen independently, using estimation (2) (Chapter I) we obtain that

$$|(\mathcal{H}_i \cup \ldots \cup \mathcal{H}_\ell) \cap \mathcal{R}| < \frac{e(\ln n)^{a/3}(k-1)^2 d\varepsilon}{i(\ln n)^{2a/3}} < \frac{e(k-1)^2 d}{i(\ln n)^{a/3}} \tag{16}$$

holds with probability greater than $1 - 2(\ln n)^{-(\ln n)^{a/3}} > 1 - 1/(n^2 \ell)$ for large n.

Thus (16) is true simultaneously for all $i = 1, 2, \ldots, \ell < \binom{k}{2} d/(\ln n)^a$ with probability at least $1 - n^{-2}$, $(n > n_0)$. Substituting (16) into (15) we obtain the inequality

$$\sum_{H \in \mathcal{H} \cap \mathcal{R}} \mu(H) < \sum_{i=1}^{\ell} \frac{1}{i} \frac{e(k-1)^2 d}{(\ln n)^{a/3}} < \frac{3(k-1)^2 d}{(\ln n)^{(a-3)/3}},$$

which holds with probability greater than $1 - n^{-2}$.

Next we have

$$P(Y_F^* = 1) \overset{3\rho}{\sim} e^{-(k-1)\varepsilon}$$

and

$$\sum_{F \in \mathcal{F}^i} Y_F^* \overset{4\rho}{\sim} e^{-(k-1)\varepsilon} |\mathcal{F}^i|$$

holds with probability greater than $1 - n^{-2}$. Taking into account (11) and (14) we obtain that

$$|\tilde{\mathcal{F}}(x)| \overset{6\rho}{\sim} e^{-(k-1)\varepsilon} d$$

holds with probability greater than $1 - 2n^{-2}$. Thus with probability greater than $1 - 2n^{-1}$ the same holds for all $x \in \tilde{\mathcal{V}}$. Therefore, (i) holds with $\tilde{d} = e^{-(k-1)\varepsilon} d$, $\tilde{\rho} = 6\rho$. Since $e^{-(k-1)\varepsilon} = O(1)$, we have that (ii) is satisfied with $\tilde{a} = a - o(1)$ when $n \to \infty$. The proof of Lemma 19 is completed. $\qquad\square$

To prove Theorem 14, consider a k-uniform hypergraph $\mathcal{H}$, which satisfies the assumptions of the theorem. Then applying Lemma 19 to hypergraph $\mathcal{F} = \mathcal{H}$ we obtain $\mathcal{R} = \mathcal{R}_1$ and $\tilde{\mathcal{F}} = \mathcal{F}_1$ with properties from the lemma. We apply the lemma now to $\mathcal{F}_1$ to obtain $\mathcal{R}_2$ and $\mathcal{F}_2$. Repeating this procedure t times, we obtain a sequence $\mathcal{R}_1, \mathcal{R}_2, \ldots, \mathcal{R}_t$ and $\mathcal{F}_1, \mathcal{F}_2, \ldots, \mathcal{F}_t$ satisfying $|\mathcal{R}_{j+1}| \overset{\varepsilon/3}{\sim} \varepsilon n e^{-j\varepsilon}/k$. After t steps we cover all but $ne^{-\varepsilon t}$ vertices with $\sum_{j=0}^{t-1} \frac{\varepsilon n}{k} e^{-j\varepsilon}$ edges. For each of these uncovered points we pick one edge containing it. Hence for $t > t_0(\varepsilon)$

$$\sum_{j=0}^{t-1} \varepsilon \frac{n}{k} e^{-j\varepsilon} + ne^{-\varepsilon t} < (1+\varepsilon)\frac{n}{k}$$

edges cover all points. More precisely, we first fix a small $\varepsilon > 0$, then choose $t = t(\varepsilon)$ and finally ρ. This proves the theorem. ∎

Now we are going to use Theorem 14 to prove one corollary, which we use in Lecture 11. First we give some definitions (caution: we change the meaning of k). Let $\mathcal{H}$ be a family of t-sets over k elements, that is, $\mathcal{H}$ is a t-uniform hypergraph on k vertices. Suppose that $\mathcal{F} \subset \binom{N}{k}$ is a family of k-sets on N, $|N| = n$ and for every $F \in \mathcal{F}$ there exists a copy $\mathcal{H}_F$ of $\mathcal{H}$ on F, that is, $\mathcal{H}_F \subset \binom{F}{t}$ and $\mathcal{H}_F \approx \mathcal{H}$. If every t-set $T \in \mathcal{H}_F$ is covered only by F (i.e., $T \not\subset F' \in \mathcal{F}$, $F' \neq F$), then we call $\mathcal{F}$ a $(k, \mathcal{H})$-packing. It is evident that

$$|\mathcal{F}| \leq \frac{\binom{n}{t}}{|\mathcal{H}|}. \tag{17}$$

Next we define a $c - (k, \mathcal{H})$-packing, which imposes some more restrictions. For F with $|F| = k$ let $\mathcal{H}_F \subset \binom{F}{t}$, $c = \binom{k}{t} - |\mathcal{H}|$ and consider a partition

$$\binom{F}{t} = \mathcal{H}_F \cup \{T_1\} \cup \ldots \cup \{T_c\}. \tag{18}$$

The family $\mathcal{F} \subset \binom{N}{k}$ is called $c - (k, \mathcal{H})$-packing if $|F \cap F'| \leq t$ for every $F, F' \in \mathcal{F}$ and there exists a partition (coloring)

$$\chi : \binom{N}{t} = \mathcal{C}_0 \cup \mathcal{C}_1 \cup \ldots \cup \mathcal{C}_c \tag{19}$$

such that for each $F \in \mathcal{F}$ the partition

$$\binom{F}{t} = \left(\binom{F}{t} \cap \mathcal{C}_0\right) \cup \left(\binom{F}{t} \cap \mathcal{C}_1\right) \cup \ldots \cup \left(\binom{F}{t} \cap \mathcal{C}_c\right)$$

is isomorphic to the partition (18) and

$$\left|\left(\binom{F}{t} \cap \mathcal{C}_0\right) \cap \left(\binom{F'}{t} \cap \mathcal{C}_0\right)\right| < t.$$

The last condition means that the parts of F that are isomorphic to $\mathcal{H}$ intersect in less than t vertices.

If we denote

$$f(n,k,\mathcal{H}) = \max_{\mathcal{F} \text{ is a } (k,\mathcal{H})\text{-packing}} |\mathcal{F}|,$$

$$f_c(n,k,\mathcal{H}) = \max_{\mathcal{F} \text{ is a } c-(k,\mathcal{H})\text{-packing}} |\mathcal{F}|,$$

then $f(n,k,\mathcal{H}) \geq f_c(n,k,\mathcal{H})$. Hence any lower bound for $f_c(n,k,\mathcal{H})$ is a lower bound also for $f(n,k,\mathcal{H})$.

Corollary 2 (Frankl, Füredi 1987) *For every given k and $\mathcal{H}$ the following bound is valid:*

$$f_c(n,k,\mathcal{H}) > (1+o(1))\frac{\binom{n}{t}}{|\mathcal{H}|} \quad (n \to \infty). \tag{20}$$

This corollary is a consequence of Theorem 14. To apply this theorem we need

Lemma 20 *Let $k > t$ and $4k\ln n \leq d < n^{2/3}$. Then there exists a family of k-sets $\mathcal{S}$ of n-element set N such that*

(i)
$$|S \cap S'| \leq t$$

holds for all distinct $S, S' \in \mathcal{S}$,

(ii)
$$|\deg_{\mathcal{S}}(T) - d| < 2\sqrt{kd\ln n} \tag{21}$$

holds for every $T \in \binom{N}{t}$.

Here $\deg_{\mathcal{S}}(T) = |\{S \in \mathcal{S} : T \subset S\}|$.

We first prove this lemma and then return to the proof of Corollary 2.
Proof. For every $F \in \binom{N}{k}$ we define the random variables Y_F:

$$P(Y_F = 1) = \frac{d}{\binom{n-t}{k-t}}, P(Y_F = 0) = 1 - P(Y_F = 1). \tag{22}$$

Let $\mathcal{F} = \left\{F \in \binom{N}{k} : Y_F = 1\right\}$ be a random family.

Let $Y_T = \sum_{F \supset T} Y_F$, $T \in \binom{N}{t}$. Then $\mathbb{E}(Y_T) = d$ and relation (1) (Chapter I) implies for every given T

$$P\left(\left|Y_T - d\right| > (2kd\ln n)^{1/2}\right) < 2n^{-k}.$$

Hence, uniformly in $T \in \binom{N}{t}$, with probability greater than

$$1 - 2\binom{n}{t}n^{-k}, \tag{23}$$

the inequality

$$|Y_T - d| \leq (2kd\ln n)^{1/2} \tag{24}$$

holds. The same simple probabilistic argument shows that for some $U \in \binom{N}{t+1}$ $\deg_{\mathcal{F}}(U) > 3k$ the probability P' satisfies the relation

$$P' < \frac{1}{t+1}. \tag{25}$$

Indeed, it is easy to calculate that

$$P' \leq \binom{n}{t+1} \binom{\binom{n-t-1}{k-t-1}}{3k} \left(\frac{d}{\binom{n-t}{k-t}} \right)^{3k} < \frac{1}{t+1}.$$

Here we used the inequality $d < n^{2/3}$. A similar proof (by a counting argument, see Exercise 1) shows the validity of the following:

Lemma 21 *If* $s = \left[0.2k^{-2.5}(d\ln n)^{0.5} \right]$, *then the probability that there exists a* $T \in \binom{N}{t}$ *and* $2s$ *elements* $F_1, \ldots, F_{2s} \in \mathcal{F}$ *such that for* $i \leq s$, $T \subset F_i$ *and* $|F_i \cap F_{s+i}| > t$, $i = 1, \ldots, s$, *is less than* k^{-2} *(for sufficiently large* $n > n_0(k)$).

Now we are ready to prove Lemma 20. Choose $\mathcal{F}$ at random as before. Then the sum of probabilities (23), (25), and the last condition is less than 1. Hence there exists a family $\mathcal{F}$ without the prescribed configurations and

$$|\deg_{\mathcal{F}}(T) - d| < \sqrt{2kd\ln n} \tag{26}$$

holds for all $T \in \binom{N}{t}$.

We call a set $F \in \mathcal{F}$ bad if there exists an $F' \in \mathcal{F}$, $F \neq F'$ with $|F \cap F'| > t$. We denote the family of bad sets by $\mathcal{B}$ and $\mathcal{S} = \mathcal{F} - \mathcal{B}$. This family $\mathcal{S}$ fulfils the constraints of Lemma 20. Indeed, (i) is satisfied and (ii) follows from the inequality

$$\deg_{\mathcal{B}}(T) < (2 - \sqrt{2})\sqrt{kd\ln n}, \ T \in \binom{N}{t}.$$

Suppose the contrary is true, that is, for some $T \in \binom{N}{t}$ we have $\deg_{\mathcal{B}}(T) > 3k^3 s$. Then by (25) there exist $B_1, \ldots, B_{sk} \in \mathcal{B}$ such that $T \subset B_i$ and $B_i \cap B_j = T$, $1 \leq i < j \leq sk$. And for each i there exist distinct $B_i' \in \mathcal{B}$ such that $|B_i' \cap B_i| > t$. Then we can choose a subsequence $B_{i_1}, \ldots, B_{i_s}$ and a sequence $B_{i_1}', \ldots, B_{i_s}'$ which contradicts Lemma 21. Lemma 20 is proved. □

Proof of Corollary 2. We suppose that $t > 1$, otherwise the statement of the corollary is trivial. To construct a colored packing we begin with the family $\mathcal{S} \subset \binom{N}{k}$, $|N| = n$, given by Lemma 20 with $d = \sqrt{n}$. We suppose that $h > 0$ is some sufficiently small number, depending only on k, and $n > n_0(k)$ is sufficiently large. Let $p = n^{-h}$ and $\left\{ Z_T : T \in \binom{N}{t} \right\}$ be a set of i.i.d random variables such that

$$P(Z_T = i) = p, \ i = 1, \ldots, c \text{ and } P(Z_T = 0) = 1 - cp.$$

Recall that $c = \binom{k}{t} - |\mathcal{H}|$. Let $C_i = \{T : Z_T = i\}$, then for $i \geq 1$, $\mathbb{E}(|C_i|) = p\binom{n}{t}$ and by (1) in Chapter I

$$P\left(\left||C_i| - p\binom{n}{t}\right| > n^{t-h/2}\right) < 2e^{-n^t/2}.$$

We call a set $S \in \mathcal{S}$ good if the restriction of coloring to $\binom{S}{t}$ is isomorphic to χ. Denote the family of good sets by $\mathcal{W}$. If g is the number of non-isomorphic imbeddings of χ into $\binom{S}{t}$, then

$$P(S \in \mathcal{W}) = gp^c(1-cp)^{\binom{k}{t}-c}.$$

Define the $|\mathcal{H}|$-uniform hypergraph $\mathcal{A}$ with vertex set C_0,

$$\mathcal{A} = \left\{\binom{S}{t} \cup C_0 : S \in \mathcal{W}\right\}.$$

Denote $d^* = gp^c(1 - pc)^{\binom{k}{t}-c-1}|\mathcal{H}|d/\binom{k}{t}$. By the choices of p and d we have $d^* > n^{0.4}$.

Let us show that if $T \in C_0$, then

$$P\left(|\deg_{\mathcal{A}}(T) - d^*| > n^{0.3}\right) < e^{-n^h}. \tag{27}$$

Consider the sets $S_i \in \mathcal{S}$ with $T \subset S_i$. As $T \in C_0$, show that

$$P(S_i \in \mathcal{W}) = d^*/d.$$

Moreover, events $S_i \in \mathcal{W}$ for different i are independent, hence by (1) (Chapter I) we have

$$P(|\deg_{\mathcal{A}}(T) - \mu| > n^h \sqrt{\mu}) < 2e^{-n^{2h}} < e^{-n^h},$$

where $\mu = \frac{d^*}{d}\deg_{\mathcal{S}}(T)$. This and (21) imply (27).

Thus there exists a choice of $\{C_0, \ldots, C_c\}$ for which

$$|C_i| < p\binom{n}{t} + n^{t-h/2} < 2n^{t-h/2}$$

holds for all $i > 0$ and for every $T \in C_0$

$$|\deg_{\mathcal{A}}(T) - d^*| < n^{0.3}.$$

We have $d^* > n^{0.4}$ and for every $T_1 \neq T_2$, $T_1, T_2 \in \binom{N}{t}$ we have $\deg_{\mathcal{A}}(T_1, T_2) \leq 1 < d^*/(\ln n)^4$. Hence we can apply Theorem 14 to $\mathcal{A}$. This shows that whenever $n \to \infty$

$$p(\mathcal{A}) \geq (1 + o(1)) \frac{|\mathcal{C}_0|}{|\mathcal{H}|} = (1 + o(1)) \frac{\binom{n}{t}}{|\mathcal{H}|}.$$

And this matching $\mathcal{M}$ in $\mathcal{A}$ gives a colored $\mathcal{H}$-cover. Corollary 2 is proved. $\qquad\square$

This corollary in turn implies another important corollary (Exercise 2).

Corollary 3 (Rödl 1985) *If* $\mathcal{H} = \binom{S}{t}$, *where* $|S| = k$, *then there exist an* $\mathcal{H}$-*cover and an* $\mathcal{H}$-*packing such that*

$$c(\mathcal{H}) = p(\mathcal{H})(1 + o(1)) = (1 + o(1)) \frac{\binom{n}{t}}{\binom{k}{t}}, \ n \to \infty.$$

In words, this corollary states that if we consider a maximal packing of t-sets by k-sets, that is, each t-set belongs to not more than one k-set, and a minimal covering of t-sets by k-sets, that is, each t-set belongs to at least one k-set, then the cardinalities of these packing and covering by k-sets have the same asymptotic growth $\frac{\binom{n}{t}}{\binom{k}{t}}$ as $n \to \infty$ and k, t are fixed. This fact was conjectured in [EH63].

Lecture 8 Covering of Products of Graphs and Hypergraphs

In this lecture we consider another problem connected with the problem of finding optimal coverings. Here we deal with products of (hyper)graphs and introduce the notion of a minimal partition of a product. We give a formula for the number of elements in a minimal partition of a product of graphs equal to the product of the numbers of elements in minimal partitions of the components. Next we note that the same problem for hypergraphs is more difficult and needs additional restrictions on the components. At the end we find the asymptotic growth of the covering number of the product of hypergraphs (in terms of the structure of components). The same problem for packing numbers is very difficult and far from being solved.

Now we introduce the notion of a partition, which is a packing and covering at the same time. Consider a finite hypergraph $\mathcal{H} = (\mathcal{V}, \mathcal{E})$. For the Cartesian products $\mathcal{V}^n = \prod_1^n \mathcal{V}$ and $\mathcal{E}^n = \prod_1^n \mathcal{E}$, let $\pi(\mathcal{H}^n)$ denote the minimal size of a partition of $\mathcal{V}^n$ into sets that are elements of $\mathcal{E}^n$, if such a partition exists. Otherwise, $\pi(\mathcal{H}^n)$ is not defined. Also consider the packing number $p(\mathcal{H}^n)$, that is, the maximal size of a system of disjoint sets from $\mathcal{E}^n$, and the covering number $c(\mathcal{H}^n)$, which is the minimal number of sets from $\mathcal{E}^n$ covering $\mathcal{V}^n$. Obviously $c(\mathcal{H}^n) \leq \pi(\mathcal{H}^n) \leq p(\mathcal{H}^n)$ if all values are well defined. We start with the case when $\mathcal{C} = (\mathcal{V}, \mathcal{E})$ is a complete graph, including all loops. We introduce the map $\sigma : \mathcal{E}^n \to \{0, 1\}^n$ by setting for $E^n = E_1 \times \cdots \times E_n$

$$s^n = \sigma(E^n) = (\log |E_1|, \dots, \log |E_n|).$$

Note that $|E^n| = 2^{w(s^n)}$, where $w(\cdot)$ is the Hamming weight. Instead of partitions we consider a packing $\mathcal{P}$ of C^n. We set

$$\mathcal{P}(i) = \{E^n \in \mathcal{P} : w(\sigma(E^n)) = i\}, \ P(i) = |\mathcal{P}(i)|$$

and associate with $\mathcal{P}$ the set of lower shadows $\mathcal{L}$ defined by

$$\mathcal{L} = \{E^n : E^n \subset F^n \text{ for some } F^n \in \mathcal{P}\},$$

$$\mathcal{L}(i) = \{E^n \in \mathcal{L} : w(\sigma(E^n)) = i\}, \ Q(i) = |\mathcal{L}(i)|.$$

Proposition 11 *For a packing $\mathcal{P}$ of C^n*

$$\sum_{i=k}^{n} 2^{i-k} \binom{i}{k} P(i) = Q(k), \tag{1}$$

$$|\mathcal{P}| = \sum_{i=0}^{n} P(i) = \sum_{k=0}^{n} (-1)^k Q(k), \tag{2}$$

$$P(0) = \sum_{k=0}^{n} (-1)^k 2^k Q(k). \tag{3}$$

If in addition $\mathcal{P}$ is a partition and $S = |\mathcal{V}|$ is odd, then

$$\sum_{k=0}^{n} (-1)^k 2^k Q(k) - 1 \geq 0. \tag{4}$$

This is proposed to be proved in Exercise 3. We now consider the hypergraph C^n with vertex set $\mathcal{V}^n = \prod_{t=1}^{n} \mathcal{V}_t$ and edge set $\mathcal{E}^n = \prod_{t=1}^{n} \mathcal{E}_t$, where $\mathcal{V}_t$ are finite sets of not necessarily equal cardinalities S_t. The factors $\mathcal{E}_t$ are such that $(\mathcal{V}_t, \mathcal{E}_t)$ is a complete graph with all loops included. We shall write, with positive integers α_t,

$$S_t = |\mathcal{V}_t| = 2\alpha_t + \varepsilon_t, \ \varepsilon_t \in \{0, 1\}.$$

An inspection shows that the sizes of factors do not affect the proofs of equalities (1), (2) and since $P(0) \geq 0$ if $\varepsilon_t = 1$, $t = 1, \ldots, n$, we have a generalization of (3):

$$\sum_{k=0}^{n} (-1)^k 2^k Q(k) - \prod_{k=1}^{n} \varepsilon_k \geq 0. \tag{5}$$

We also have the following inequality:

Lemma 22 *For a partition $\mathcal{P}$ of C^n*

$$\binom{n}{m} \prod_{i=1}^{n} S_i + \sum_{k=1}^{m} (-1)^k \binom{n-k}{m-k} 2^k Q(k) - \sum_{I:|I|=m} \prod_{i \in I} \varepsilon_i \prod_{j \in I^c} S_j \geq 0. \tag{6}$$

Proof. We shall use (5) by applying it to classes of subhypergraphs, which we now define. For any $I \subset \{1, \ldots, n\}$ and any specification $(v_j)_{j \in I^c}$ with $v_j \in \mathcal{V}_j$, we set

$$C^n(I,(v_j)_{j\in I^c}) = \left(\prod_{i=1}^n \mathcal{V}_i, \prod_{i=1}^n \mathcal{F}_i\right) = (\mathcal{V}^n, \mathcal{F}^n). \tag{7}$$

Here

$$\mathcal{V}_i = \begin{cases} \mathcal{V}_i, & i\in I, \\ \{v_i\}, & i\in I^c, \end{cases}$$

$$\mathcal{F}_i = \begin{cases} \mathcal{E}-i, & i\in I, \\ \{v_i\}, & i\in I^c. \end{cases}$$

Clearly, for a partition $\mathcal{P}$ of C^n and the shadow $\mathcal{L}$ of $\mathcal{P}$, the set $\mathcal{L}(I,(v_j)_{j\in I^c}) = \mathcal{L}\cap\mathcal{F}^n$ is a downset and the map

$$\chi : \mathcal{F}^n \to \prod_{i\in I}\mathcal{E}_i, \quad \chi\left(\prod_{i=1}^n E_i\right) = \prod_{i\in I} E_i$$

is a bijection. Write $\tilde{\mathcal{L}} = \chi(\mathcal{L}\cap\mathcal{F}^n)$ and let $\tilde{\mathcal{L}}(i,I,(v_j)_{j\in I^c})$ count the members of $\tilde{\mathcal{L}}$ of weight i. Since $\tilde{\mathcal{L}}$ is a downset in $\prod_{i\in I}\mathcal{E}_i$ and its maximal elements form a partition of $\prod_{i\in I}\mathcal{V}_i$, we know that $\tilde{\mathcal{L}}(0,I,(v_j)_{j\in I^c}) = \prod_{i\in I} S_i$. This fact and (5) yield

$$\prod_{i\in I} S_i + \sum_{k=1}^n (-1)^k 2^k \tilde{\mathcal{L}}(k,I,(v_j)_{j\in I^c}) - \prod_{i\in I}\mathcal{E}_i \geq 0. \tag{8}$$

Next, the map χ preserves inclusions and weights. Each $E^n \in \mathcal{L}$ with $w(E^n) = k$ is contained in exactly $\binom{n-k}{m-k}$ sets of the form $\mathcal{L}(I,(v_j)_{j\in I^c})$ and thus for the sets of weight k

$$\binom{n-k}{m-k}Q(k) = \sum_{(I,(v_j)_{j\in I^c}),\ |I|=m} \tilde{\mathcal{L}}(k,I,(v_j)_{j\in I^c}).$$

We have one inequality of the form (5) for each pair $(I,(v_j)_{j\in I^c})$. Summation of their LHS over these pairs gives (6). $\qquad\square$

Theorem 15 (Ahlswede and Cai 1993) *For a partition $\mathcal{P}$ of C^n*

$$|\mathcal{P}| \geq \prod_{i=1}^n \left\lceil \frac{S_i}{2} \right\rceil. \tag{9}$$

Proof. Summing up the expressions on the LHS in (7) for $m = 1,\ldots,n$ results in

$$0 \leq \sum_{m=1}^n \binom{n}{m}\prod_{i=1}^n S_i + \sum_{m=1}^n \sum_{k=1}^m \binom{n-k}{m-k}(-1)^k 2^k Q(k)$$

$$- \sum_{m=1}^n \sum_{I:|I|=m} \prod_{i\in I}\mathcal{E}_i \prod_{j\in I^c} S_j$$

$$= (2^n - 1) \prod_{i=1}^{n} S_i + \sum_{k=1}^{n} (-1)^k 2^k Q(k) \sum_{m=k}^{n} \binom{n-k}{m-k} - \sum_{I \neq \emptyset} \prod_{i \in I} \varepsilon_i \prod_{j \in I^c} S_i$$

$$= 2^n \left[\prod_{i=1}^{n} + \sum_{k=1}^{n} (-1)^k Q(k) \right] - \sum_{I} \prod_{i \in I} \varepsilon_i \prod_{j \in I^c} S_j$$

or

$$|\mathcal{P}| \geq 2^{-n} \sum_{I} \prod_{i \in I} \varepsilon_i \prod_{j \in I^c} S_j.$$

Now note that

$$\sum_{I} \prod_{i \in I} \varepsilon_i \prod_{j \in I^c} S_j = \prod_{j=1}^{n} (S_j + \varepsilon_j).$$

From this the theorem follows.   ∎

It is easy to see that the bound from the theorem is achievable and hence the partition number equals $\prod_{j=1}^{n} \lceil \frac{S_j}{2} \rceil$.

Theorem 15 can also be proved for arbitrary (not necessarily complete) graphs $\mathcal{G}_t = (\mathcal{V}_t, \mathcal{E}_t)$, $t = 1, \ldots, n$, with all loops included. Obviously, we have for the partition number

$$\pi(\mathcal{G}_t) = |\mathcal{V}_t| - \nu(\mathcal{G}_t), \tag{10}$$

where $\nu(\mathcal{G}_t)$ is the matching number of $\mathcal{G}_t$.

Lemma 23 For the product $\mathcal{G}^n = \prod_{t=1}^{n} \mathcal{G}_t$,

$$\pi(\mathcal{G}^n) = \prod_{t=1}^{n} \pi(\mathcal{G}_t).$$

Here only the inequality

$$\pi(\mathcal{G}^n) \geq \prod_{t=1}^{n} \pi(\mathcal{G}_t) \tag{11}$$

is nontrivial. We make use of a fact from matching theory [G63].

Theorem 16 (Gallai 1963) If a graph $\mathcal{G} = (\mathcal{V}, \mathcal{E})$ is connected and, for all $v \in \mathcal{V}$, $\nu(\mathcal{G} - v) = \nu(\mathcal{G})$, then for all $v \in \mathcal{V}$, $\mathcal{G} - v$ has a perfect matching.

Proof. Let $\mathcal{G} = (\mathcal{V}, \mathcal{E})$ be a graph and D be the set of those vertices of $\mathcal{G}$, which are missed by some maximum matching. Then by the hypothesis of the theorem $D = \mathcal{V}$. Now consider the equivalence relation $\sim$ defined as follows (see Exercise 4). For $u, v \in \mathcal{V}$ we write $u \sim v$ iff $u = v$ or no maximum matching misses both u and v.

Obviously, any two adjacent vertices are in relation $\sim$, since a matching missing both of them can be augmented by the edge connecting them and so cannot be maximum. By the connectivity of $\mathcal{G}$, any two points of $\mathcal{G}$ must be equivalent. But this means that no maximum matching can miss more than one point, or

$$\nu(\mathcal{G}) \geq \frac{1}{2}(|\mathcal{V}| - 1).$$

On the other hand

$$v(\mathcal{G}) = v(\mathcal{G} - v) \le \frac{1}{2}(|\mathcal{V}| - 1).$$

These two inequalities prove the theorem. ∎

Next, for every $t \in \{1,\dots,n\}$ we modify $\mathcal{G}_t$ as follows: remove any vertex $v \in \mathcal{V}_t$ with $v(\mathcal{G}_t - v) < v(\mathcal{G}_t)$ and iterate this until a graph $\overline{\mathcal{G}}_t$ with $v(\overline{\mathcal{G}}_t - v) = v(\overline{\mathcal{G}}_t)$ for all $v \in \overline{\mathcal{V}}_t$ is obtained.

Notice that (10) ensures that

$$\pi(\mathcal{G}_t) = \pi(\overline{\mathcal{G}}_t). \tag{12}$$

Denote the set of the connected components of $\overline{\mathcal{G}}_t$ by $\{\overline{\mathcal{G}}_t^j\}_{j \in J_t}$. Clearly,

$$\pi(\overline{\mathcal{G}}_t) = \sum_{j \in J_t} \pi(\overline{\mathcal{G}}_t^j). \tag{13}$$

Moreover, by Theorem 16, each component $\overline{\mathcal{G}}_t^j$ has a vertex set $\overline{\mathcal{V}}_t^j$ of odd size and

$$v(\overline{\mathcal{G}}_t^j) = \frac{|\overline{\mathcal{V}}_t^j| - 1}{2} \triangleq \alpha_t^j.$$

Thus

$$\pi(\overline{\mathcal{G}}_t) = \sum_j (\alpha_t^j + 1).$$

Now, for $\overline{\mathcal{G}}^n = \prod_{t=1}^n \overline{\mathcal{G}}_t$ we have

$$\pi(\mathcal{G}^n) \ge \pi(\overline{\mathcal{G}}^n), \tag{14}$$

because the modifications described above transform a partition of $\mathcal{G}^n$ into a partition of $\overline{\mathcal{G}}^n$ with not more parts.

Finally, by Theorem 15, we have for the product $\mathcal{C}^n$ of complete graphs with vertex sets $\overline{\mathcal{V}}_t^j$ that

$$\pi\left(\prod_{m=1}^n \overline{\mathcal{G}}_m^{j_m}\right) \ge \pi(\mathcal{C}^n) = \prod_{m=1}^n (\alpha_m^{j_m} + 1).$$

Therefore

$$\pi(\overline{\mathcal{G}}^n) = \sum_{j_1 \in J_1,\dots,j_n \in J_n} \pi\left(\prod_{m=1}^n \overline{\mathcal{G}}_m^{j_m}\right)$$

$$\ge \sum_{(j_1,\dots,j_n)} \prod_{m=1}^n (\alpha_m^{j_m} + 1) = \prod_{t=1}^n \sum_{j \in J_t} (\alpha_t^j + 1) = \prod_{t=1}^n \pi(\overline{\mathcal{G}}_t) = \prod_{t=1}^n \pi(\mathcal{G}_t).$$

This and (14) imply (11).

Note that it is easy to show that the analog of Lemma 23 is not valid if instead of graphs one considers hypergraphs $\mathcal{G}_t$. One can find a generalization of Theorem 15 for uniform hypergraphs in [AC94]. However, that generalization imposes some additional condition on the sizes of $\mathcal{V}_t$, which is essentially restrictive when one considers hypergraphs which are not graphs.

Here is a note about the covering problem for Cartesian products of (hyper)graphs. The result below was proved in 1971 and finally printed in [G] with some historical comments. Let $\mathcal{H} = (\mathcal{X}, \mathcal{E})$ be a finite hypergraph with the property $\bigcup_{E \in \mathcal{E}} E = \mathcal{X}$. For any natural $n \in \mathbb{N}$ we have the Cartesian product of spaces $\mathcal{X}^n = \prod_{i=1}^n \mathcal{X}$ and $\mathcal{E}^n = \prod_{i=1}^n \mathcal{E}$. The elements of $\mathcal{E}^n$ are subsets of $\mathcal{X}^n$.

As usual, we say that $\mathcal{F}^n \subset \mathcal{E}^n$ covers $\mathcal{X}^n$, if $\mathcal{X}^n = \bigcup_{F^n \in \mathcal{F}^n} F^n$. We are interested in obtaining bounds on the numbers $c(\mathcal{F}^n)$ defined by

$$c(\mathcal{H}^n) = \min_{\mathcal{F}^n \text{ covers } \mathcal{X}^n} |\mathcal{F}^n|.$$

Clearly $c(\mathcal{H}^{n_1 + n_2}) \leq c(\mathcal{H}^{n_1}) c(\mathcal{H}^{n_2})$.

Denote by Q the set of all probability distributions on $\mathcal{E}$, denote by $I_E(\cdot)$ the indicator function of a set E, and define K by

$$K = \max_{q \in Q} \min_{x \in \mathcal{X}} \sum_{E \in \mathcal{E}} I_E(x) q(E). \tag{15}$$

Theorem 17 (Ahlswede 1971) *With* $C = -\ln K$ *we have*

$$\lim_{n \to \infty} \frac{\ln c(\mathcal{H}^n)}{n} = C.$$

Proof. We prove first that

$$c(n) \geq \exp\{Cn\}. \tag{16}$$

Let us assume that $\mathcal{F}^{n+1}$ covers $\mathcal{X}^{n+1}$ and that $|\mathcal{F}^{n+1}| = c(n+1)$. Write an element F^{n+1} of $\mathcal{F}^{n+1}$ as $F_1 \times F_2 \times \ldots \times F_{n+1}$ and denote by $\mathcal{X}^{n+1}(x)$ the set of all those elements of $\mathcal{X}^{n+1}$ that have x as their first component. Finally, define a probability distribution q^* on $\mathcal{E}$ by

$$q^*(E) = |\{F^{n+1} \in \mathcal{F}^{n+1}, F_1 = E\}| c^{-1}(\mathcal{H}^{n+1}), \ E \in \mathcal{E}.$$

To cover the set $\mathcal{X}^{n+1}(x)$ we need at least $c(\mathcal{H}^n)$ elements of $\mathcal{F}^{n+1}$. This and the definition of q^* yield

$$c(\mathcal{H}^{n+1}) \sum_{E \in \mathcal{E}} I_E(x) q^*(E) \geq c(\mathcal{H}^n). \tag{17}$$

Since (17) holds for all $x \in \mathcal{X}$, we obtain

$$c(\mathcal{H}^{n+1}) \min_{x \in \mathcal{X}} \sum_{E \in \mathcal{E}} I_E(x) q^*(E) \geq c(\mathcal{H}^n)$$

and therefore also

$$c(\mathcal{H}^{n+1}) \max_{q \in Q} \min_{x \in \mathcal{X}} \sum_{E \in \mathcal{E}} I_E(x) q(E) \geq c(\mathcal{H}^n).$$

Relation (16) follows from the last inequality. Next we show by simple random choice that

$$c(\mathcal{H}^n) \leq \exp\{Cn + \ln n + \ln \ln |\mathcal{X}|\} + 1. \tag{18}$$

This inequality and inequality (16) yield Theorem 17. Let r be an element of Q on which the maximum in (15) is assumed. Denote by r^n the product probability distribution on $\mathcal{E}^n$

$$r^n(E^n) = \prod_{t=1}^{n} r(E_t), \ E^n = E_1 \times \ldots \times E_n \in \mathcal{E}^n.$$

Let N be a number specified later. Select now N elements $E_1^n, \ldots, E_N^n$ of $\mathcal{E}^n$ independently according to the random experiment $(\mathcal{E}^n, r^n)$. Let $x^n = (x_1, \ldots, x_n)$ be any element of $\mathcal{X}^n$. Let also

$$\mathcal{E}(x^n) = \{E^n \in \mathcal{E}^n : x^n \in E^n\}.$$

Clearly, $\mathcal{E}(x^n) = \prod_{t=1}^{n} \{E \in \mathcal{E} : x_t \in E\}$ and

$$r^n(\mathcal{E}(x^n)) = \prod_{t=1}^{n} \left(\sum_E I_E(x_t) r(E) \right) \geq K^n.$$

Hence x_n is not contained in any one of the selected sets with probability smaller than $(1 - K^n)^N$ and therefore $\mathcal{X}^n$ is not covered by those sets with probability smaller than $|X|^n (1 - K^n)^N$. Thus there exists a covering of cardinality N for all N satisfying

$$|\mathcal{X}|^n (1 - K^n)^N.$$

Now using inequality $(1 - K^n)^N \leq \exp\{-K^n N\}$ one can easily derive inequality (18). ∎

Lecture 9 Multiple Packing

Multiple packing is an area of investigation in Coding Theory, which can be considered as a part of Combinatorial Extremal Theory. The main problem in Coding Theory is to find the maximal number of points in a metric space such that their pairwise distance is not less than some given number d. This set of points is called code and d is called the distance of the code. The terms code and packing are synonyms and the word packing is often used in Coding Theory instead of code.

Slightly modifying the packing problem we come to the notion of multiple packing. A code with distance d can be considered as a set of nonintersecting balls of radius $\lfloor (d-1)/2 \rfloor$ centered in the points of the code. Now we allow the balls to be intersecting, but in such a way, that any point of the space belongs to (is covered by) not more than L balls for a given positive integer L. The set of centers of the balls with this property is called L-packing or multiple packing with multiplicity L. It is obvious that the maximal cardinality of an L-packing is a nondecreasing function of L. To find maximal L-packings for a given radius of the balls is in general a hopeless task, but it is extremely interesting to find asymptotic bounds for the cardinality of optimal L-packings. For $L = 1$ this problem becomes the classical one.

Consider the Hamming space $\mathcal{H}_q^n$ with Hamming metric d_H. Denote $B(x^n, r) = \{z^n \in [q]^n : d_H(x^n, z^n) \le r\}$ the ball in $\mathcal{H}_q^n$ with center x^n and radius r. A code $\mathcal{C}_n \subset \mathcal{H}_q^n$ is a 1-packing of $\mathcal{H}_q^n$ by balls of radius t iff for all $x^n \in \mathcal{H}_q^n$ the ball $B(x^n, t)$ contains not more than one point of $\mathcal{C}_n$:

$$|\mathcal{C}_n \cap B(x^n, t)| \le 1.$$

It is obvious that this formulation is equivalent to the requirement that the balls $B(x^n, t)$, $B(y^n, t)$ do not intersect for any $x^n \ne y^n \in \mathcal{C}_n$. In case $L = 1$ the lower bound of Varshamov and Gilbert for the rate $R = \log_q(|\mathcal{C}_n|)/n$ of the code is known when $\tau = t/n$ is given. It says that there exists a sequence of codes $\mathcal{C}_n$, $n \to \infty$, such that the following asymptotic relation is true:

$$R = 1 - H_q(2\tau) + o(1), \ n \to \infty, \tag{1}$$

where $H_q(\tau) = -\tau \log_q \tau - (1 - \tau) \log_q (1 - \tau) + \tau \log_q (q - 1)$. This bound is a particular case of our general lower bound for arbitrary L. It is known that the Varshamov–Gilbert bound can be improved for large q by using algebraic methods. We are not going to present here those considerations and refer the interested reader to [TV91].

One of the known upper bounds on the code rate is the Bassalygo–Elias bound:

$$R \le 1 - H_q \left(\frac{q-1}{q} - \frac{q-1}{q} \sqrt{1 - \frac{2q\tau}{q-1}} \right). \tag{2}$$

This bound can be improved by using the linear programming method, but we are not going to describe it here and refer to [MRRW77].

Next in this lecture we will show how to extend the lower bound (1) and the upper bound (2) to the case of arbitrary L. These extensions are nontrivial and need notions and ideas, which at first sight seem to lie far from Coding Theory.

The problem of finding bounds for the rate of an L-packing arose immediately after the works of Elias and Wonenkraft (for references see [E91]) where they formulate the concept of an L-packing (or list-of-L decoding codes, see the next lecture), and only in 1984 this problem was solved in the binary case ($q = 2$) and published in [B86a]. Here we present the solution of this problem for arbitrary q. It appeared in [B05b]. While the lower bound for arbitrary L and q is a consequence of the method called random choice with expurgation, the main difficulty here is to find a characteristic for the centers of $L+1$ balls of a given radius, which would tell whether they have a point in common or not. This leads to the notion of the average radius, which is a prototype of the moment inertia in Physics. The average radius r can be easily calculated for given $L+1$ points in $\mathcal{H}_q^n$ and has the property that $L+1$ balls of radius $t \leq r$ do not have any point in common. Hence if we derive a lower bound for the rate $R(\rho)$, when the average radius $r = \lceil \rho n \rceil$ is given, then this bound also has to be a bound for the rate $R(\tau)$ with substitution $\tau \leftrightarrow \rho$.

The upper bound for the rate of a multiple packing involves rich technique. First we obtain a so-called Plotkin-type bound for the rate when the minimal average radius $r = \lceil \rho n \rceil$ is given. Then we use Ramsey's Theorem and Komlós' Lemma to show that the same bound is true if we replace ρ by τ, where $t = \lceil \tau n \rceil$ is the radius of the balls whose centers constitute the L-packing set. The upper bound (20), which we obtain in this way, is a natural generalization of the Bassalygo–Elias bound (2) to the case of multiple packing. The key lemma here is that the upper bound (20) is true if we substitute τ for ρ.

Now we come to precise definitions. We say that the set (code) $\mathcal{C}_{q,n,L}(t)$ is an L-packing of $\mathcal{H}_q^n$ by balls of radius t if

$$\max_{x^n \in \mathcal{H}_q^n} |\mathcal{C}_{q,n,L}(t) \cap B(x^n,t)| \leq L.$$

In the case $L = 1$ we have the usual definition of a t-error correcting code. An L-packing $\mathcal{C}_{q,n,L}(t)$ has the property that any ball of radius t in $\mathcal{H}_q^n$ contains fewer than $L+1$ points of $\mathcal{C}_{q,n,L}(t)$ or, equivalently, the multiplicity of covering of any point from $\mathcal{H}_q^n$ by the balls of radius t with centers in the points of $\mathcal{C}_{q,n,L}(t)$ does not exceed L.

We set

$$R_{q,L}(\tau) = \limsup_{n \to \infty} \frac{\max \log_q |\mathcal{C}_{q,n,L}(\lceil \tau n \rceil)|}{n}, \quad \tau \in (0,1).$$

§1 A Lower Bound

We are going to introduce a scheme for obtaining a lower bound for the value $R_{q,L}(\tau)$. We will show that at the zero rate our bound is tight, that is, we find the exact value of $\tau_0 = \sup_{R_{q,L}(\tau)=0} \tau$. Also we obtain an upper bound for $R_{q,L}(\tau)$ for arbitrary τ; however, our proof is incomplete: to complete it, it is necessary to prove the convexity of an explicitly given function.

Define the following value, which depends on j vectors $x_1^n, \ldots, x_j^n \in \mathcal{H}_q^n$:

$$r_j(x_1^n, \ldots, x_j^n) = \frac{1}{j} \min_{y^n \in \mathcal{H}_q^n} \sum_{i=1}^{j} d_H(y^n, x_i^n).$$

We call it the average radius of the points $x_1^n, \ldots, x_j^n \in H_q^n$. It has the meaning of moment inertia of j points of unit mass in the Hamming space.

It is clear that

$$r_j(x_1^n,\ldots,x_j^n) = \frac{1}{j} \sum_{m=1}^{n} \min_{y_m \in [q]} \sum_{i=1}^{j} d_H(x_{im}, y_m) = \frac{1}{j} \sum_{m=1}^{n} (j - s(x_{1m},\ldots,x_{jm})),$$

where $s(x_{1m},\ldots,x_{jm})$ is the number of occurrences of a most frequent element in the sequence $(x_{im})_{i=1}^{j}$.

An n-tuple $y^n = (y_1,\ldots,y_n) \in \mathcal{H}_q^n$ at which the average radius r achieves its minimum is called the center of inertia.

Define

$$R_{q,L}(\rho) = \limsup_{n \to \infty} \frac{\max \log_q |\mathcal{C}_{q,n,L}(\rho)|}{n},$$

where max is taken over all codes $\mathcal{C}_{q,n,L}(\rho) \subset \mathcal{H}_q^n$ such that

$$\min_{\{x_1^n,\ldots,x_{L+1}^n\} \subset \mathcal{C}_{q,n,L}(\rho)} r_{L+1}(x_1^n,\ldots,x_{L+1}^n) \geq \rho n.$$

Let $\omega_{m_1,\ldots,m_j}(x_1^n,\ldots,x_j^n)$ be the joint type of the n-tuples $x_1^n,\ldots,x_j^n$:

$$\omega_{m_1,\ldots,m_j}(x_1^n,\ldots,x_j^n) = \frac{|i : (x_{1i}^n,\ldots,x_{ji}^n) = (m_1,\ldots,m_j)|}{n}.$$

It is easy to establish the following equalities:

$$\rho_j(x_1^n,\ldots,x_j^n) \overset{\Delta}{=} \frac{r_j(x_1^n,\ldots,x_j^n)}{n} \tag{3}$$

$$= \sum_{(m_1,\ldots,m_j) \in [q]^j} \left(1 - \frac{s(m_1,\ldots,m_j)}{j}\right) \omega_{m_1,\ldots,m_j}(x_1^n,\ldots,x_j^n),$$

$$\delta_j(x_1^n,\ldots,x_j^n;i) \overset{\Delta}{=} \frac{d_H(y^n,x_i^n)}{n} \tag{4}$$

$$= \sum_{(m_1,\ldots,m_i,\ldots,m_j) \in [q]^j : m_i \neq f(m_1,\ldots,m_j)} \omega_{m_1,\ldots,m_j}(x_1^n,\ldots,x_j^n),$$

where function $f(m_1,\ldots,m_j)$ is defined as follows. Let $m_{p_1},\ldots,m_{p_s'}$ be the elements that most frequently occur in the sequence $(m_1,\ldots,m_j)$. To every set $\{m_1,\ldots,m_j\}$ uniquely corresponds the set $\{m_{p_1},\ldots,m_{p_s'}\}$. We pick up an element $m \in \{m_{p_1},\ldots,m_{p_s'}\}$ and put $f(m_1,\ldots,m_j) = m$. We set $y_m = f(x_{1m},\ldots,x_{jm})$. Considering the values ρ, δ as functions of type ω, we see that these functions are linear and satisfy the following relations:

$$\rho_j(x_1^n,\ldots,x_j^n) \leq \tau_j(x_1^n,\ldots,x_j^n) + \frac{1}{n} \leq \max_i \delta_j(x_1^n,\ldots,x_j^n;i), \tag{5}$$

where $\tau_j(x_1^n,\ldots,x_j^n) = t_j(x_1^n,\ldots,x_j^n)/n$ and

$$t_j(x_1^n,\ldots,x_j^n) = \max\{t : |B(x^n,t) \cap \{x_1^n,\ldots,x_j^n\}| < j, \ x^n \in \mathcal{H}_q^n\}$$

is the maximal radius of a ball, which for any choice of the center x^n does not contain all j vectors $\{x_1^n, \ldots, x_j^n\}$. In other words, $t_j(x_1^n, \ldots, x_j^n)$ is the maximal radius of a ball such that the code $\{x_1^n, \ldots, x_j^n\}$ is a packing (by these balls) of multiplicity $j-1$. Let $\pi(m_1, \ldots, m_j)$ be an arbitrary permutation of the numbers $m_1, \ldots, m_j$. Note that if for $\pi(m_1, \ldots, m_j)$ the following relation is valid:

$$\omega_{m_1, \ldots, m_j}(x_1^n, \ldots, x_j^n) = \omega_{\pi(m_1, \ldots, m_j)}(x_1^n, \ldots, x_j^n),$$

then for all i the following equalities hold:

$$r_j(x_1^n, \ldots, x_j^n) = d_H(y^n, x_i^n)$$

and consequently in this case we have

$$r_j(x_1^n, \ldots, x_j^n) = t_j(x_1^n, \ldots, x_j^n) + 1.$$

First we obtain a bound for the value $R_{q,L}(\rho)$. From relations (5) it follows that any lower bound for $R_{q,L}(\rho) : R_{q,L}(\rho) \geq f(\rho)$ is also a lower bound for $R_{q,L}(\tau) : R_{q,L}(\tau) \geq f(\tau)$, when $\rho = \tau$. For upper bounds this is not true in general, but using the original method we will prove that the upper bound on $R_{q,L}(\rho) : R_{q,L}(\rho) \leq \varphi(\rho)$, which we will show is also valid for $R_{q,L}(\tau)$ when $\rho = \tau : R_{q,L}(\tau) \leq \varphi(\tau)$.

We obtain the lower bound for $R_{q,L}(\rho)$ by using the method of random choice with expurgation. Choosing the coordinates of vectors $x_1^n, \ldots, x_{L+1}^n$ from the alphabet $[q]$ independently with uniform distribution, it is easy to see that for the average value we have

$$\mathbb{E}(\rho) = \frac{1}{q^{L+1}} \sum_{\{j_i\}:\sum_{i=1}^q j_i = L+1} \left(1 - \frac{\max\{j_1, \ldots, j_q\}}{L+1}\right) \binom{L+1}{j_1, \ldots, j_q}. \quad (6)$$

At the same time the moment generating function $\mathbb{E}(q^{h\rho})$ satisfies the relation

$$\mathbb{E}(q^{h\rho}) = \left[\frac{1}{q^L} \sum_{\{j_i\}:\sum_{i=1}^q j_i = L+1} \binom{L+1}{j_1, \ldots, j_q} q^{\left(1 - \frac{\max\{j_1, \ldots, j_q\}}{L+1}\right)h}\right]^n. \quad (7)$$

Using the Chernoff inequality we obtain

$$\frac{\log_q P(\mathbb{E}(\rho) - \rho_{L+1} \geq \alpha)}{n} \leq \frac{\log_q \mathbb{E}(q^{-hr_{L+1}})}{n} + h(\mathbb{E}(\rho) - \alpha) \stackrel{\Delta}{=} \varepsilon(h, \alpha), \ h \geq 0. \quad (8)$$

Hence, using the additive bound for the probability of the union of events, we obtain the existence of a code $\mathcal{C}_n$ of size M for which the number of families of $L+1$ words $\{x_1^n, \ldots, x_{L+1}^n\}$ with $\rho(x_1^n, \ldots, x_{L+1}^n) < \mathbb{E}(\rho) - \alpha$ ("bad" families) does not exceed

$$M^{L+1} q^{n\varepsilon(h, \alpha)}.$$

If we impose the following restriction:

$$M^{L+1} q^{n\varepsilon(h,\alpha)} < \frac{M}{2}, \tag{9}$$

then there exists a code C_n with the number of "bad" families less than $M/2$. Expurgating one codeword from each "bad" family from C_n, we obtain a code of cardinality greater than $M/2$, in which all families of $L+1$ words $\{x_1^n, \ldots, x_{L+1}^n\}$ have large values of ρ_{L+1} : $\rho_{L+1}(x_1^n, \ldots, x_{L+1}^n) > \mathbb{E}(\rho) - \alpha$. Taking the $\log_q$ on both sides in (9) and making obvious transformations we obtain the asymptotic lower bound

$$R_{q,L}(\rho) \geq -\frac{1}{L}\varepsilon(h, \mathbb{E}(\rho) - \rho) + o(1). \tag{10}$$

Substituting (7) into (10) and optimizing over h (by setting the derivative in h of the LHS of (10) equal to zero) we obtain the parametric relations

$$R_{q,L}(\rho) = 1 - \tag{11}$$

$$\frac{1}{L}\left(\log_q\left(\frac{1}{q}\sum_{\{j_i\}:\sum_{i=1}^q j_i=L+1}\binom{L+1}{j_1,\ldots,j_q}q^{-h\left(1-\frac{\max\{j_1,\ldots,j_q\}}{L+1}\right)}\right) + h\rho\right),$$

$$\rho = \frac{\sum_{\{j_i\}:\sum_{i=1}^q j_i=L+1}\binom{L+1}{j_1,\ldots,j_q}\left(1-\frac{\max\{j_1,\ldots,j_q\}}{L+1}\right)q^{-\left(1-\frac{\max\{j_1,\ldots,j_q\}}{L+1}\right)h}}{\sum_{\{j_i\}:\sum_{i=1}^q j_i=L+1}\binom{L+1}{j_1,\ldots,j_q}q^{-\left(1-\frac{\max\{j_1,\ldots,j_q\}}{L+1}\right)h}}.$$

This is the final lower bound for $R_{q,L}(\rho)$. As we mentioned before, this bound is also valid for $R_{q,L}(\tau)$ when $\tau = \rho$.

One can find an interesting approach to establish a lower bound for multiple packing in the case where this multiple packing is at the same time a linear subspace of the binary Hamming space in [B00].

§2 An Upper Bound for $R_{q,L}(\rho)$

First to obtain an upper bound for $R_{q,L}(\rho)$, we use a Plotkin-type bound and the asymptotic Johnson bound (see end of the chapter for the proof)

$$R_{q,L}(\rho) \leq 1 - H_q(\lambda(\rho)) + o(1), \tag{12}$$

where

$$H_q(x) = -x\log_q x - (1-x)\log_q(1-x) + x\log_q(q-1),$$

$\lambda(\rho) = \omega/n$, and $\omega \in \{1, \ldots, \lfloor n/2 \rfloor\}$, with the property that for an arbitrary fixed $\varepsilon > 0$ the number of words in H_q^n of weight $\omega - \varepsilon n$ and with given minimal value of ρ is bounded uniformly in n.

From the above it follows that it is enough to obtain an explicit expression of the function $\lambda(\rho)$. For this we use a Plotkin-type bound.

Now we explain what we mean by Plotkin-type bound. We will use this bound several times in the proofs in this and the next lectures. Let us have some code $C_n \subset \mathcal{H}_q^n$. We enumerate its vectors by the numbers $1,\ldots,|C_n|$ and arrange them in that order in the rows of a matrix of dimension $|C_n| \times n$. Let $K_j = \{q_1,\ldots,q_{|C_n|}\}$ be the sequence of symbols in the jth column of this matrix and $K_{i,j}$ be the number of occurrences of symbol i, $0 \le i \le q-1$, in the jth column. We want to estimate the minimum of the value $r(x_{i_1}^n,\ldots,x_{i_{L+1}}^n)$ over the choice of $L+1$ vectors from the code $C_n \subset \mathcal{H}_q^n$ and use

$$r(x_{i_1}^n,\ldots,x_{i_{L+1}}^n) = \sum_{j=1}^{n} \xi(x_{i_1,j},\ldots,x_{i_{L+1},j}).$$

Sometimes we assume that ξ depends only on the number of entries of the symbols $\{0,\ldots,q-1\}$ in the same component of x_{i_k}. We want to find an upper bound on

$$r_m = \min_{(x_{i_1}^n,\ldots,x_{i_{L+1}}^n) \subset C_n,\, i_p \ne i_t,\, p \ne t} r(x_{i_1}^n,\ldots,x_{i_{L+1}}^n).$$

It is easy to see that this minimum does not exceed the average $\bar{r}$ of $r(x_{i_1}^n,\ldots,x_{i_{L+1}}^n)$ over the choice of different vectors from the code. This gives the bound

$$r_m \le \bar{r} = \frac{\sum_{(x_{i_1}^n,\ldots,x_{i_{L+1}}^n) \subset C_n,\, i_p < i_t,\, i < t} r(x_{i_1}^n,\ldots,x_{i_{L+1}}^n)}{\binom{|C_n|}{L+1}}$$

$$= \frac{\sum_{(x_{i_1}^n,\ldots,x_{i_{L+1}}^n) \subset C_n,\, i_p < i_t,\, i < t} \sum_{j=1}^{n} \xi(x_{i_1,j},\ldots,x_{i_{L+1},j})}{\binom{|C_n|}{L+1}}$$

$$= \frac{\sum_{j=1}^{n} \sum_{(x_{i_1},\ldots,x_{i_{L+1}}) \subset K_j,\, i_p < i_t,\, p < t} \xi(x_{i_1},\ldots,x_{i_{L+1}})}{\binom{|C_n|}{L+1}}$$

$$= \frac{\sum_{j=1}^{n} f(K_{1,j},\ldots,K_{q,j})}{\binom{|C_n|}{L+1}},$$

where $f(K_{1,j},\ldots,K_{q,j})$ is a function which is the same for all $j = 1,\ldots,n$. Sometimes, when we do not care about the order of the code vectors, we sum in the expressions for $\bar{r}$ over all $(c_{i_1}^n,\ldots,c_{i_{L+1}}^n) \subset C_n$, $i_p \ne i_t$, $p \ne t$ and divide the sum by $|C_n|(|C_n|-1)\cdots(|C_n|-L)$. The last step of the estimation of r_m is different in different problems and consists in finding the asymptotic behavior of the RHS of the last chain of relations and then optimizing it over the choice of $\{K_{i,j}\}$.

Now we return to our problem. In this problem we use a Plotkin-type bound with $\xi = 1 - \frac{\max\{j_1,\ldots,j_q\}}{L+1}$. Consider M_n words from $\mathcal{H}_q^n$ (the code $C_n \subset \mathcal{H}_q^n$, $|C_n| = M_n$ consists of vectors of Hamming weight $\omega = [\lambda n]$) and let $k_\nu(i)$ be the number of symbols $i \in [q]$ in the νth column of the code matrix of size $M_n \times n$.

The average value $\overline{\rho}_{L+1}(x_{i_1}^n,\ldots,x_{i_{L+1}}^n)$ over choosing families of different $L+1$ words $\{x_{i_1}^n,\ldots,x_{i_{L+1}}^n\} \subset C_n$ satisfies the equality

$$\overline{\rho} = \frac{\sum_{v=1}^n \sum_{\{j_i\}:\sum_{i=1}^q j_i = L+1} \left(1 - \frac{\max\{j_1,\ldots,j_q\}}{L+1}\right) \prod_{i=1}^q \binom{k_v(i)}{j_i}}{n\binom{M_n}{L+1}}. \tag{13}$$

When $M_n \to \infty$, (13) implies the asymptotical inequality

$$\overline{\rho} \leq \frac{1}{n}\sum_{v=1}^n \sum_{\{j_i\}:\ \sum_{i=1}^q j_i = L+1} \left(1 - \frac{\max\{j_1,\ldots,j_q\}}{L+1}\right)\binom{L+1}{j_1,\ldots,j_q}\prod_{i=1}^q \kappa_v^{j_i}(i)(1+o(1)),$$

where $\kappa_v(i) = \frac{k_v(i)}{M}$ is the type of the vth column of the code matrix and $0^0 = 1$. Here we used the relations

$$\binom{a}{b} \leq \frac{a^b}{b!}, \quad \binom{a}{b} = \frac{a^b}{b!}(1+o(1)), \ b = const, \ a \to \infty.$$

Denote

$$\gamma(\{j_i\};\{x(i)\}) = \sum_{\pi(\{j_i\})}\prod_{i=1}^q x^{j_i}(i),$$

where $j_i \geq 0$, $\sum_{i=1}^q j_i = L+1$, and the sum is taken over all permutations π of the set $\{j_1,\ldots,j_q\}$. Function γ is symmetric in variables $x(1),\ldots,x(q)$. Denote also

$$\alpha(\{x(i)\}) = \sum_{\{j_i\}}^* \left(1 - \frac{\max\{j_1,\ldots,j_q\}}{L+1}\right)\binom{L+1}{j_1,\ldots j_q}\gamma(\{j_i\};\{x(i)\}),$$

where the sum $\sum^*$ is taken over the sets $(j_1,\ldots,j_q)$, $\sum_{i=1}^q j_i = L+1$, which are not equivalent under permutations. It is easy to see that up to $o(1)$ the following inequality is valid:

$$\overline{\rho} \leq \frac{1}{n}\sum_{v=1}^n \alpha(\{\kappa_v(i)\}). \tag{14}$$

Suppose that $x(i) \geq 0$ and $\sum_{i=1}^{q-1} x(i) = \lambda$. Exercise 5 asks to prove that α achieves its maximum for given λ, when $x(i) = \frac{\lambda}{q-1}$, $i = 1,\ldots,q-1$, and when $\lambda \in [0,1]$ can vary, α achieves its maximum at $\lambda = (q-1)/q$, $x(i) = 1/q$. Hence if $\sum_{i=1}^{q-1} \kappa_v(i) = \lambda_v$, then

$$\overline{\rho} \leq \tag{15}$$

$$\frac{1}{n}\sum_{v=1}^n \sum_{\{j_i\}:\Sigma_{i=1}^q j_i = L+1}\left(1 - \frac{\max\{j_1,\ldots,j_q\}}{L+1}\right)\binom{L+1}{j_1,\ldots,j_q}$$

$$\times \left(\frac{\lambda_v}{q-1}\right)^{L+1-j_q}(1-\lambda_v)^{j_q}(1+o(1)).$$

From the above considerations also follows that we can put $\lambda_v = (q-1)/q$ in the last formula and obtain a universal bound, which is an upper bound over other values of λ_v:

$$\bar{\rho} \leq \tag{16}$$
$$\frac{1}{q^{L+1}} \sum_{\{j_i\}:\, \sum_{i=1}^{q} j_i = L+1} \left(1 - \frac{\max\{j_1,\ldots,j_q\}}{L+1}\right) \binom{L+1}{j_1,\ldots,j_q} (1+o(1)).$$

It can be proved that the function

$$f_L(\lambda_v) = \tag{17}$$
$$\sum_{\{j_i\}:\, \sum_{i=1}^{q} j_i = L+1} \left(1 - \frac{\max\{j_1,\ldots,j_q\}}{L+1}\right) \binom{L+1}{j_1,\ldots,j_q}$$
$$\times \left(\frac{\lambda_v}{q-1}\right)^{\sum_{i=1}^{q-1} j_i} (1-\lambda_v)^{j_q}$$

satisfies the inequality[1]

$$\frac{1}{n} \sum_{v=1}^{n} f_L(\lambda_v) \leq f_L\left(\frac{\sum_{i=1}^{v} \lambda_v}{n}\right) = f_L(\lambda). \tag{18}$$

Recall now that $\omega = [\lambda n]$ is the weight of the codewords from C_n and $\sum_{v=1}^{n} \lambda_v = \lambda$. Suppose that (18) is valid and if for given λ, volume M of the code C_n tends to infinity, then the minimal value ρ_{min} of ρ while choosing $L+1$ different words from C_n is less than the average value $\bar{\rho}$. Hence

$$\rho_{min} \leq f_L(\lambda) + o(1). \tag{19}$$

We reformulate the previous idea by saying that if $\rho_{min} > f_L(\lambda) + o(1)$, then M_n is bounded uniformly in n. In this case we can use the asymptotic Johnson bound, which says that if λ is given and M_n is uniformly bounded, then the value of the rate R of an arbitrary code in the whole space $\mathcal{H}_q^n$ satisfies (12). For the proof of the Johnson bound see the comments at the end of this chapter. This considerations give the following upper bound on $R_{q,L}(\rho)$ in parametric form:

$$R_{q,L}(\rho) \leq 1 - H_q(\lambda), \tag{20}$$
$$\rho = f_L(\lambda) =$$
$$\sum_{\{j_i\}:\, \sum_{i=1}^{q} j_i = L+1} \left(1 - \frac{\max\{j_1,\ldots,j_{L+1}\}}{L+1}\right) \binom{L+1}{j_1,\ldots,j_q} \left(\frac{\lambda}{q-1}\right)^{\sum_{i=1}^{q-1} j_i} (1-\lambda)^{j_q}.$$

---

[1] The proof has not been included, because it was found by Blinovsky while the book was already in print. The reader can give a proof himself by doing Exercise 15.

This is our upper bound. At zero rate (when $R_{q,L}(\rho) = 0$) we have $\lambda = (q-1)/q$ and, as it was shown in this case,

$$f_L((q-1)/q)) = \max_{\lambda \in [0,1]} f_L(\lambda)$$

(Exercise 5) and we have

$$\rho = \frac{1}{q^{L+1}} \sum_{\{j_i\}:\sum_{i=1}^{q} j_i = L+1} \left(1 - \frac{\max\{j_1,\ldots,j_{L+1}\}}{L+1}\right)\binom{L+1}{j_1,\ldots,j_q}. \quad (21)$$

In Exercise 6 it is asked to prove that this bound coincides with the lower bound (11) at zero rate (when still $\log_q |C_{q,n,L}(\rho)| \to \infty$, it is $o(n)$). Hence, at zero rate the bound (21) is tight.

§3 An Upper Bound for $R_{q,L}(\tau)$

We have the following result:

Theorem 18 (Blinovsky 2005) *Bounds (20) are still valid after substituting $\rho \leftrightarrow \tau$.*

This part is devoted to the sophisticated proof of this theorem.

We need some auxiliary results formulated in the following two lemmas. Let $X_1, X_2, \ldots, X_M$ be a sequence of random variables taking values in $[q]$. We assume later that $\alpha_m(i) = I_{X_m=i}$.

Lemma 24 (Komlós 1990) *Let $\alpha_i(m)$, $1 \le i \le M$, $m \in [q]$ be square integrable functions, and let*

$$\max_{m,i} ||\alpha_i(m)|| \le 1.$$

Define the averages $\bar{\alpha}_i(m) = \frac{1}{i}(\alpha_1(m) + \ldots + \alpha_i(m))$, $1 \le i \le M$. If for some number r and for all $i < j$,

$$\left|\int \alpha_i(m)\alpha_j(m)dP - r\right| < \delta,$$

then for all $i < j$

$$\int (\bar{\alpha}_i(m) - \bar{\alpha}_j(m))^2 dP < \frac{2+\delta}{i}\left(1 - \frac{i}{j}\right) + 4\delta\left(1 - \frac{i}{j}\right)^2.$$

Proof. We have

$$(\bar{\alpha}_i(m) - \bar{\alpha}_j(m))^2 = \frac{1}{i^2 j^2}\left[(j-i)\sum_{k=1}^{i}\alpha_k(m) - i\sum_{k=i+1}^{j}\alpha_k(m)\right]^2.$$

Expanding the square on the RHS, we get the expression

$$(j-i)^2 \sum_{k=1}^{i} \alpha_k^2(m) + i^2 \sum_{k=i+1}^{j} \alpha_k^2(m) + 2(j-i)^2 \sum_{1 \le k < k' \le i} \alpha_k(m) \alpha_{k'}(m)$$
$$+ 2i^2 \sum_{i < k < k' \le j} \alpha_k(m) \alpha_{k'}(m) - 2i(j-i) \sum_{1 \le k \le i < k' \le j} \alpha_k(m) \alpha_{k'}(m).$$

Multiplying it by $i^{-2} j^{-2}$, integrating it term by term, and using inequalities $\int \alpha_k^2 dP \le K^2$ and $r - \delta < \int \alpha_k(m) \alpha_{k'}(m) dP < r + \delta$ $(k \ne k')$, we get the upper bound

$$(1-r) \frac{j-i}{ij} + 4\delta \frac{(j-i)^2}{j^2}.$$

This gives the desired bound, since the inequality $|r| \le 1 + \delta$ can be easily obtained by applying the Cauchy–Schwartz inequality $\left| \int \alpha_k(m) \alpha_{k'}(m) dP \right| \le \| \alpha_k(m) \| \| \alpha_{k'}(m) \| \le 1.$ □

Lemma 25 (Komlós 1990) *Let $\alpha_i(m)$, $1 \le i \le M$, $M \ge 2$, $m \in [q]$ be square integrable functions satisfying the conditions of the previous lemma (with possibly different values of r). If for some numbers r', r'', for some $m, m' \in [q]$, and for all i, j*

$$\left| \int \alpha_i(m) \alpha_j(m') dP - r' \right| < \delta, \quad \left| \int \alpha_j(m) \alpha_i(m') dP - r'' \right| < \delta, \tag{22}$$

then

$$|r' - r''| < \frac{6}{\sqrt{M-1}} + 6\sqrt{\delta} + 2\delta.$$

Proof. We assume $2|M$, otherwise we consider $M - 1$ instead of M. By the previous lemma, for the averages

$$a = \bar{\alpha}_{M/2}(m), \ A = 2\bar{\alpha}_M(m) - \bar{\alpha}_{M/2}(m) = \frac{2}{M}(\alpha_{M/2+1}(m) + \ldots + \alpha_M(m))$$

we have

$$\|a - A\| = 2\|\bar{\alpha}_{M/2}(m) - \bar{\alpha}_M(m)\| < 2\sqrt{\frac{2+\delta}{M}} + \delta < \frac{3}{\sqrt{M}} + 3\sqrt{\delta} \stackrel{\Delta}{=} \Delta.$$

Similarly, for the averages

$$b = \frac{2}{m} \sum_{k=1}^{M/2} \alpha_k(m'), \ B = \frac{2}{M} \sum_{k=M/2+1}^{M} \alpha_k(m')$$

we have

$$\|b - B\| < \Delta.$$

By the conditions of the lemma, for $n > M/2$,

$$\left|\int a\alpha_n(m')dP - r'\right| < \delta, \quad \left|\int b\alpha_n(m)dP - r''\right| < \delta$$

and hence

$$\left|\int aBdP - r'\right| < \delta, \quad \left|\int bAdP - r''\right| < \delta.$$

Now

$$\left|\int aBdP - \int bAdP\right| \leq ||a-A||||B|| + ||b-B||||A|| < 2\Delta$$

and the lemma follows.   □

We resume in the statements of the previous two lemmas the following one, which we use in our next considerations.

Lemma 26 *Let $\alpha_i(m)$, $1 \leq i \leq M$, $M \geq 2$, $m \in [q]$ be square integrable functions and $\int \alpha_i^2(m)dP \leq 1$. Suppose that for some function $r(i,j)$, $1 \leq i,j \leq M$, and all $m_1, m_2 \in [q]$*

$$\left|\int \alpha_i(m_1)\alpha_j(m_2)dP - r(i,j)\right| < \delta.$$

Then

$$|r(i,j) - r(j,i)| \leq \frac{6}{\sqrt{M}} + 6\sqrt{\delta} + 2\delta.$$

As an application of Lemma 26, consider a code C_n, $|C_n| = M_n$, and the matrix of the code C_n of dimension $M_n \times n$. Recall that the rows of this matrix are filled by codewords $x_1^n, \ldots, x_{M_n}^n \in C_n$ in some order. Let $\alpha_i(m)$ be the indicator of the set of positions, where the ith codeword x_i^n has the symbol m. Then $\alpha_i(m_1)\alpha_j(m_2)$ is the indicator function of the set of positions, where the word x_i^n has the symbol m_1 and the word x_j^n has the symbol m_2. If we consider the uniform distribution on the set of positions $\{1, \ldots, n\}$, then

$$\omega_m(x_i^n) = \int \alpha_i(m)dP$$

is the type of the word x_i^n and

$$\omega_{m_1,m_2}(x_i^n, x_j^n) = \int \alpha_i(m_1)\alpha_j(m_2)dP$$

is the joint type of the words x_i^n, x_j^n. Then Lemma 26 says that if for $i < j$

$$\omega_{m_1,m_2}(x_i^n, x_j^n) = \varphi_2(m_1, m_2) + o(1), \quad o(1) \overset{M_n \to \infty}{\to} \infty,$$

then the joint type $\omega_{m_1,m_2}(x_i^n, x_j^n)$ does not depend (asymptotically) on the choice of the ordered pair of codewords (x_i^n, x_j^n) and

$$|\varphi_2(m_1, m_2) - \varphi_2(m_2, m_1)| = o(1),$$

that is, asymptotically the joint type is a symmetric function of elements m_1, m_2 and hence also $\omega_{m_1,m_2}(x_i^n, x_j^n)$ is an asymptotically symmetric function of (i, j) (or (m_1, m_2)).

Next we need a similar property for the joint type of more than two words

$$\omega_{m_1,\ldots,m_\ell}(x_{i_1}^n, \ldots, x_{i_\ell}^n) = \int \alpha_{i_1}(m_1) \ldots \alpha_{i_\ell}(m_\ell) dP.$$

We will show that if

$$\omega_{m_1,\ldots,m_\ell}(x_{i_1}^n, \ldots, x_{i_\ell}^n) = \varphi_\ell(m_1, \ldots, m_\ell) + o(1) \; (M_n \to \infty), \; i_1 < \ldots < i_\ell, \quad (23)$$

that is, the joint type $\omega_{m_1,\ldots,m_\ell}(x_{i_1}^n, \ldots, x_{i_\ell}^n)$ asymptotically does not depend on the choice of the ordered set of code words $(x_{i_1}^n, \ldots, x_{i_\ell}^n)$ (ℓ is fixed), then

$$|\varphi_\ell(m_1, \ldots, m_\ell) - \varphi_\ell(\pi(m_1), \ldots, \pi(m_\ell))| = o(1) \; (M_n \to \infty), \quad (24)$$

where π is a permutation of $(m_1, \ldots, m_\ell)$, that is, $\varphi_\ell(m_1, \ldots, m_\ell)$ is an asymptotically symmetric function and hence $\omega_{m_1,\ldots,m_\ell}(x_{i_1}^n, \ldots, x_{i_\ell}^n)$ is an asymptotically symmetric function of $(i_1, \ldots, i_\ell)$ (and $(m_1, \ldots, m_\ell)$).

We observe that to prove (24) it is enough to show that

$$\varphi_\ell(m_1, \ldots, m_{p-1}, m_p, m_{p+1}, m_{p+2}, \ldots, m_\ell) \quad (25)$$
$$= \varphi_\ell(m_1, \ldots, m_{p-1}, m_{p+1}, m_p, m_{p+2}, \ldots, m_\ell) + o(1).$$

Note that if (23) is true for fixed ℓ, then it is true for arbitrary $\ell' < \ell$. Indeed,

$$\omega_{m_1,\ldots,m_{\ell'}}(x_1^n, \ldots, x_p^n) = \sum_{(m_{\ell'+1},\ldots,m_\ell) \in [q]^{\ell-\ell'}} \omega_{m_1,\ldots,m_\ell}(x_1^n, \ldots, x_\ell^n)$$

$$= \sum_{(m_{\ell'+1},\ldots,m_\ell) \in [q]^{\ell-\ell'}} \varphi_\ell(m_1, \ldots, m_\ell) + o(1) = \varphi_{\ell'}(m_1, \ldots, m_{\ell'}) + o(1),$$

where

$$\varphi_{\ell'}(m_1, \ldots, m_{\ell'}) = \sum_{(m_{\ell'+1},\ldots,m_\ell) \in [q]^{\ell-\ell'}} \varphi_\ell(m_1, \ldots, m_\ell).$$

Thus, if $\varphi_{\ell'}(m_1, \ldots, m_{\ell'}) = 0$ for some $(m_1, \ldots, m_{\ell'}) \in [q]^{\ell'}$, then

$$\varphi_\ell(m_1, \ldots, m_{\ell'}, \ldots, m_\ell) = 0$$

for arbitrary $(m_{\ell'+1}, \ldots, m_\ell) \in [q]^{\ell-\ell'}$. The same is true for an arbitrary given set of ℓ' variables (not only when $(m_1, \ldots, m_{\ell'})$ are the first ℓ' variables of the function φ_ℓ).

Next assume that $\varphi_{\ell-2}(m_1,\ldots,m_{p-1},m_{p+2},\ldots,m_\ell) \neq 0$. We can consider the conditional types defined as follows

$$\omega_{m_p,m_{p+1}|m_1,\ldots,m_{p-1},m_{p+2},\ldots,m_\ell}(x_1^n,\ldots,x_\ell^n)$$

$$= \frac{\omega_{m_1,\ldots,m_\ell}(x_1^n,\ldots,x_\ell^n)}{\omega_{m_1,\ldots,m_{p-1},m_{p+2},m_\ell}(x_1^n,\ldots,x_{p-1}^n,x_{p+2}^n,\ldots,x_\ell)}$$

$$= \frac{\varphi_\ell(m_1,\ldots,m_p,m_{p+1},\ldots,m_\ell)}{\varphi_{\ell-2}(m_1,\ldots,m_{p-1},m_{p+2},\ldots,m_\ell)} + o(1).$$

Let $x_{i_1}^n = x_1^n,\ldots,x_{i_{p-1}}^n = x_{p-1}^n, x_{i_{p+2}}^n = x_{M_n-\ell+p+2}^n,\ldots,x_{i_\ell}^n = x_{M_n}^n$, that is, we set the first $p-1$ words from the family $(x_{i_1}^n,\ldots,x_{i_\ell}^n)$ to be the first $p-1$ words from the code C_n and the last $\ell-p-1$ words from this family to be the last $\ell-p-1$ words from the code. The ordered words $x_{i_p}^n, x_{i_{p+1}}^n$ we choose from the rest of the $M_n-\ell+2$ words from C_n.

Now introduce some random variables: let $\beta_{i_p}(m)$ be the indicator of the set Ω of the positions k, where word $x_{i_p}^n$ has symbol $m \in [q]$ and the ordered set of coordinates $(x_{1k}^n,\ldots,x_{(p-1)k}^n,x_{(M_n-\ell+p+2)k}^n,\ldots,x_{M_nk}^n)$ of n-tuples

$$(x_1^n,\ldots,x_{p-1}^n,x_{M_n-\ell+p+2}^n,\ldots,x_{M_n}^n) \tag{26}$$

is $(m_1,\ldots,m_{p-1},m_{p+2},\ldots,m_\ell)$. Let $\beta_{i_{p+1}}(m)$ have the same meaning for word $x_{i_{p+1}}^n$. Let also ω be the uniform distribution on the set Ω. Then we have for the conditional type

$$\omega_{m_p,m_{p+1}|m_1,\ldots,m_{p-1},m_{p+2},\ldots,m_\ell}(x_1^n,\ldots,x_{p-1}^n,x_{i_p}^n,x_{i_{p+1}}^n,x_{M_n-\ell+p+2}^n,\ldots,x_{M_n}^n)$$

$$= \int \beta_{i_{p+1}}(m_{p+1})\beta_{i_p}(m_p)d\omega \tag{27}$$

$$= \frac{\omega_{m_1,\ldots,m_\ell}(x_1^n,\ldots,\ldots,x_{p-1}^n,x_{i_p}^n,x_{i_{p+1}}^n,x_{M_n-\ell+p+2}^n,\ldots,x_{M_n}^n)}{\omega_{m_1,\ldots,m_{p-1},m_{p+1},\ldots,m_\ell}(x_1^n,\ldots,x_{p-1}^n,x_{M_n-\ell+p+2}^n,\ldots,x_{M_n}^n)}$$

$$= \frac{\varphi_\ell(m_1,\ldots,m_\ell)}{\varphi_{\ell-2}(m_1,\ldots,m_{p-1},m_{p+2},\ldots,m_\ell)} + o(1) =$$

$$= \varphi_{\ell,2}(m_p,m_{p+1}|m_1,\ldots,m_{p-1},m_{p+2},\ldots,m_\ell) + o(1),$$

where

$$\varphi_{\ell,2}(m_p,m_{p+1}|m_1,\ldots,m_{p-1},m_{p+2},\ldots,m_\ell) = \frac{\varphi_\ell(m_1,\ldots,m_\ell)}{\varphi_{\ell-2}(m_1,\ldots,m_{p-1},m_{p+2},\ldots,m_\ell)}.$$

Since $M_n-\ell+2 \to \infty$, we have from Lemma 26 that if

$$\omega_{m_p,m_{p+1}|m_1,\ldots,m_{p-1},m_{p+2},\ldots,m_\ell}(x_1^n,\ldots,x_{p-1}^n,x_{i_p}^n,x_{i_{p+1}}^n,x_{M_n-\ell+p+2}^n,\ldots,x_{M_n}^n)$$

(asymptotically) does not depend on the choice of ordered pair $x_{i_p}^n, x_{i_{p+1}}^n$, that is, $(p-1 < i_p' < i_{p+1}' < M_n-\ell+p+2$

$$\omega_{m_p,m_{p+1}|m_1,\ldots,m_{p-1},m_{p+2},\ldots,m_\ell}(x_1^n,\ldots,x_{p-1}^n,x_{i_p}^n,x_{i_{p+1}}^n,x_{M_n-\ell+p+2}^n,\ldots,x_{M_n}^n)$$
$$= \omega_{m_p,m_{p+1}|m_1,\ldots,m_{p-1},m_{p+2},\ldots,m_\ell}(x_1^n,\ldots,x_{p-1}^n,x_{i_p}^n,x_{i_{p+1}}^n,x_{M_n-\ell+p+2}^n,\ldots,x_{M_n}^n)$$
$$+o(1), \; M_n \to \infty,$$

then it is asymptotically symmetric in variables m_p, m_{p+1}:

$$\omega_{m_p,m_{p+1}|m_1,\ldots,m_{p-1},m_{p+2},\ldots,m_\ell}(x_1^n,\ldots,x_{p-1}^n,x_{i_p}^n,x_{i_{p+1}}^n,x_{M_n-\ell+p+2}^n,\ldots,x_{M_n}^n)$$
$$= \omega_{m_{p+1},m_p|m_1,\ldots,m_{p-1},m_{p+2},\ldots,m_\ell}(x_1^n,\ldots,x_{p-1}^n,x_{i_p}^n,x_{i_{p+1}}^n,x_{M_n-\ell+p+2}^n,\ldots,x_{M_n}^n)$$
$$+o(1). \tag{28}$$

Hence from (27) and (28) it follows that

$$\varphi_\ell(m_1,\ldots,m_{p-1},m_p,m_{p+1},m_{p+2},\ldots,m_\ell) \tag{29}$$
$$= \varphi_\ell(m_1,\ldots,m_{p-1},m_{p+1},m_p,m_{p+2},\ldots,m_\ell)+o(1).$$

This completes the proof of (24). □

Now we are ready to formulate a key lemma, which says that we can replace ρ by τ in the upper bounds (20) without their violation.

Lemma 27 *From an arbitrary sequence of codes C_n, $|C_n| = M_n \to \infty$, $n \to \infty$ we can extract a sequence of subcodes C'_n, $|C'_n| = M'_n \to \infty$ in such a way that for an arbitrary family of $L+1$ different codewords $\mathcal{F} = \{x_{m_1}^n,\ldots,x_{m_{L+1}}^n\} \subset C'_n$ the following relations hold*

$$\rho_{L+1}(x_{i_1}^n,\ldots,x_{i_{L+1}}^n) = \delta(y^n,x_{i_j}^n)+o(1) \tag{30}$$
$$= \tau_{L+1}(x_{i_1}^n,\ldots,x_{i_{L+1}}^n)+o(1),$$
$$j = 1,\ldots,L+1, \; n \to \infty.$$

To prove this result we need one auxiliary lemma.

Lemma 28 *From an arbitrary sequence of sets $C_n = \{x_i^n\}$ consisting of M_n words $x_i^n \in [q]^n$, $M_n \to \infty$, a subsequence $C'_n = \{z_1^n,\ldots,z_{M'_n}^n\}$, $M'_n \to \infty$, can be selected in such a way that any ordered set $\{z_{i_1}^n,\ldots,z_{i_{L+1}}^n, \; i_j < i_{j+1}\} \subset C'_n$ has the property*

$$\omega_{m_1,\ldots,m_{L+1}}(x_{i_1}^n,\ldots,x_{i_{L+1}}^n) = \varphi_{L+1}(m_1,\ldots,m_{L+1})+o(1).$$

The latter means that the types $\omega_{m_1,\ldots,m_{L+1}}(x_{i_1}^n,\ldots,x_{i_{L+1}}^n)$ depend (asymptotically) on the ordered set $(m_1,\ldots,m_{L+1}) \in [q]^{L+1}$ only.

Proof. The proof of Lemma 28 employs Ramsey's Theorem for a uniform hypergraph.

Consider the natural partition of a unit cube $K = [0,1]^{q^{L+1}}$ into $s^{-q^{L+1}}$ subcubes with dimensions $1/s$. Enumerate all subcubes by different natural numbers in an arbitrary way. Next, to each $L+1$-tuple $(m_1,\ldots,m_{L+1}) \in [q]^{L+1}$ assign an edge of K

in one-to-one manner. Then to a vector with coordinates $\omega_{m_1,\ldots,m_{L+1}}(x_{i_1}^n,\ldots,x_{i_{L+1}}^n)$ (with given i_j such that $i_{j_1} < i_{j_2}$ if $j_1 < j_2$) corresponds a point in K and a natural number, which is the number of the subcube in which this point lies.

Now consider a hypergraph on the ordered set of vertices C_n, where the edges are all possible collections of $L+1$ vertices from C_n with the order of vertices induced by the order of vertices in C_n. Assign to each edge $\{x_{i_1}^n,\ldots,x_{i_{L+1}}^n\}$ of the hypergraph a number equal to the number of the subcube that contains the vector with coordinates $\omega_{m_1,\ldots,m_{L+1}}(x_{i_1}^n,\ldots,x_{i_{L+1}}^n)$. It is clear that the number of possible numbers (for a given s) is finite. Therefore, Ramsey's Theorem implies that for any code C_n a subcode C_n' of cardinality M_n' can be extracted such that, for any ordered collection $\{x_{i_1}^n,\ldots,x_{i_{L+1}}^n\} \subset C_n'$, the vector with coordinates $\omega_{m_1,\ldots,m_{L+1}}(x_{i_1}^n,\ldots,x_{i_{L+1}}^n)$ belongs to one subcube and $M_n' \to \infty$. Note that such a subcube can change as n changes.

Finally, letting $s \to \infty$ slowly enough, we can get that all the above conditions are valid and each coordinate $\omega_{m_1,\ldots,m_{L+1}}(i_1,\ldots,i_{L+1})$ (for given $m_1,\ldots,m_{L+1}$) for any choice of $i_1,\ldots,i_{L+1}$ assumes the same value to within $1/s = o(1)$. The lemma is proved.   □

Proof of Lemma 27. From Lemma 28 follows that from the sequence of codes C_n, $M_n = |C_n| \to \infty$ we can extract a sequence of subcodes C_n', $M_n' = |C_n'| \to \infty$ such that for an arbitrary family of different words $\{x_{i_1}^n,\ldots,x_{i_{L+1}}^n\}$ the following relation holds:

$$\omega_{m_1,\ldots,m_{L+1}}(x_{i_1}^n,\ldots,x_{i_{L+1}}^n) = \varphi_{L+1}(m_1,\ldots,m_{L+1}) + o(1) \tag{31}$$

and φ_{L+1} is an asymptotically symmetric function. From here, relation (4), and the choice of function f in relation (4) it follows that $d_H(y^n, x_{i_j}^n)$ asymptotically does not depend on the choice of the word $x_{i_j}^n \in \{x_{i_1}^n,\ldots,x_{i_{L+1}}^n\}$ and hence

$$r_{L+1}(x_{i_1}^n,\ldots,x_{i_{L+1}}^n) = \frac{1}{L+1} \sum_{j=1}^{L+1} d_H(y^n, x_{i_j}^n) = d_H(y^n, x_{i_j}^n) + o(n)$$

$$= \max_{j=1,\ldots,L+1} d_H(y^n, x_{i_j}^n) + o(n).$$

The relations (30) follow.   □

Now we are ready to complete the proof of Theorem 18. Let us have the sequence of codes C_n with codewords of equal weight $\omega = [\lambda n]$. As we have proved before, if $M_n = |C_n| \to \infty$, then the minimal average radius r_{L+1} in the code C_n satisfies relation (20). At the same time, we can extract from the sequence of codes C_n a sequence of subcodes C_n' also of increasing cardinality $M_n' = |C_n'| \to \infty$ such that relations (30) are valid. It means that the maximal average radius r_{L+1}' in the subcode C_n' coincides (asymptotically) with the maximal radius of the L-packing ball t_{L+1}'. Actually we proved more: for an arbitrary set of different codewords $\{x_{i_1}^n,\ldots,x_{i_{L+1}}^n\} \subset C_n'$, the average radius $r_{L+1}(x_{i_1}^n,\ldots,x_{i_{L+1}}^n)$ and the radius of the L-packing ball $t_{L+1}(x_{i_1}^n,\ldots,x_{i_{L+1}}^n)$ asymptotically coincide. We can apply the upper bound (19) to the sequence of codes C_n'. Hence the upper bound (20) for the value of

ρ'_{L+1} for the sequence of codes C'_n is also an upper bound for the value τ'_{L+1} for this sequence of codes. Since $C'_n \subset C_n$, the value of the minimal radius of the L-packing ball $t_{L+1} = \lceil \tau_{L+1} n \rceil$ calculated for the code C_n can be only smaller than the value $t'_{L+1} = \lceil \tau'_{L+1}, n \rceil$ calculated for the code C'_n. Hence the bound (20) is also a bound for τ_{L+1}. This completes the proof of the theorem.   ∎

Lecture 10 List Decoding

In this lecture we consider a problem that has interpretations in both Extremal Combinatorics and Information Theory. It is closely related to the material of the previous lecture. The problem is to obtain bounds on the exponent of the list-of-L decoding error probability. List decoding and multiple packing are closely related notions. List codes were introduced by Elias ([E57]).

First we explain the notion of the decoding. For a given code $C_n \subset \mathcal{X}^n$ decoding means that to each vector $y^n \in \mathcal{Y}^n$ one assigns a unique code vector $x^n \in C_n$. The notion "decoding" comes from Information Theory where a channel causes distortion of the transmitted code vector $x^n \in C_n$, such that the output of the channel is some vector y^n with probability $p(y^n|x^n)$. We consider so-called memoryless channels, which means that the probability $p(y^n|x^n)$ for each $y^n = (y_1,\dots,y_n)$, $x^n = (x_1,\dots,x_n)$ can be decomposed into the product $p(y^n|x^n) = \prod_{i=1}^n p(y_i|x_i)$. The set $\{p(\cdot|x), x \in [q]\}$ is the set of probability distributions on $[q]$. A decoding algorithm assigns to each output y^n of the channel a codeword $x^n \in C_n$ as the result of the decoding. One decoding algorithm, which has the most transparent combinatorial interpretation is the following: to a given output of the channel y^n assign $x^n \in C_n$ such that the distance $d_H(x^n, y^n)$ (in this case $\mathcal{X}^n = \mathcal{Y}^n = \mathcal{H}_q^n$) is minimal over the choice of x^n (minimal distance decoding). This is an optimal decoding (with minimal average probability of the decoding error) for the symmetric channel, that is, the channel generated by the transition probabilities,

$$p(y|x) = \left(\frac{p}{q-1}\right)^{1-\delta_{x,y}} (1-p)^{\delta_{x,y}}, \; x,y \in \mathcal{X} = \mathcal{Y},$$

(the definition of the average decoding error probability comes later). If $d_H(x^n, y^n) < t$, where $d = 2t+1$ is the minimal distance of the code C_n, then the minimal distance decoding procedure guarantees that the result of the decoding will be the input vector x^n. Otherwise in some (but possibly not all) cases an error may occur: the result of decoding may differ from the channel input.

Next we consider a generalization of the previous scheme to the case when the input (code) sequences take values from $C_n \subset \mathcal{X}^n$ and the outputs of the channel are n-tuples from the set $\mathcal{Y}^n$. Consider finite sets $\mathcal{X} = \{1,2,\dots,|\mathcal{X}|\}$ and $\mathcal{Y} = \{1,2,\dots,|\mathcal{Y}|\}$ and a set $\{p(\cdot|j), j \in \mathcal{X}\}$ of probability distributions on $\mathcal{Y}$. For each given $x^n = (x_1,\dots,x_n) \in \mathcal{X}^n$ this set generates a distribution on $\mathcal{Y}^n$:

$$p(y^n|x^n) = \prod_{i=1}^n p(y_i|x_i), \tag{1}$$

thus defining a discrete memoryless channel. Any subset $C_n \subset \mathcal{X}^n$ we call a code. We assume that the elements of C_n (codewords) are indexed by numbers from $1, 2, \ldots, |C_n|$. For a given code $C_n \subset \mathcal{X}^n$ the maximum likelihood decoding is defined as follows: to each vector $y^n \in \mathcal{Y}^n$ a code vector $x^n \in \mathcal{X}^n$ is assigned,

$$x^n = \arg \max_{z^n \in C_n} p(y^n | z^n). \tag{2}$$

If there are several words satisfying (2), we take any one of them. For general channels the maximum likelihood decoding algorithm is an analog of the minimum distance decoding. In the case of symmetric (memoryless) channels these algorithms coincide.

In general, a list-of-L decoding algorithm assigns to each output of the channel $y^n \in \mathcal{Y}^n$ a set of L code sequences $\{x_1^n, \ldots, x_L^n\} \subset C_n$. Let us also define the list-of-L maximum likelihood decoding: that is, for each $y^n \in \mathcal{Y}^n$ a list of L codewords is chosen, $\varphi(y^n) = \{x_1^n, \ldots, x_L^n\} \subset C_n$, such that

$$x_i^n = \arg \max_{z^n \notin \{x_1^n, \ldots, x_{i-1}^n\}} p(y^n | z^n). \tag{3}$$

We use this decoding rule in the last section of the lecture where we derive the expurgation bound.

Below we consider the situation when a codeword $x^n \in C_n$ is transmitted over a discrete memoryless channel and the channel output is the word $y^n \in \mathcal{Y}^n$ with probability $p(y^n | x^n)$. The decoder assigns to each output word $y^n \in \mathcal{Y}^n$ a list of L codewords from C_n according to (3). We say that a decoding error occurs if the transmitted vector x^n is not contained in the list $\varphi(y^n) = \{x_1^n, \ldots, x_L^n\}$. Thus the whole set $\mathcal{Y}^n$ is covered by decoding regions in such a way that each point is L-fold covered. This means that the list decoding rule generates a covering $\{B_i, i = 1, \ldots, |C_n|\}$ of $\mathcal{Y}^n$ such that $y^n \in B_i$ iff $x_i^n \in \varphi(y^n)$. It is more precise to speak about an L-packing of the space $\mathcal{Y}^n$ by decoding regions. This is a generalization of the situation from the previous lecture, where we considered L-packings of the whole space by balls of equal radius. Actually, there we considered the case where every point is covered by not more than L balls and here we consider the case where every point is covered by exactly L regions.

The average list-of-L decoding error probability $\bar{P}_{n,L}(C_n)$ is defined as

$$\bar{P}_{n,L}(C_n) = \frac{1}{|C_n|} \sum_{i=1}^{|C_n|} \sum_{y^n \notin B_i} p(y^n | x_i^n). \tag{4}$$

It can be easily seen that the list-of-L maximum likelihood decoding algorithm (3) minimizes the average list-of-L decoding error probability over the choice of decoding algorithms.

The smaller the average error probability, the better is the choice of the code and decoding algorithm. We also introduce another measure of effectiveness of the decoding (and coding): the maximum of list-of-L decoding error probability $P_{n,L}^m(C_n)$, which is defined as follows

$$P_{n,L}^m(C_n) = \max_i \sum_{y^n \notin B_i} p(y^n|x_i^n). \tag{5}$$

By foregoing remarks these values are indeed probabilities of error of list-of-L decoding. We show that asymptotically these two measures of effectiveness of the decoding are equivalent (see (12), (13)). Define also the following parameters:

$$\bar{\pi}_{n,L}(M) = \min_{C_n:\, |C_n|=M} \bar{P}_{n,L}(C_n), \tag{6}$$

$$\pi_{n,L}^m(M) = \min_{C_n:\, |C_n|=M} P_{n,L}^m(C_n), \; M=1,2,\ldots, \tag{7}$$

$$\bar{P}_{n,L}(R) = \min_{C_n:\, |C_n|\geq |\mathcal{X}|^{nR}} \bar{P}_{n,L}(C_n), \tag{8}$$

$$P_{n,L}^m(R) = \min_{C_n:\, |C_n|\geq |\mathcal{X}|^{nR}} P_{n,L}^m(C_n), \tag{9}$$

$$\bar{E}_L(R) = \limsup_{n\to\infty} -\frac{\ln \bar{P}_{n,L}(R)}{n}, \tag{10}$$

$$E_L^m(R) = \limsup_{n\to\infty} -\frac{\ln P_{n,L}^m(R)}{n}. \tag{11}$$

The parameter $\bar{E}_L(R)$ $(E_L^m(R))$ is the average (maximum) list-of-L decoding error probability exponent.

From the inequalities (see Exercise 7)

$$2\bar{P}_{n,L}\left(R+\frac{\log_{|\mathcal{X}|}2}{n}\right) \geq P_{n,L}^m(R) \geq \bar{P}_{n,L}(R), \tag{12}$$

it follows that

$$\bar{E}_L(R) = E_L^m(R) \overset{\Delta}{=} E_L(R). \tag{13}$$

The form of the function for arbitrary R is not known in general and only upper and lower bounds on $E_L(R)$ are known. Lower bounds are the random-coding bound $E_L^r(R)$ and the expurgation bound $E_L^{ex}(R)$. Usually, there exist $R_1 \leq R_2$ such that, if $R \leq R_1$, then $E_L^{ex}(R) > E_L^r(R)$. Also, if $R_1 < R < R_2$, then $E_L^{ex}(R) = E_L^r(R) = -RL\ln|\mathcal{X}| + \beta$ (for some $\beta > 0$ such that this line is tangent to the sphere-packing bound (14)) and if $R_2 \leq R \leq K$, then $E_L^{ex}(R) < E_L^r(R)$, and $E_L(R) = E_L^r(R) = E^{sp}(R)$, where

$$K = \max_{p_\mathcal{X}} I(X;Y)$$

is the channel capacity and $I(X;Y)$ is the mutual information between the channel input and output:

$$I(X;Y) = \sum_{x\in\mathcal{X},\, y\in\mathcal{Y}} p_\mathcal{X}(x)p(y|x)\log_{|\mathcal{X}|}\frac{p(y|x)}{p_\mathcal{X}(x)}.$$

Here $p_{\mathcal{X}}$ is a probability distribution on $\mathcal{X}$. If $R \in [R_2, K]$ then, as we have already mentioned, $E_L(R)$ coincides with the sphere-packing bound $E^{sp}(R)$ and $E_L^r(R)$, that is,

$$E_L(R) = E^{sp}(R),$$

where $E^{sp}(R)$ satisfies the relation

$$E^{sp}(R) = \sup_{\rho \geq 0}[E_0(\rho) - \rho R \ln|\mathcal{X}|]. \tag{14}$$

Here,

$$E_0(\rho) = -\min_{p_{\mathcal{X}}} \ln \sum_{j=1}^{|\mathcal{Y}|} \left[\sum_{k=1}^{|\mathcal{X}|} p_k (p(j|k))^{1/(1+\rho)} \right]^{1+\rho}$$

and $p_{\mathcal{X}} = (p_1, \ldots, p_{|\mathcal{X}|})$ is a distribution on $\mathcal{X}$. At low rates R this bound can be improved considerably. We have the lower bound (expurgation bound)

$$E_L^{ex}(R) = \sup_{\rho \geq 1}(E_{0,ex}(\rho) - \rho R L \ln|\mathcal{X}|), \tag{15}$$

where

$$E_{0,ex}(\rho) = \tag{16}$$

$$-\rho \min_{p_{\mathcal{X}}} \ln \left(\sum_{(i_1,\ldots,i_{L+1}) \in \mathcal{X}^{L+1}} p_{i_1} \cdots p_{i_{L+1}} \left(\sum_{j=1}^{|\mathcal{Y}|} (p(j|i_1) \cdots p(j|i_{L+1}))^{\frac{1}{L+1}} \right)^{1/\rho} \right).$$

At zero rate (when $|\mathcal{C}_n| = 2^{o(n)} \overset{n \to \infty}{\to} \infty$ or $R = 0$), this bound looks as follows (see Exercise 8)

$$E_L^{ex}(0) = \tag{17}$$

$$-\min_{p_{i_j}} \sum_{(i_1,\ldots,i_{L+1}) \in \mathcal{X}^{L+1}} p_{i_1} \cdots p_{i_{L+1}} \ln \left(\sum_{j=1}^{|\mathcal{Y}|} (p(j|i_1) \cdots p(j|i_{L+1}))^{\frac{1}{L+1}} \right).$$

In Figure 1 the typical behavior of these bounds is shown. One can find more information about the behavior of the exponent of probability of list-of-L decoding error with proofs in [SGB67]. Only the exponent of probability of list-of-L decoding error at low rates is not introduced in that work. While the lower bound $E_L^{ex}(R)$ can be derived for arbitrary L in a similar way as in the case $L = 1$, we introduce this derivation in the last section, the obtaining of the upper bound for $E_L(0)$ involves a lot of new technique. We prove that $E_L^{ex}(0)$ is the true value of $E_L(0)$. The knowledge of $E_L(0)$ allows to construct the so-called straight-line upper bound for $E_L(R)$ at low rates (see the Notes at the end of the chapter).

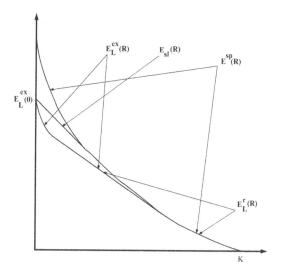

Fig. 1 Bounds on the Error Exponent

§1 The Exponent of List-of-L Decoding Error Probability at Zero Rate

In this section we prove that for a discrete memoryless channel $E_L(0) = E_L^{ex}(0)$. Since $E_L^{ex}(0)$ is a lower bound for $E_L(0)$ (we reproduce the proof of this fact at the end of the lecture), all we need to prove is the upper bound $E_L(0) \leq E_L^{ex}(0)$. In the case $L = 1$ we present the original proof, which allows to find an estimation including not only the main asymptotic term $E_L^{ex}(0)$ but also the rest term.

We formulate the following main theorem:

Theorem 19 (Blinovsky 2001) *For an arbitrary discrete memoryless channel with finite input and output alphabets, we have the following relation:*

$$E_L(0) = E_L^{ex}(0). \tag{18}$$

Proof. The proof of this theorem uses some rather usual technique from Information Theory and also the original technique, which was developed in the previous lecture when we proved the upper bound for the cardinality of a multiple packing.

In the sequel, for any $x_1, \ldots, x_{L+1} \in \mathcal{X}$, we assume that there exists a $y \in \mathcal{Y}$ such that

$$p(y|x_1) \ldots p(y|x_{L+1}) > 0.$$

Otherwise, calculation shows that $E_L^{ex}(0) = \infty$ and the inequality $E_L(0) \leq E_L^{ex}(0)$ is obvious.

Fix $x_1^n, \ldots, x_{L+1}^n \in \mathcal{C}_n$ and define

$$\mu(\lambda) = \ln \sum_{y^n \in \mathcal{Y}^n} p^{\lambda_1}(y^n|x_1^n) \ldots p^{\lambda_{L+1}}(y^n|x_{L+1}^n),$$

where $\lambda = (\lambda_1, \ldots, \lambda_{L+1})$, $\lambda_i \in (0,1)$, and $\sum_i \lambda_i = 1$. In the following reasoning, it suffices to assume that the sums are taken over vectors $y^n \in \mathcal{Y}^n$ such that $p(y^n|x_1^n) \ldots p(y^n|x_{L+1}^n) > 0$ only. The continuous function $\mu(\lambda)$ achieves its minimum on the set $\lambda_i \geq 0$, $\sum_i \lambda_i = 1$ in some point $\lambda^* = (\lambda_1^*, \ldots, \lambda_{L+1}^*)$. W.l.o.g. we assume that $\lambda_{L+1}^* > 0$. For a memoryless channel, we have

$$\mu(\lambda) = \sum_{i=1}^{n} \mu_i(\lambda), \tag{19}$$

where

$$\mu_i(\lambda) = \ln \sum_{j=1}^{|\mathcal{Y}|} \prod_{k=1}^{L+1} p^{\lambda_k}(j|x_{k,i}), \tag{20}$$

$x_k^n = (x_{k,1}, \ldots, x_{k,n})$.

Let

$$T_\lambda(y^n) = \frac{\prod_{k=1}^{L+1} p^{\lambda_k}(y^n|x_k^n)}{\sum_{y^{n\prime} \in \mathcal{Y}^n} \prod_{k=1}^{L+1} p^{\lambda_k}(y^{n\prime}|x_k^n)}$$

and

$$D_k(y^n) = \ln \frac{p(y^n|x_k^n)}{p(y^n|x_{L+1}^n)}, \quad k = 1, \ldots, L+1.$$

For the points y^n for which $p(y^n|x_1^n) \ldots p(y^n|x_{L+1}^n) = 0$ we set $T_\lambda(y^n) = 0$ and they can be disregarded in further considerations. Then

$$p(y^n|x_k^n) = e^{\mu(\lambda) - (\lambda(k), D(y^n))} T_\lambda(y^n), \quad k = 1, \ldots, L,$$

$$p(y^n|x_{L+1}^n) = e^{\mu(\lambda) - (\lambda, D(y^n))} T_\lambda(y^n),$$

where $(\lambda, D(y^n))$ is the scalar product, $\lambda(k) = (\lambda_1, \ldots, \lambda_{k-1}, \lambda_k - 1, \lambda_{k+1}, \ldots, \lambda_{L+1})$. Next, putting $\lambda_{L+1} = 1 - \lambda_1 - \ldots - \lambda_L$, we have

$$\mu_{\lambda_i}'(\lambda) = \sum_{y^n \in \mathcal{Y}^n} \left(\frac{\prod_{k=1}^{L+1} p^{\lambda_k}(y^n|x_k^n)}{\sum_{y^{n\prime} \in \mathcal{Y}^n} \prod_{k=1}^{L+1} p^{\lambda_k}(y^{n\prime}|x_k^n)} \ln \frac{p(y^n|x_i^n)}{p(y^n|x_{L+1}^n)} \right),$$

$$\mu_{\lambda_i}''(\lambda) = \sum_{y^n \in \mathcal{Y}^n} \left(\frac{\prod_{k=1}^{L+1} p^{\lambda_k}(y^n|x_k^n)}{\sum_{y^{n\prime} \in \mathcal{Y}^n} \prod_{k=1}^{L+1} p^{\lambda_k}(y^{n\prime}|x_k^n)} \left(\ln \frac{p(y^n|x_i^n)}{p(y^n|x_{L+1}^n)} \right)^2 \right) - \left(\mu_{\lambda_i}'(\lambda) \right)^2,$$

$i = 1, \ldots, L$. Define the sets $\mathcal{Y}_\lambda^k \subset \mathcal{Y}^n$, $k = 1, \ldots, L$, as follows

$$\mathcal{Y}_\lambda^k = \left\{ y^n \in \mathcal{Y}^n : \left| D_k(y^n) - \mu_{\lambda_k}'(\lambda) \right| \leq \sqrt{(L+1)\mu_{\lambda_k}''(\lambda)} \right\}.$$

Using the Chebyshev inequality

$$T_\lambda \left(\left| R(y^n) - \mathbb{E}(R(\cdot)) \right| \geq A\sqrt{Var(R(\cdot))} \right) \leq \frac{1}{A^2}, \quad A > 0,$$

where $\mathbb{E}$ and *Var* are, respectively, the mathematical expectation and the variance of the random variable $R(y^n)$ according to the distribution T_λ on $\mathcal{Y}^n$, we estimate the probability of the event $\mathcal{Y}^n - \mathcal{Y}_\lambda^k$ (according to distribution T_λ the value $\mu'_{\lambda_k}(\lambda)$ is the expectation and $\mu''_{\lambda_i}(\lambda)$ is the variance of random variable $D_k(y^n)$) and find

$$\sum_{y^n \in \mathcal{Y}_\lambda^k} T_\lambda(y^n) > 1 - \frac{1}{L+1}.$$

Thus, for $\mathcal{Y}_\lambda = \cap_{k=1}^{L} \mathcal{Y}_\lambda^k$, we have

$$\sum_{y^n \in \mathcal{Y}_\lambda} T_\lambda(y^n) > \frac{1}{L+1}. \tag{21}$$

The decoding rule divides the set $\mathcal{Y}^n$ into $L+1$ regions $\mathcal{Y}_i$ such that

$$\bigcap_{i=1}^{L+1} \mathcal{Y}_i = \emptyset \tag{22}$$

and any vector $y^n \in \mathcal{Y}^n$ belongs to exactly L of the subsets $\mathcal{Y}_i$. If $y^n \in \mathcal{Y}_{i_1} \cap \ldots \cap \mathcal{Y}_{i_L}$, then the result of the decoding is a list of code vectors $\{x_{i_1}^n, \ldots, x_{i_L}^n\} \subset \mathcal{C}_n$.

Then we have the following estimates for the decoding error probability $P_{e,i}$ in the case when vector x_i^n was transmitted over the channel:

$$P_{e,i} = \sum_{y^n \in \mathcal{Y}_i^c} p(y^n | x_i^n) \geq \sum_{y^n \in \mathcal{Y}_i^c \cap \mathcal{Y}_\lambda} p(y^n | x_i^n)$$

$$\geq \exp\left(\mu(\lambda) - (\lambda(i), \mu'_\lambda(\lambda)) - \left(\lambda(i), \sqrt{(L+1)\mu''_\lambda(\lambda)}\right)\right.$$

$$\left. + 2(\lambda_i - 1)\sqrt{(L+1)\mu''_{\lambda_i}(\lambda)}\right) \sum_{y^n \in \mathcal{Y}_i^c \cap \mathcal{Y}_\lambda} T_\lambda(y^n), \ i = 1, \ldots, L,$$

$$P_{e,L+1} = \sum_{y^n \in \mathcal{Y}_{L+1}^c} p(y^n | x_i^n) \geq \sum_{y^n \in \mathcal{Y}_{L+1}^c \cap \mathcal{Y}_\lambda} p(y^n | x_i^n)$$

$$\geq \exp\left(\mu(\lambda) - (\lambda, \mu'_\lambda(\lambda)) - \left(\lambda, \sqrt{(L+1)\mu''_\lambda(\lambda)}\right)\right) \sum_{y^n \in \mathcal{Y}_{L+1}^c \cap \mathcal{Y}_\lambda} T_\lambda(y^n).$$

Here

$$\mu'_\lambda(\lambda) = \left(\mu'_{\lambda_1}(\lambda), \ldots, \mu'_{\lambda_L}(\lambda), 0\right),$$

$$\sqrt{\mu''_\lambda(\lambda)} = \left(\sqrt{\mu''_{\lambda_1}(\lambda)}, \ldots, \sqrt{\mu''_{\lambda_L}(\lambda)}, 0\right).$$

Estimate (21) implies the validity of at least one of the following $L+1$ inequalities:

$$\sum_{y^n \in \mathcal{Y}_i^c \cap \mathcal{Y}_\lambda} T_\lambda(y^n) \geq \frac{1}{(L+1)^2}, \ i = 1, \ldots, L+1.$$

Thus, we have proved the following result:

Lemma 29 *At least one of the inequalities*

$$P_{e,k} \geq \frac{1}{(L+1)^2} exp\left\{ \mu(\lambda) - (\lambda(k), \mu'_\lambda(\lambda)) - \left(\lambda(k), \sqrt{(L+1)\mu''_\lambda(\lambda)}\right) \right.$$

$$\left. + 2(\lambda_k - 1)\sqrt{(L+1)\mu''_{\lambda_k}(\lambda)} \right\},$$

$$k = 1, \ldots, L,$$

$$P_{e,L+1} \geq \frac{1}{(L+1)^2} exp\left\{ \mu(\lambda) - (\lambda, \mu'_\lambda(\lambda)) - \left(\lambda, \sqrt{(L+1)\mu''_\lambda(\lambda)}\right) \right\}$$

holds.

Using (19) and (20) and taking the derivative, we obtain the relations

$$\mu''_{\lambda_i}(\lambda) = \sum_{m=1}^{n} \sum_{j' \in \mathcal{Y}} \left(\frac{\prod_{k=1}^{L+1} p^{\lambda_k}(j'|x_{k,m})}{\sum_{j \in \mathcal{Y}} \prod_{k=1}^{L+1} p^{\lambda_k}(j|x_{k,m})} \left(\ln \frac{p(j'|x_{i,m})}{p(j'|x_{(L+1),m})} \right)^2 \right) \quad (23)$$

$$- \left(\sum_{m=1}^{n} \sum_{j' \in \mathcal{Y}} \frac{\prod_{k=1}^{L+1} p^{\lambda_k}(j'|x_{k,m})}{\sum_{j \in \mathcal{Y}} \prod_{k=1}^{L+1} p^{\lambda_k}(j|x_{k,m})} \ln \frac{p(j'|x_{im})}{p(j'|x_{(L+1),m})} \right)^2 \quad (24)$$

$$\leq n(\ln P_{min})^2,$$

where $\prod_{k=1}^{L+1} p^{\lambda_k}(j|x_{k,m}) = 0$ if $\prod_{k=1}^{L+1} p(j|x_{k,m}) = 0$. Here we used the inequality

$$\left| \ln \frac{p(j|x_{i,k})}{p(j|x_{(L+1),k})} \right| \leq -\ln P_{min},$$

where

$$P_{min} = \min_{j \in \mathcal{Y}, \, x \in \mathcal{X}: \, p(j|x) > 0} p(j|x).$$

At first we concentrate our attention on the case $L = 1$. In this case we take care about the rest, asymptotically vanished terms and derive the precise lower bound (30) for the maximal probability of list-of-L decoding error (which is valid for an arbitrary code and decoding algorithm). Then in the case of arbitrary L we find only the main term $E_L^{ex}(0)$ of the asymptotics of the lower bound of the probability of error. Caution: The upper bound for the exponent of the probability of decoding error is the asymptotics of the lower bound for the probability of the decoding error!

Let

$$\mu_{m,m'}(s) = \ln \sum_j p^s(j|m) p^{1-s}(j|m'), \quad s \in [0,1], \quad m, m' \in \mathcal{X}.$$

For a pair of code vectors $x_i^n, x_k^n \subset C_n$, let $\omega_{m,m'}(i,k)$ be their joint type. Denote

$$D(i,k) = -\min_{s \in [0,1]} \mu(s,i,k),$$

where

$$\mu(s,i,k) = \sum_{m,m'} \omega_{m,m'}(i,k)\mu_{m,m'}(s).$$

Recall that λ^* is the point where $\mu(\lambda)$ achieves its minimum. In the one-dimensional case $L = 1$, if $\mu(\lambda_1) \overset{\Delta}{=} \mu(\lambda_1, 1 - \lambda_1)$ achieves its minimum in the point $\lambda_1 \in (0,1)$, then $\mu'_{\lambda_1} = 0$; otherwise if $\lambda_1^* = 0$, we have

$$\lambda_1^* \mu'_{\lambda_1}(\lambda_1^*) = 0, \ (\lambda_1^* - 1)\mu'_{\lambda_1}(\lambda_1^*) \leq 0.$$

We avoid the case $\lambda_1^* = 1$, because we have the assumption that $\lambda_{L+1}^* > 0$, which means in our case that $1 - \lambda_1^* > 0$. Hence we can omit terms with μ' in the estimation of $P_{e,k}$, $1 \leq k \leq L+1$ from Lemma 29. Taking into account this fact, estimations (23), and the definition of $D(i,k)$, we have the following estimation for maximal probability of the decoding error $P_e(\mathcal{C}_n)$ of the code $\mathcal{C}_n$

$$\ln P_e(\mathcal{C}_n) \geq -n\left(D_{\min} + \sqrt{\frac{2}{n}}\ln\frac{1}{P_{\min}} + \frac{2\ln 2}{n}\right), \tag{25}$$

where

$$D_{\min} = \min_{i \neq k} D(i,k).$$

Since the function $\mu(s,i,k)$ is $\cup$-convex ($\mu_s''(s,i,k) \geq 0$, $s \in [0,1]$), we have

$$\mu(s,i,k) \geq \mu(1/2,i,k) + (s - 1/2)\mu_s'(1/2,i,k).$$

Under the condition

$$|\omega_{m,m'}(i,k) - \omega_{m,m'}(k,i)| < \gamma$$

the following inequality holds:

$$|\mu_s'(1/2,i,k) - \mu_s'(1/2,k,i)| < \gamma,$$

and since

$$\mu_s'(1/2,i,k) = -\mu_s'(1/2,k,i)$$

we have

$$|\mu_s'(1/2,i,k)| < \gamma/2.$$

From here it follows that

$$\mu(s,i,k) \geq \mu(1/2,i,k) - \gamma/4$$

and thus

$$D(i,k) \leq -\mu(1/2,i,k) + \gamma/4.$$

To estimate the value $D_{\min}$ we use a Plotkin-type bound:

$$D_{\min} \leq \overline{D} = \frac{\sum_{i \neq k} D(i,k)}{M(M-1)} \tag{26}$$

$$\leq -\frac{1}{M(M-1)} \sum_{i \neq k} \mu(1/2,i,k) + \gamma/4$$

$$\leq -\frac{1}{M(M-1)} \sum_{i \neq k} \sum_{m,m' \in \mathcal{X}} \omega_{m,m'}(i,k)\mu_{m,m'}(1/2) + \gamma/4$$

$$\leq -\frac{1}{n} \sum_{t=1}^{n} q_m^t q_{m'}^t \mu_{m,m'}(1/2) + \gamma/4.$$

Here in the last inequality we used the equality

$$\frac{1}{M(M-1)} \sum_{i \neq k} \omega_{m,m'}(i,k) = \frac{1}{n} \sum_{t=1}^{n} q_m^t q_{m'}^t,$$

where $\{q_m^t\}$ is the type of the tth column of the code matrix of dimension $M_n \times n$. Optimization of the RHS of (26) over $q = \{q_m^t\}$ leads to the inequality

$$D_{\min} \leq -\min_q \sum_{m,m'} q_m q_{m'} \mu_{m,m'}(1/2) + \gamma/4 = E_L^{ex}(0) + \gamma/4.$$

Hence, we obtain the inequality

$$\ln P_e(\mathcal{C}) \geq -n \left(E_1^{ex}(0) + \gamma/4 + \sqrt{\frac{2}{n}} \ln \frac{1}{P_{\min}} + \frac{2\ln 2}{n} \right). \tag{27}$$

Now we use Ramsey's Theorem, which tells us about the magnitude of the Ramsey numbers. We have that from a sequence of codes $\mathcal{C}_n$ with $M_n = |\mathcal{C}_n| \to \infty$ we can extract a sequence of subcodes $\mathcal{C}_n'$ with $|\mathcal{C}_n'| = M_n' \geq \left\lfloor \delta^{|\mathcal{X}|^2} \log_{\delta^{-|\mathcal{X}|^2}} \frac{M_n}{\delta^{-|\mathcal{X}|^2}+3} \right\rfloor$, $\delta \in (0,1)$, where $1/\delta$ is an integer such that for arbitrary $m, m' \in \mathcal{X}$ and $i < k$

$$|\omega_{m,m'}(i,k) - \varphi_2(m,m')| < \delta, \tag{28}$$

where $\varphi_2(m,m')$ is some function of m and m' (see Exercise 9).

Then by Lemma 25, if (28) is valid, then we have

$$|\omega_{m,m'}(i,k) - \omega_{m,m'}(k,i)| < |\omega_{m,m'}(i,k) - \varphi_2(m,m')| \tag{29}$$
$$+ |\omega_{m,m'}(k,i) - \varphi_2(m',m)| + |\varphi_2(m,m') - \varphi_2(m',m)|$$
$$< 2\delta + \frac{6}{\sqrt{M_n'}} + 6\sqrt{\delta} + 2\delta$$

$$= 4\delta + 6\sqrt{\delta} + \cfrac{6}{\sqrt{\left\lceil \delta|\mathcal{X}|^2 \log_{\delta-|\mathcal{X}|^2} \frac{M_n}{\delta-|\mathcal{X}|^2+3}\right\rceil - 1}} = \gamma$$

Substituting (29) into (27) we obtain the bound

$$\ln P_e(\mathcal{C}_n) \geq -n\left(E_1^{ex}(0) + \delta + \frac{3}{2}\sqrt{\delta} \right. \tag{30}$$

$$+ \frac{3}{2}\cfrac{6}{\sqrt{\left\lceil \delta|\mathcal{X}|^2 \log_{\delta-|\mathcal{X}|^2} \frac{M_n}{\delta-|\mathcal{X}|^2+3}\right\rceil - 1}} + \sqrt{\frac{2}{n}}\ln\frac{1}{P_{\min}} + \left. \frac{2\ln 2}{n} \right).$$

This is our final result in the case $L = 1$.

Now we continue with the general case of arbitrary L. We write all rest terms in different expressions as $o(n)$, not taking care of their explicit expressions.

Lemma 29 and estimate (23) imply the validity of at least one of the following estimates:

$$\ln P_{e,k} \geq \mu(\lambda) - (\lambda(k), \mu'_\lambda(\lambda)) + o(n) \tag{31}$$

for some $k = 1, \ldots, L$ or

$$\ln P_{e,L+1} \geq \mu(\lambda) - (\lambda, \mu'_\lambda(\lambda)) + o(n).$$

Note that if

$$\lambda^* = (\lambda_1^*, \ldots, \lambda_{L+1}^*) = \arg\min_\lambda \mu(\lambda),$$

then (31) implies that for some $i = 1, \ldots, L+1$ we have

$$\ln P_{e,i} \geq \mu(\lambda^*) + o(n). \tag{32}$$

This is because if λ^* is an interior point of the set

$$\{\lambda = (\lambda_1, \ldots, \lambda_{L+1}) : \sum_i \lambda_i = 1, \ \lambda_i \geq 0\},$$

then $\mu'_\lambda(\lambda^*) = 0$. If λ^* belongs to the boundary set and $\lambda_{m_1}^* = \lambda_{m_2}^* = \cdots = \lambda_{m_\ell}^* = 0$, $\lambda_m \in (0,1)$, $\mu'_{\lambda_m} = 0$ for $m \notin \{m_1, \ldots, m_\ell\}$, then

$$(\lambda^*, \mu'_\lambda(\lambda^*)) = 0,$$
$$(\lambda^*(k), \mu'_\lambda(\lambda^*)) \leq 0.$$

Hence (32) follows.

Let $\omega_{k_1,\ldots,k_{L+1}}(i_1,\ldots,i_{L+1})$ be the joint type of an ordered sequence $x_{i_1}^n, \ldots, x_{i_{L+1}}^n \in \mathcal{C}_n$ of vectors. Then we have

$$\mu(\lambda) = n \sum_{(k_1,\ldots,k_{L+1})\in\mathcal{X}^{L+1}} \mu_{k_1,\ldots,k_{L+1}}(\lambda)\omega_{k_1,\ldots,k_{L+1}}(i_1,\ldots,i_{L+1}),$$

where

$$\mu_{k_1,\ldots,k_{L+1}}(\lambda) = \ln\left(\sum_{j\in\mathcal{Y}}\prod_{m=1}^{L+1}p^{\lambda_m}(j|k_m)\right).$$

Next we are going to prove that for an arbitrary sequence of codes $\mathcal{C}_n \subset \mathcal{X}^n$ such that $|\mathcal{C}_n| \overset{n\to\infty}{\to} \infty$, there exists a set of code vectors $\{x_{i_1}^n,\ldots,x_{i_{L+1}}^n\} \subset \mathcal{C}_n$ such that

$$\lim_{n\to\infty}\frac{\mu(\lambda^*)}{n} \geq \min_{p_{\mathcal{X}}}\sum_{(k_1,\ldots,k_{L+1})\in\mathcal{X}^{L+1}}p_{k_1}\cdots p_{k_{L+1}}\mu_{k_1,\ldots,k_{L+1}}(\lambda^{**}) = -E_L^{ex}(0). \quad (33)$$

Here $p_k \geq 0$, $\sum_{k\in\mathcal{X}}p_k = 1$, and $\lambda^{**} = (1/(L+1),\ldots,1/(L+1))$.
    Thus

$$\lim_{n\to\infty}\frac{\ln P_{n,L}^m(\mathcal{C}_n)}{n} \geq -E_L^{ex}(0).$$

Relation (31) is the key in the proof of the theorem.
    We call a code having the property (31) almost good. Since the function $\mu(\lambda)$ is $\cup$-convex (see Exercise 10), we have

$$\mu(\lambda^*) \geq \mu(\lambda^{**}) + (\lambda^* - \lambda^{**}, \mu_\lambda'(\lambda^{**})).$$

Now consider a sequence of codes $\mathcal{C}_n$ of cardinality $M_n \to \infty$. As relation (31) demonstrates, there exists a sequence of subcodes $\mathcal{C}_n'$ of cardinality $M_n' \to \infty$ such that for any sequence $\{z_{j_1}^n,\ldots,z_{j_{L+1}}^n\} \subset \mathcal{C}_n'$, $j_p \neq j_q$, $p \neq q$ and for some constants c_i, $|c_i| < 1$, $i = 1,\ldots,L$ (which depend on $\mathcal{C}_n'$ only), we have

$$\mu(\lambda^*) \geq \xi(\lambda^{**}) + \sum_{i=1}^{L}c_i\xi_{\lambda_i}'(\lambda^{**}) + o(n), \quad (34)$$

where

$$\xi(\lambda) = n\sum_{(k_1,\ldots,k_{L+1})\in\mathcal{X}^{L+1}}\varphi_{L+1}(k_1,\ldots,k_{L+1})\mu_{k_1,\ldots,k_{L+1}}(\lambda)$$

$$\mu_{k_1,\ldots,k_{L+1}}(\lambda) = \ln\sum_{j=1}^{|\mathcal{Y}|}\prod_{i=1}^{L+1}p^{\lambda_i}(j|k_i)$$

and $\varphi_{L+1}(k_1,\ldots,k_{L+1})$ is a symmetric function. Indeed, λ^* depends only on types

$$\omega_{k_1,\ldots,k_{L+1}}(x_1^n,\ldots,x_{L+1}^n) = \varphi_{L+1}(k_1,\ldots,k_{L+1}) + o(1).$$

which, in turn, asymptotically depend only on the nonordered set $(k_1,\ldots,k_{L+1}) \in \mathcal{X}^{L+1}$ and hence λ^* is (asymptotically) the same for all sets of code vectors $(z_{j_1}^n,\ldots,z_{j_{L+1}}^n) \subset \mathcal{C}_n'$, $j_p \neq j_q$, $p \neq q$. From this follows that

$$c_i = \lambda^* - \lambda^{**} = const,$$

where *const* means that c_i does not depend on the choice of $(z_{j_1}^n,\ldots z_{j_{L+1}}^n) \subset \mathcal{C}_n'$, $j_p \neq j_q$, $n \neq q$. The maximum list-of-L decoding error probability for $\mathcal{C}_n$ is greater than the same probability for $\mathcal{C}_n'$, as calculation shows. Thus,

$$\ln P_{n,L}^m(\mathcal{C}_n) \geq \ln P_{n,L}^m(\mathcal{C}_n') \geq -\left(-\xi(\lambda^{**}) - \sum_{i=1}^{L} c_i \xi_{\lambda_i}'(\lambda^{**})\right) + o(n). \qquad (35)$$

Finally note that since $\varphi_{L+1}(k_1, \ldots, k_{L+1})$ is a symmetric function, we have

$$\xi_{\lambda_i}'(\lambda^{**}) = 0, \; i = 1, \ldots, L. \qquad (36)$$

Next we use a Plotkin-type bound:

$$-\min_{\{x_1^n, \ldots, x_{L+1}^n\} \subset \mathcal{C}_n'} \mu(\lambda^{**}) \leq -\xi(\lambda^{**}) + o(n)$$

$$= -\sum_{\{x_1^n, \ldots, x_{L+1}^n\} \subset \mathcal{C}_n'} \frac{\xi(\lambda^{**})}{M_n'^{(L+1)}} + o(n)$$

$$= -n \sum_{\{x_1^n, \ldots, x_{L+1}^n\} \subset \mathcal{C}_n'} \sum_{(k_1, \ldots, k_{L+1}) \in \mathcal{X}^{L+1}} \frac{\varphi_{L+1}(k_1, \ldots, k_{L+1}) \mu_{k_1, \ldots, k_{L+1}}(\lambda^{**})}{M_n'^{(L+1)}}$$
$$+ o(n)$$

$$= -n \sum_{\{x_1^n, \ldots, x_{L+1}^n\} \subset \mathcal{C}_n'} \sum_{(k_1, \ldots, k_{L+1}) \in \mathcal{X}^{L+1}} \frac{\omega_{k_1, \ldots, k_{L+1}}(x_1^n, \ldots, x_{L+1}^n) \mu_{k_1, \ldots, k_{L+1}}(\lambda^{**})}{M_n'^{(L+1)}}$$
$$+ o(n)$$

$$= -\sum_{m=1}^{n} \sum_{(k_1, \ldots, k_{L+1}) \in \mathcal{X}^{L+1}} p_{k_1}(m) \ldots p_{k_{L+1}}(m) \mu_{k_1, \ldots, k_{L+1}}(\lambda^{**}) + o(n),$$

where $p_k(m)$ is the type of the mth column of the matrix of code $\mathcal{C}_n'$. In the last equality we used the relations

$$\sum_{\{x_1^n, \ldots, x_{L+1}^n\} \subset \mathcal{C}_n'} \frac{\omega_{k_1, \ldots, k_{L+1}}(x_1^n, \ldots, x_{L+1}^n)}{M_n'^{(L+1)}} = \frac{1}{n} \sum_{m=1}^{n} p_{k_1}(m) \ldots p_{k_{L+1}}(m) + o(1).$$

Our last step in the proof of the theorem follows from (35):

$$\ln P_{n,L}^m(\mathcal{C}_n) \geq \ln P_{n,L}^m(\mathcal{C}_n') \geq \xi(\lambda^{**}) + o(n) \qquad (37)$$

$$\geq \min_{\{p_k(m)\}} \sum_{m=1}^{n} \sum_{(k_1, \ldots, k_{L+1}) \in \mathcal{X}^{L+1}} p_{k_1}(m) \ldots p_{k_{L+1}}(m) \mu_{k_1, \ldots, k_{L+1}}(\lambda^{**}) + o(n)$$

$$\geq \sum_{m=1}^{n} \min_{\{p_k(m)\}} \sum_{(k_1, \ldots, k_{L+1}) \in \mathcal{X}^{L+1}} p_{k_1}(m) \ldots p_{k_{L+1}}(m) \mu_{k_1, \ldots, k_{L+1}}(\lambda^{**}) + o(n)$$

$$= -n E_L^{ex}(0) + o(n).$$

This completes the proof of the theorem. ∎

§2 The Expurgation Bound

To complete this lecture we show how to prove the expurgation bound for the reliability function $E_L(R)$.

Let us fix some $L+1$ code vectors $x_{i_1}^n, \ldots, x_{i_{L+1}}^n \in C_n$. If $\mathcal{Y}_{i_j}$ is the decoding region for the vector $x_{i_j}^n$, then the probability P'_{e,i_s} that $p(y^n|x_{i_j}^n) \geq p(y^n|x_{i_s}^n)$ for some $j \in [L+1] \setminus m$ satisfies the inequality

$$P'_{e,i_s} \leq \sum_{y^n \in \mathcal{Y}_{i_s}^c} \prod_{j \geq 1,\, j \neq s}^{L} \left(\frac{p(y^n|x_{i_j}^n)}{p(y^n|x_{i_s}^n)} \right)^{\frac{1}{L+1}} p(y|x_{i_s}^n) \leq \sum_{y^n \in \mathcal{Y}^n} \prod_{j=1}^{L+1} (p(y^n|x_{i_j}^n))^{\frac{1}{L+1}}.$$

Here we used the fact that $p(y^n|x_{i_j}^n) \geq p(y^n|x_{i_s}^n)$ in the area $y^n \in \mathcal{Y}_{i_s}$.

Then, using the additive bound for the probability of the union of events, we see that the probability of error of the list-of-L decoding when the code vector $x_s \in C_n$ was transmitted satisfies the inequality

$$P_{e,s} \leq \sum_{(s_1,\ldots,s_L) \in [|C_n|]^L,\, s_i \neq s} \sum_{y^n} \left(p(y^n|x_s^n) \prod_{j=1}^{L} p(y^n|x_{s_j}^n) \right)^{1/(L+1)}.$$

If $\gamma \in (0,1]$, then on the basis of $(\sum a_i)^\gamma \leq \sum a_i^\gamma$, $a_i \geq 0$, we have

$$P_{e,s}^\gamma \leq \sum_{(s_1,\ldots,s_L) \in [|C_n|]^L,\, s_i \neq s} \left(\sum_{y^n} \left(p(y^n|x_s^n) \prod_{j=1}^{L} p(y^n|x_{s_j}^n) \right)^{1/(L+1)} \right)^\gamma.$$

Consider an ensemble of codes in which each code vector is selected independently with probability $Q(x^n)$. Then

$$\mathbb{E}(P_{e,s}^\gamma) \leq \qquad\qquad\qquad\qquad\qquad\qquad\qquad\qquad (38)$$

$$\sum_{\substack{(s_1,\ldots,s_L) \in [|C_n|]^L \\ s_i \neq s}} \left[\sum_{x_1^n,\ldots,x_{L+1}^n \in \mathcal{X}^n} \prod_{k=1}^{L+1} Q(x_k^n) \left(\sum_{y^n} (p(y^n|x_s^n) \prod_{j=1}^{L} p(y^n|x_{s_j}^n))^{\frac{1}{L+1}} \right)^\gamma \right].$$

Now we start with an ensemble of codes each with $M'_n = 2M_n - 1$ codewords. Then we show that, for at least one code in the ensemble, there are at least M codewords for which $P_{e,s}$ satisfies a given bound. This bound will be derived in terms of an arbitrary parameter, $\gamma > 0$, which can be optimized. Over the ensemble of codes, $P_{e,s}^\gamma$ is a random variable for each $s \in [M'_n]$. Applying the Chebyshev inequality to this random variable we obtain

$$P(P_{e,s}^\gamma \geq 2\mathbb{E}(P_{e,s}^\gamma)) \leq \frac{1}{2}. \qquad\qquad\qquad\qquad\qquad (39)$$

Then for any $\gamma > 0$, there is at least one code in the ensemble of codes with $M_n' = 2M_n - 1$ code vectors for which at least M code vectors satisfy the inequality

$$P_{e,s} < 2^{1/\gamma} \mathbb{E}^{1/\gamma}(P_{e,s}^\gamma). \tag{40}$$

Indeed, let for each s, φ_s be an indicator random variable. Let $\varphi_s = 1$ for the codes in which (40) is satisfied and $\varphi_s = 0$ otherwise. From (39), the probability in (40) is at least $1/2$ and thus $\mathbb{E}\varphi_s \geq 1/2$. The number of code vectors in a randomly selected code satisfying (40) is equal to $\sum_{s=1}^{M_n'} \varphi_s$.

The expectation of the number of words that satisfy (40) is thus

$$\sum_{s=1}^{M_n'} \mathbb{E}(\varphi_s) \geq \frac{M_n'}{2}.$$

It follows that there is at least one code for which $\sum_s \varphi_s \geq M_n'/2$ holds and for such a code $\sum_s \varphi_s \geq M_n$. If all but M words satisfying (40) are expurgated from the code, then the decoding regions of the remaining code vectors cannot decrease, and thus we have constructed a code with M_n code vectors that satisfy (40).

Now, returning to (38), we see that x_s^n is a dummy variable of summation, the term in square brackets is independent of s and we have $M_n' = 2(M_n - 1)$ identical terms. Substituting (38) with modifications into (40), we have

$$P_{e,s} <$$

$$2^\rho \left\{ 2(M_n - 1) \sum_{x_1^n, \ldots, x_{L+1}^n \in \mathcal{X}^n} \prod_{j=1}^{L+1} Q(x_j^n) \left[\sum_{y^n \in \mathcal{Y}^n} \left(\prod_{j=1}^{L+1} p(y^n | x_j^n) \right)^{\frac{1}{L+1}} \right]^{1/\rho} \right\}^\rho,$$

where $\rho \geq 1$. We restrict $Q(x^n)$ to being a product distribution,

$$Q(x^n) = \prod_{j=1}^n Q(x_j).$$

Then using (1) we obtain the inequality

$$P_{e,s} <$$

$$(4(M_n - 1))^\rho \left\{ \sum_{(k_1, \ldots, k_{L+1}) \in \mathcal{X}^{L+1}} \prod_{i=1}^{L+1} Q(k_i) \left[\sum_{j \in \mathcal{Y}} \left(\prod_{i=1}^{L+1} p(j|k_i) \right)^{1/(L+1)} \right]^{1/\rho} \right\}^{\rho n}.$$

Taking the natural logarithm on both sides of this inequality, then dividing the expression by $n \to \infty$, and optimizing the RHS over the distribution Q and parameter ρ, we obtain that the expurgation exponent (15) is a lower bound for $E_L(R)$.

Notes to Chapter III

Theorem 14 was proved by Frankl and Rödl in [FR85]. Corollary 2 was proved by Frankl and Füredi in [FF87]. The important Corollary 3 was proved as the first in this list by Rödl in [R85].

The results of Lecture 9, in particular the upper bound (20) and Theorem 18, were obtained by Blinovsky in [B05b]. Extensions to general sum-type metric spaces can be found in [AB08]. Lemmas 24 and 25 were obtained by Komlós in [K90]. In the binary case one can avoid applying the Komlós Lemma, because the joint types $\omega_{m_1,\ldots,m_{L+1}}(x_1^n,\ldots,x_{L+1}^n)$ in that case can be expressed as linear combinations of types $\omega_{1,\ldots,1}(x_{j_1}^n,\ldots,x_{j_\ell}^n)$, $(j_1,\ldots,j_\ell) \subset (1,\ldots,L+1)$. Hence applying Ramsey's Theorem as before we extract a subcode such that types $\omega_{1,\ldots,1}(x_{j_1}^n,\ldots,x_{j_\ell}^n) = \omega_\ell$ do not depend on the choice of $(x_{j_1}^n,\ldots,x_{j_\ell}^n)$ (the order here does not matter, because $\omega_{1,\ldots,1}(x_{j_1}^n,\ldots,x_{j_\ell}^n)$ is automatically symmetric). Then $\omega_{m_1,\ldots,m_{L+1}}(x_1^n,\ldots,x_{L+1}^n) = \chi(\{\omega_\ell\})$ is also a symmetric function of $(m_1,\ldots,m_{L+1})$, because the coefficients of the linear form

$$f(\{\omega_\ell\}) = \sum_{i=1}^{L+1} c_\ell \omega_\ell$$

depend only on the number of ones in the sequence $(i_1,\ldots,i_{L+1})$. More about the binary case the reader can find in [B97] (see also Exercise 11).

Next we give a proof of the Johnson bound. Let $\mathcal{C}_n \subset [q]^n$ be an arbitrary code. Then (Exercise 12)

$$q^n \max_{z \in [q]^n} |\mathcal{C}_n \cap B(z,\omega)| \geq \sum_{z \in [q]^n} |\mathcal{C}_n \cap B(z,\omega)| = |\mathcal{C}_n||B(x,\omega)|. \tag{41}$$

On the other hand, if $\mathcal{C}_n$ is a code with given minimal average radius $[\rho_{L+1}n]$ (L-packing radius $[n\tau_{L+1}]$) then $|\mathcal{C}_n \cap B(z,\omega)| \leq |\mathcal{C}_{n,\omega}|$, where $\mathcal{C}_{n,\omega}$ is a code of maximal volume consisting of words of weight ω with given minimal average radius $[\rho_{L+1}n]$ (L-packing radius $[n\tau_{L+1}]$). Then from (41) we have

$$\frac{\log_q |\mathcal{C}_n|}{n} \leq 1 - \frac{\log_q B(x,\omega)}{n} + \frac{\log_q |\mathcal{C}_{n,\omega}|}{n}. \tag{42}$$

Now if $|\mathcal{C}_{n,\omega}| = 2^{o(n)}$, using (42) and the inequality from Exercise 13 we have the Johnson bound

$$R = \frac{\log_q |\mathcal{C}_n|}{n} \leq 1 - H_q\left(\frac{\omega}{n}\right) + o(1), \ n \to \infty.$$

The proof of Theorem 19, which we presented here, was obtained by Blinovsky [B01a]. There exists also another proof of this theorem in [B01b].

In conclusion, we give without proof a result that establishes an upper bound for the reliability function $E_L(R)$ at low rates.

The following inequality holds:

$$\bar{\pi}_{(n_1+n_2),L_2}(M) \geq \bar{\pi}_{n_1,L_1}(M)\pi_{n_2,L_2}^m(L_1+1).$$

Assume for simplicity that $E_L(0) < \infty$. Then from the last inequality, the sphere-packing bound, and the proved theorem, one can get that the following function is an upper bound (straight line bound) for the reliability function:

$$E_{s\ell}(R) = \lambda E^{sp}(R_0) + (1-\lambda)E_L^{ex}(0), \tag{43}$$

where $\lambda = R/R_0$ and R_0 is the tangent point of the functions $E_{s\ell}(R)$ and $E^{sp}(R)$. The proofs of these facts can be found in [SGB67] or in [G68].

Exercises

1. Using counting arguments show the additive upper bound for the probability from Lemma 21 and prove the following: That probability is upper bounded by the value

$$\binom{n}{t}\binom{\binom{n-t}{k-t}}{s}\left(\binom{n-t-1}{k-t-1}\binom{k}{t+1}\right)^s \left(\frac{d}{\binom{n-t}{k-t}}\right)^{2s}.$$

2. Using Lemma 20 prove Corollary 3.
3. Prove Proposition 11. For the proof of (2) and (3) use (1) (Lecture 8) and the inversion formulas

$$\sum_{i=k}^{n}\binom{i}{k}a_i c_{ik} = b_k,$$

$$\sum_{k=0}^{n}(-1)^k b_k = \sum_{i=0}^{n}\sum_{k=0}^{i}a_i\binom{i}{k}(-1)^k c_{ik}.$$

To prove (4) use the fact that $P(0) \geq 1$.
4. For a graph $\mathcal{G} = (\mathcal{V}, \mathcal{E})$ and all $u, v \in \mathcal{V}$ define $u \sim v$ iff $u = v$ or no maximum matching misses both u and v. Prove that $\sim$ is an equivalence. First prove that if $A, B \subset \mathcal{V}$, $|A| < |B|$ and there exists a maximum matching that avoids A and one that avoids B, then there exists a maximum matching that avoids A and at least one point of $B \setminus A$.
   This fact and Gallai's Theorem we took from [LP86].
5. Suppose that $x(1) + x(2) = c$, $x(i) \geq 0$. Then

$$\gamma(\{j_i\}; \{x(i)\}) = \sum_{\pi(\{j_i\})} x^{j_1}(1)(c - x(1))^{j_2}\prod_{i=3}^{q} x^{j_i}(i).$$

Prove that the derivative of this function in $x(1)$ equals zero when $x(1) = c/2$. Calculate the second derivative $\alpha''(\{x(i)\})$ in $x(1)$, when $x(1) + x(2) = c$ and the variables $x(i), i = 3, \ldots, q$, are fixed, and prove that it is nonpositive. To do this, consider the summands in $\alpha''_{x(1)}$, which have given degrees a of $x(1)$ and b of $x(2)$ and the degrees of the other variables are the same. Such summands arise from the summands of α with the same degrees of variables $x(i), i \geq 3$ and pair degrees $(a+2,b)$, $(a,b+2)$, $(a+1,b+1)$ of the pair $(x(1),x(2))$. Next consider cases: first, when the set $\{j_i\}$ satisfies the inequality $\max\{j_3,\ldots,j_q\} \geq \max\{a+2,b+2\}$. In this case the multiple $\left(1 - \frac{\max\{j_1,\ldots,j_q\}}{L+1}\right)$ is the same and we can omit it. The other cases are $\max\{j_3,,\ldots,j_q\} = b+2 \leq a$, $\max\{j_3,,\ldots,j_q\} = b+2$, $a = b+1$, and the same cases with only transposition $a \leftrightarrow b$. Conclude that α achieves its maximum for given λ when $x(i) = \frac{\lambda}{q-1}$, $i = 1,\ldots,q-1$, and when $\lambda \in [0,1]$ can vary, then the maximum of α is achieved, when all $x(i)$ are equal, that is, $x(i) = 1/q$ and $\lambda = (q-1)/q$.

6. Put $h = 0$ in (11) and prove that the lower bound (11) coincides with the upper bound (21) at zero rate (Lecture 9).

7. Prove inequalities (12) (Lecture 10). *Hint:* inequality

$$P_{n,L}^m(R) \geq \bar{P}_{n,L}(R)$$

follows from the definitions. To prove inequality

$$2\bar{P}_{n,L}\left(R + \frac{\log_{|\mathcal{X}|} 2}{n}\right) \geq P_{n,L}^m(R)$$

choose a code $\mathcal{K}_n$ with $2|\mathcal{C}_n|$ words on which the probability of decoding error $P\left(R + \frac{\log_{|\mathcal{X}|} 2}{n}\right)$ is achieved (we set $\log_{|\mathcal{X}|} |\mathcal{C}_n| \geq R$)

$$\bar{P}_{n,L}(\mathcal{K}_n) = \frac{1}{2|\mathcal{C}_n|} \sum_{i=1}^{2|\mathcal{C}_n|} P_i = \bar{P}_{n,L}\left(R + \frac{\log_{|\mathcal{X}|} 2}{n}\right),$$

where

$$P_i = \sum_{y^n \notin B_i} p(y^n | x_i^n),$$

and remove from this code all words x_i^n for which

$$P_i \geq 2\bar{P}_{n,L}\left(R + \frac{\log_{|\mathcal{X}|} 2}{n}\right).$$

It is easy to see that the number of such code words does not exceed $|\mathcal{C}_n|$. Then the maximum probability of error for the rest $|\mathcal{C}_n|$ codewords, which form the code, has the desired property.

8. Using (15) and (16) prove (17) (Lecture 10). *Hint:*

$$E_L^{ex}(0) = \lim_{\rho \to \infty} E_{0,ex}(\rho),$$

where the limit can be obtained by using L'Hospital's rule.

9. Using the proof of Ramsey's Theorem (see Chapter I), estimate the Ramsey number (maximal clique) in the graph whose edges are colored by q colors. Show that the maximal monochromatic clique has a number of vertices of at least $\lfloor m/q \rfloor$, where m is the number of times we can apply the operation

$$M \rightarrow \left\lfloor \frac{M-1}{q} \right\rfloor$$

starting from $M = M_n$, the number of vertices in the initial graph, such that the resulting number at the first time becomes less than $q+1$. This consideration gives the estimation of m:

$$q+1 > \frac{M_n - 2(1+q+\ldots+q^{m-1})}{q^m} > \frac{M_n}{q^m} - \frac{2}{q-1} > \frac{M_n}{q^m} - 2.$$

Thus

$$m > \log_q \frac{M_n}{q+3}.$$

10. Prove that the function $\mu(\lambda)$ is $\cup$-convex. *Hint:* use Hölder's inequality [KF61]: for nonnegative random variables $X, Y > 0$ the following inequality is valid:

$$\mathbb{E}XY \leq \left(\mathbb{E}X^{1/\theta}\right)^{\theta} \left(\mathbb{E}Y^{1/(1-\theta)}\right)^{1-\theta}, \quad \theta \in (0,1),$$

where $\mathbb{E}(\cdot)$ is the mathematical expectation.

11. Prove by induction on $k = \lfloor L/2 \rfloor$ that, when $q = 2$,

$$f_L(\lambda) = \sum_{i=1}^{k} K_{i-1}(\lambda(1-\lambda))^i,$$

where $C_i = \frac{\binom{2i}{i}}{i+1}$ are the Catalan numbers.
Using induction on k, prove that

$$f_L''(\lambda) = -k\binom{2k}{k}(\lambda(1-\lambda))^k.$$

This proves that in the binary case $f_L(\lambda)$ is $\cap$-convex.

12. Prove relation (41) (Lecture 10).

13. Using the Chernoff bound (1) (Chapter I), prove that for $\lambda = \omega/n \in (0,1)$ we have

$$\frac{\log_q |B(x, \omega)|}{n} \leq H_q(\lambda).$$

Prove, using the Stirling formula that

$$\binom{n}{\omega} \geq \frac{1}{\sqrt{8\omega(1-\lambda)}} 2^{nH_2(\lambda)}.$$

Conclude from this that

$$\frac{\log_q |B(x,\omega)|}{n} \geq H_q(\lambda) + o(1).$$

14. Prove directly that if $p(j|x_1)\ldots p(j|x_{L+1}) = 0$ for some $x_1,\ldots,x_{L+1} \in \mathcal{X}$ and all $j \in \mathcal{Y}$, then $E_L(0) = E_L^{ex}(0) = \infty$. *Hint:* arrange $L+1$ sequences $x_1^n,\ldots,x_{L+1}^n \in \mathcal{X}^n$ as the rows of a matrix of dimension $(L+1) \times n$ and choose symbols from $\mathcal{X}$ as the elements of this matrix independently with equal probability $1/|\mathcal{X}|$. Then the probability that a given string $x_1,\ldots,x_{L+1}$ does not occur as a column of the matrix is

$$\left(1 - \frac{1}{|\mathcal{X}|^{L+1}}\right)^n.$$

Then the probability that among $\binom{M_n}{L+1}$ choices of $L+1$ code sequences that the event occurs is estimated from above by

$$\left(1 - \frac{1}{|\mathcal{X}|^{L+1}}\right)^n \binom{M^n}{L+1} \leq M_n^{L+1} e^{-\frac{n}{|\mathcal{X}|^{L+1}}}.$$

15. Show that the function $f_L(\lambda)$ is $\cap$-convex and increasing when $\lambda \in [0,(q-1)/q]$, and is decreasing when $\lambda \in [(q-1)/q,1]$. Prove that from these facts follows inequality (18).

Chapter IV
Higher Level and Dimension Constrained Extremal Problems

Lecture 11 Higher Level Extremal Problems

Here we consider extremal problems of set systems. "Higher level" means that points in our consideration are sets, and sets of points are set systems or families of sets. Let $\mathcal{S} = \{\mathcal{A}_1, \ldots, \mathcal{A}_m\}$ be a system of families of sets, $\mathcal{A}_i \subset 2^{[n]}$ with some condition imposed on the relations between the pairs of sets from different families $\mathcal{A}_i$. The typical extremal problem is to find the asymptotic growth (when $n \to \infty$) of the maximal value of m. The methods of proofs in the case of higher level extremal problems are quite different in comparison with ordinary extremal problems.

First we consider the problem when all sets $A_j \in \mathcal{A}_i$ have the same cardinality: $A_j \in \binom{[n]}{k}$. Let $\mathcal{A} = \{\mathcal{A}_1, \ldots, \mathcal{A}_m\}$ be a system of pairwise disjoint families of k-element subsets of $[n]$. Further, $\mathcal{A}$ is a disjoint system of type $(\exists, \forall, n, k)$ if for every ordered pair $(\mathcal{A}_i, \mathcal{A}_j)$ of distinct families from $\mathcal{A}$ there exists an $A \in \mathcal{A}_i$ that does not intersect any element of $\mathcal{A}_j$. $\mathcal{A}$ is a disjoint system of type $(\forall, \exists, n, k)$ if for every ordered pair $(\mathcal{A}_i, \mathcal{A}_j)$ of distinct families from $\mathcal{A}$ for every $A \in \mathcal{A}_i$ there exists a $B \in \mathcal{A}_j$ that does not intersect A. Finally, $\mathcal{A}$ is a disjoint system of type $(\exists, \exists, n, k)$ if for every ordered pair $(\mathcal{A}_i, \mathcal{A}_j)$ of distinct families from $\mathcal{A}$ there exist $A \in \mathcal{A}_i$ and $B \in \mathcal{A}_j$ that do not intersect. We denote the corresponding maximal cardinalities of such set systems by $D_n(\exists, \forall, k)$, $D_n(\forall, \exists, k)$, and $D_n(\exists, \exists, k)$. This lecture consists of two sections. In the first one we show how to find the exact asymptotics of these numbers (the value of $D_n(\forall, \forall, k)$ can be obtained in a trivial way). The main result of the first section is stated in Lemma 30. The proof of this lemma is an interesting interaction between Corollary 2 and Vizing's Theorem.

In the second section we consider a number of relations between the sets whose cardinalities are not fixed and show some ways of studying the asymptotic behavior for the number of elements in the corresponding set systems in the most interesting cases. The methods of the second section are in some sense more straightforward than those in the first section, but still powerful enough and allow to find the complete solutions.

§1 The Asymptotics of the Numbers $D_n(\exists,\forall,\mathbf{k})$, $D_n(\forall,\exists,\mathbf{k})$ and $D_n(\exists,\exists,\mathbf{k})$

First note that it is evident that

$$D_n(\exists,\forall,1) = D_n(\forall,\exists,1) = D_n(\exists,\exists,1) = n.$$

Therefore, in the sequel we will suppose that $k > 1$.

Since any system of type $(\exists,\forall,n,k)$ is also a system of type $(\forall,\exists,n,k)$ and the last type of systems are in turn $(\exists,\exists,n,k)$ systems, we have

$$D_n(\exists,\forall,k) \leq D_n(\forall,\exists,k) \leq D_n(\exists,\exists,k).$$

We derive tight asymptotics of these three values.

Lemma 30 *For every $k \geq 2$ and $n \to \infty$,*

$$D_n(\exists,\forall,k) = (1+o(1))\frac{\binom{n}{k}}{k+1}, \tag{1}$$

$$D_n(\forall,\exists,k) = (1+o(1))\frac{\binom{n}{k}}{2}, \tag{2}$$

$$D_n(\exists,\exists,k) = (1+o(1))\frac{\binom{n}{k}}{2}. \tag{3}$$

Proof. To prove (1) we need Corollary 2 from Lecture 7. We reformulate it in a new notation for convenient further usage. Let $\mathcal{H} = (V,\mathcal{E})$ be a k-uniform hypergraph. Then we can embed

$$r = (1+o(1))\frac{\binom{n}{k}}{|\mathcal{E}|}$$

number of copies of $\mathcal{H}$: $\mathcal{H}_1 = (V_1,\mathcal{E}_1),\ldots,\mathcal{H}_r = (V_r,\mathcal{E}_r)$ into a complete k-uniform hypergraph on n vertices, such that $|V_i \cap V_j| \leq k$, $i \neq j$ and if $|V_i \cap V_j| = k$ and $V_i \cap V_j = B$, then $B \notin \mathcal{E}_i \cup \mathcal{E}_j$.

This corollary directly implies the bound

$$D_n(\exists,\forall,k) \geq (1+o(1))\frac{\binom{n}{k}}{k+1}.$$

Indeed, let $(V,\mathcal{E})$ be a k-uniform hypergraph consisting of $k+1$ pairwise disjoint edges. By Corollary 2 we can place $(1+o(1))\binom{n}{k}/(k+1)$ edge disjoint copies of this graph into a complete k-uniform hypergraph on n vertices. If we take the edges of each copy as a family of k-sets, we obtain $(1+o(1))\binom{n}{k}/(k+1)$ pairwise disjoint families. For two arbitrary families $\mathcal{H}_1 = (V_1,\mathcal{E}_1)$ and $\mathcal{H}_2 = (V_2,\mathcal{E}_2)$, we have $|V_1 \cap V_2| \leq k$ and since $\mathcal{H}_1$ consists of $k+1$ nonintersecting k-sets, there exists $A \in \mathcal{A}_1$, which does not intersect any set from $\mathcal{A}_2$. Hence the constructed disjoint family is an $(\exists,\forall,n,k)$-system.

Next we show that the proved lower bound for $D_n(\exists, \forall, k)$ is also asymptotically an upper bound for this value.

Let $\mathcal{S} = \{\mathcal{A}_1, \ldots, \mathcal{A}_m\}$ be a disjoint system of type $(\exists, \forall, n, k)$. Denote by $\mathcal{S}_1$ the families from $\mathcal{A}$ that contain only one set, by $\mathcal{S}_2$ the families that contain from 2 up to k elements, and by $\mathcal{S}_3$ the families containing more than k elements. Any two one-set families are disjoint and we have $|\mathcal{S}_1| \leq n/k$. Let $\mathcal{A}_i = \{A_1, \ldots, A_t\} \in \mathcal{S}_2$, $t \in \{2, 3, \ldots, k\}$. Since we consider an $(\exists, \forall, n, k)$-system, we have that an arbitrary set $B, |B| = k$ with the property

$$B \subset \bigcup_{j=1}^{t} A_j, \ (B \cap A_j) \neq \emptyset \tag{4}$$

could not be a member of any other family. Choose (not necessarily distinct) points $a_j \in A_j$, $j = 2, 3, \ldots, t$ such that $a_j \notin A_1$ and set $L = \{a_2, \ldots, a_t\}$. We have $|L| \leq k - 1$. Let $\mathcal{L}$ be the family of all sets of the form $L_p = L \cup J_p$, where J_p runs over all $(k - |L|)$-element subsets of A_1. Each L_p satisfies (4), $L_p \neq A_1$, and $|\mathcal{L}| \geq k$. At the same time each k-set can satisfy the relation (4) only for one choice of $\mathcal{A}_i$. If this was false, then there would exist a set B with properties (4) for two families $\mathcal{A} = \{A_1, \ldots, A_t\}$ and $\mathcal{B} = (B_1, \ldots, B_r\}$. This would imply the existence of a set $A_i \in \mathcal{A}_1$ such that $A_i \cap B_j = \emptyset, j = 1, 2, \ldots, r$ and as $B \subset \bigcup_{j=1}^{r} B_j$, we would have $B \cap A_i = \emptyset$, which contradicts (4).

Hence with each family $\mathcal{A}_i = \{A_1^i, \ldots, A_t^i\} \in \mathcal{S}_2$ we can associate $k + 1$ sets (k-sets of the form L_p together with the first element A_1^i) that cannot be associated with any other family and are not members of any other family. Each family from $\mathcal{S}_3$ contains at least $k + 1$ points. Hence we have

$$(k+1)|\mathcal{S}_2| + (k+1)|\mathcal{S}_3| \leq \binom{n}{k}$$

and

$$D_n(\exists, \forall, k) = |\mathcal{S}_1| + |\mathcal{S}_2| + |\mathcal{S}_3| \leq \frac{n}{k} + \frac{\binom{n}{k}}{k+1} = (1 + o(1)) \frac{\binom{n}{k}}{k+1}.$$

This proves relation (1).

Next we prove (2). The same reasoning as above shows that $|\mathcal{S}_1| \leq n/k$ is true. Since all the other sets contain at least two elements, we have

$$2(|\mathcal{S}_2| + |\mathcal{S}_3|) \leq \binom{n}{k}$$

and hence

$$D_n(\forall, \exists, k) \leq |\mathcal{S}_1| + |\mathcal{S}_2| + |\mathcal{S}_3| \leq \frac{n}{k} + \frac{\binom{n}{k}}{2} = (1 + o(1)) \frac{\binom{n}{k}}{2}.$$

To prove the reverse inequality, consider the random graph $G(n, p)$ on n vertices with probability $p \approx 1$ of an edge connecting two given vertices. For every realization of $G(n,p)$ we construct a new graph G' whose vertices are k-cliques from $G(n,p)$, and two vertices in G' share an edge iff the corresponding k-cliques in $G(n,p)$ are vertex disjoint and there are no edges between them. By using simple probabilistic arguments we prove that for any given $\varepsilon, \delta > 0$ with high probability G' has $(1 - \varepsilon/3)\binom{n}{k}$ vertices and all degrees $deg(X)$ of vertices X in G' are almost the same:

$$|deg(X) - d| < \delta d.$$

Assume that $d \overset{n \to \infty}{\longrightarrow} \infty$. Suppose for a moment that $G' = (V,E)$ satisfies these conditions, then the chromatic number $\chi(G') \le \max_X deg(X) + 1 < d(\delta + 1) + 1$ (Vizing's Theorem) and the number of edges in G' is at least $(1 - \delta)d|V|/2$. Therefore, there exists a matching in G', which contains

$$\frac{(1-\delta)d|V|}{2\chi(G')} \ge \frac{1-\delta}{2(1+\delta)}\left(1 - \frac{\varepsilon}{2}\right)\binom{n}{k}$$

edges. Taking each pair of vertices connected by an edge in this matching as a family we obtain a disjoint system of size at least $\frac{1}{2}(1 - \varepsilon)\binom{n}{k}$ (for sufficiently small $\delta > 0$). Let us show that this system is of type $(\forall, \exists, n, k)$. Assume that this is false. Then there exist two families $\mathcal{A} = \{A_1, A_2\}$ and $\mathcal{B} = \{B_1, B_2\}$ such that $A_1 \cap B_i \ne \emptyset$, $i = 1, 2$. We choose $a_1 \in A_1 \cap B_1$ and $a_2 \in A_1 \cap B_2$. As $a_1, a_2 \in A_1$, they are adjacent in $G(n,p)$. However, $a_1 \in B_1$ and $a_2 \in B_2$, but this contradicts the fact that the subgraph of $G(n,p)$ induced on $B_1 \cup B_2$ has no edges between B_1 and B_2. Hence the system is indeed of type $(\forall, \exists, n, k)$ and

$$D_n(\forall, \exists, k) > (1 - \varepsilon)\frac{\binom{n}{k}}{2}.$$

It is left to the reader to prove the existence of a graph G' with the required properties.

We note that with high probability for every set X of k vertices of $G(n,p)$ the number of vertices that do not have any neighbor in X is

$$(1 + o(1))(1 - p)^k (n - k). \tag{5}$$

This follows from inequality (1) (Chapter I) and is to be proved in Exercise 1. We also need the following:

Lemma 31 *For given $c > 0$ with high probability for every set Y of n_1 vertices from $G(n,p)$ $(cn \le n_1 \le n)$, the number of k-cliques of the induced subgraph of $G(n,p)$ on Y is*

$$(1 + o(1))\binom{n_1}{k}p^{\binom{k}{2}}.$$

This is to be proved in Exercise 2.

Next consider a k-clique X in $G(n,p)$. The degree d of X as a vertex of G' is the number of k-cliques in the induced subgraph of G on the set of all vertices that have no neighbors in X. Using (5) and Lemma 31 we obtain

$$d = (1 + o(1)) \binom{n_1}{k} p^{\binom{k}{2}},$$

where $n_1 = (1 + o(1))(1 - p)^k(n - k)$. Therefore, G' is with high probability almost regular and $d \overset{n \to \infty}{\to} \infty$. Also from Lemma 31 follows that with high probability the number of k-cliques (the number of vertices in G') is

$$(1 + o(1)) \binom{n}{k} p^{\binom{k}{2}}.$$

Fix $p < 1$, $p^{\binom{k}{2}} > 1 - \varepsilon/4$. Then again with high probability the number of vertices in G' is greater than $(1 - \varepsilon/2) \binom{n}{k}$. This completes the proof of (2).

To prove (3) note that as before the number of one-element families $|\mathcal{S}_1| < n/k$ and therefore $D_n(\exists, \exists, k) \le \frac{n}{k} + \frac{1}{2} \binom{n}{k} = (1 + o(1)) \frac{\binom{n}{k}}{2}$. The reverse inequality follows from the inequality $D_n(\forall, \exists, k) \le D_n(\exists, \exists, k)$. This completes the proof of Lemma 30. $\qquad\square$

§2 Other Cases of Higher Level Extremal Problems

We continue to investigate the maximum cardinality of a set system of families of sets. In this section we consider the general case of families of unrestricted subsets of $\{1, 2, \ldots, n\}$. This means that the sets can be of different cardinalities. We consider the situation of a system $\{\mathcal{A}_1, \ldots, \mathcal{A}_m\}$ with disjoint families $\mathcal{A}_i$ of nonempty subsets of $[n]$. We impose some relations between subsets from different families $\mathcal{A}_i$. For instance, for two sets $A \in \mathcal{A}_i$, $B \in \mathcal{A}_j$, $i \ne j$ we consider properties like being comparable (and reserve letter C for this property), disjoint (letter D we have already used for it before), and intersecting (letter I). As before, we use different combinations of $\forall$ and $\exists$. For example, $I_n(\forall, \exists)$ denotes the maximal number of disjoint families of subsets of $[n]$ such that if we take two distinct families $\mathcal{A}_i, \mathcal{A}_j$ $i \ne j$, then for any subset $A \in \mathcal{A}_i$ there exists a subset $B \in \mathcal{A}_j$, which intersects A. The reader can easily formulate the other possibilities such as $D_n(\forall, \exists)$, $C_n(\exists, \forall)$, etc.

At first we find the asymptotic behavior of the values C_n, I_n, D_n. Note also that as before, everywhere here we consider the case of disjoint families, except in the last Theorem 23, where (for the values $N_n(\cdot, \cdot)$) we allow the families to intersect.

Now we assume that we have the graph $\mathcal{G} = (\mathcal{V}, \mathcal{E})$ with vertex set $\mathcal{V}$ colored with m colors from $[m]$. Also we have a coloring of the directed edges of the graph

$$\mathcal{E}^* = \{(a, b), (b, a) : \{a, b\} \in \mathcal{E}\} = \{e_1, \ldots, e_N\} \text{ with } N = 2|\mathcal{E}|$$

by pairs of colors (i, j). Let also the vertices of the graph be indexed by subsets of $[n]$. Suppose also that two vertices are connected by an edge $e = \{a, b\} \in \mathcal{E}$ iff a and b as subsets of $[n]$ are a. intersecting (I), b. comparable (C), and c. disjoint (D).

First we introduce a simple argument enabling to find the asymptotics of $C_n(\exists,\exists)$, $I_n(\exists,\exists)$, and $D_n(\exists,\exists)$. In this case any coloring of graph $\mathcal{G}$ generates a partition of it. We impose the property on this partition that for arbitrary two colors (i,j) there exists an edge colored by (i,j). The number of colors m in that case is the desired value. The next simple result from [ACZ92b] gives the tools to find the necessary asymptotic growth.

Lemma 32 *For any graph $\mathcal{G} = (\mathcal{V}, \mathcal{E})$ we have with $N = 2|\mathcal{E}|$*

$$m \leq N^{1/2} + 1. \tag{6}$$

Moreover, if

$$\Delta \overset{\Delta}{=} \max_{x \in \mathcal{V}} \deg(x) \leq \left(\frac{N}{e^5 \ln N} \right)^{1/2}, \tag{7}$$

then for sufficiently large N

$$m \geq \left(\frac{N}{e^5 \ln N} \right)^{1/2}. \tag{8}$$

Proof. Obviously $\binom{m}{2} \leq |\mathcal{E}|$, hence

$$m \leq \left(2|\mathcal{E}| + \frac{1}{4} \right)^{1/2} + \frac{1}{2} \leq N^{1/2} + 1$$

and (6) follows. The proof of (7) uses the probabilistic argument. Let $Z_1, \ldots, Z_n$ with $[n] = \mathcal{V}$ be independent, identically distributed random variables assuming values in $\{1, \ldots, m\}$ with equal probabilities. Then $Z^n = (Z_1, \ldots, Z_n)$ defines a random coloring of the vertices. We consider the events

$$E_{ij} = \{\text{no edge in } \mathcal{E}^* \text{ is colored by } (i,j)\},$$

whose probability is by symmetry independent of (i,j), and for the random variable

$$S = |\{(i,j) : \text{ no edge in } \mathcal{E}^* \text{ is colored by } (i,j)\}|$$

we have

$$\mathbb{E}(S) = \sum_{i,j} P(E_{ij}) \leq m^2 P(E_{ij}),$$

and if $\mathbb{E}(S) < 1$, then, with positive probability, every pair of colors occurs.

We derive now conditions under which $\mathbb{E}(S) < 1$ holds. For this we define first $E_{ij}^{(s)} = \{e_s \text{ is not colored by } (i,j)\}$, $E_{ij}^s = \{e_1, \ldots, e_s \text{ are not colored by } (i,j)\} = \bigcap_{t=1}^s E_{ij}^t$, and write

$$P(E_{ij}) = P(E_{ij}^{(1)})P(E_{ij}^{(2)}|E_{ij}^1)\dots P(E_{ij}^{(k)}|E_{ij}^{k-1})\dots P(E_{ij}^{(N)}|E_{ij}^{N-1}). \qquad (9)$$

We now estimate $P(E_{ij}^{(k)}|E_{ij}^{k-1})$ from above. With $\bar{E}_{ij}^{(k)} = \{E_k \text{ is colored by } (i,j)\}$ we have

$$P(E_{ij}^{(k)}|E_{ij}^{k-1}) = 1 - \frac{P(\bar{E}_{ij}^{(k)} \cap E_{ij}^{k-1})}{P(E_{ij}^{k-1})}.$$

First we estimate $P(\bar{E}_{ij}^{(n)} \cap E_{ij}^{n-1})$ from below. Write $e_t = (a_t, b_t)$, $1 \le t \le N$, and set

$$V_k' = \{a_t : 1 \le t < k, \, b_t = b_k\}, V_k'' = \{b_t : 1 \le t < k, \, a_t = a_k\}.$$

Consider the edges

$$\mathcal{E}_k^* = \{e_t : 1 \le t \le k-1, \{a_t, b_t\} \cap (V_k' \cup V_k'' \cup \{a_k, b_k\}) = \emptyset\}$$

and the events

$$E_{ij}^* = \{\text{no } E_t \in \mathcal{E}_k^* \text{ is colored by } (i,j)\}.$$

Then it holds

$$P(\bar{E}_{ij}^{(k)} \cap E_{ij}^{k-1}) = P(E_{ij}^{k-1}|\bar{E}_{ij}^{(k)})P(\bar{E}_{ij}^{(k)}) = \frac{1}{m^2}P(E_{ij}^{k-1}|\bar{E}_{ij}^{(k)}). \qquad (10)$$

Now E_{ij}^{k-1} clearly occurs if all vertices in $V_k' \cup V_k''$ are not colored by either i or j and if E_{ij}^* occurs. The last two events and $\bar{E}_{ij}^{(k)}$ are mutually independent, the first event has probability $\left(1 - \frac{2}{m}\right)^{|V_k' \cup V_k''|}$ and obviously

$$P(E_{ij}^*) \ge P(E_{ij}^{k-1}).$$

We conclude from (10) that

$$P(\bar{E}_{ij}^{(k)} \cap E_{ij}^{k-1}) \ge \frac{1}{m^2}\left(1 - \frac{2}{m}\right)^{|V_k' \cup V_k''|} P(E_{ij}^*).$$

Therefore,

$$P(E_{ij}^{(k)}|E_{ij}^{k-1}) \le 1 - \frac{1}{m^2}\left(1 - \frac{2}{m}\right)^{|V_k' \cup V_k''|} \le 1 - \frac{1}{m^2}\left(1 - \frac{2}{m}\right)^{2D}$$

and thus by (9)

$$P(E_{ij}) \le \left(1 - \frac{1}{m^2}\left(1 - \frac{2}{m}\right)^{2D}\right)^N.$$

Hence

$$\mathbb{E}(S) \leq m^2 \left(1 - \frac{1}{m^2}\left(1 - \frac{2}{m}\right)^{2D}\right)^N$$

$$= m^2 \exp\left\{N\ln\left(1 - \frac{1}{m^2}\left(1 - \frac{2}{m}\right)^{2D}\right)\right\} \leq m^2 \exp\left\{-\frac{N}{m^2}\left(1 - \frac{2}{m}\right)^{2D}\right\}.$$

Here in the last inequality we used the inequality $\ln(1-x) \leq -x$, $x \in [0,1)$. Next we choose m, D, N in such a way that $\mathbb{E}(S) < 1$. We can choose $m \geq D$. Then from the last relations for sufficiently large N we have

$$\mathbb{E}(S) \leq m^2 \exp\left\{-\frac{N}{m^2}e^{-5}\right\},$$

where we use the relation

$$\left(1 - \frac{2}{m}\right)^{2D} > e^{-5},$$

which is valid for sufficiently large N and the choices of m and D specified in the lemma. From here we obtain a sufficient condition for $\mathbb{E}(S) < 1$:

$$2\ln m < \frac{N}{m^2}e^{-5}. \tag{11}$$

If we choose $m = \left(\frac{N}{e^5 \ln N}\right)^{1/2}$, then

$$2\ln m < \ln N = \frac{N}{m^2}e^{-5}$$

and condition (11) holds. The proof of the lemma is completed.   □

We turn now to the determination of $C_n(\exists, \exists)$ and $D_n(\exists, \exists)$. The exact values of $I_n(\cdot, \cdot)$ are determined for all cases (see Theorem 21 below).

Theorem 20 (Ahlswede, Cai, and Zhang 1986) *The following relations are valid:*

$$\lim_{n \to \infty} \frac{1}{n}\ln C_n(\exists, \exists) = \frac{1}{2}\ln 3, \tag{12}$$

$$\lim_{n \to \infty} \frac{1}{n}\ln D_n(\exists, \exists) = \frac{1}{2}\ln 3. \tag{13}$$

Proof. We give the proof of this result in the case of the property "comparable." Define the graph $\mathcal{G} = (\mathcal{V}, \mathcal{E})$ as follows:

$$\mathcal{V} = [n], \quad \mathcal{E} = \{\{a, b\} : a, b \in \mathcal{V}, a \neq b, a \subset b\}.$$

Then

$$N = 2|\mathcal{E}| = 2 \sum_{k=1}^{n} \binom{n}{k} (2^{n-k} - 1) = 2(3^n - 2^{n+1} + 1)$$

and by Lemma 32

$$C_n(\exists, \exists) \leq (2 \cdot 3^n)^{1/2} + 1.$$

To prove the lower bound we choose the graph

$$\mathcal{G}_{k,\ell} = (\mathcal{V}_{k,\ell}, \mathcal{E}_{k,\ell}),$$

where

$$\mathcal{V}_{k,\ell} = \binom{[n]}{k} \cup \binom{[n]}{k+\ell}$$

for suitable $k, \ell : k = \ell = \frac{n}{3}$. We have

$$N = 2\binom{n}{k}\binom{n-k}{\ell}, \quad \Delta = \max_{x \in \mathcal{V}_{kl}} \deg(x) = \max\left\{ \binom{n-k}{\ell}, \binom{k+\ell}{k} \right\}.$$

Then we have

$$N = 2\binom{n}{\frac{n}{3}}\binom{\frac{2n}{3}}{\frac{n}{3}} = 2\frac{n!}{\left(\left(\frac{n}{3}\right)!\right)^3} \geq \frac{2}{n} 3^n,$$

$$\Delta = \binom{\frac{2n}{3}}{\frac{n}{3}} \leq 2^{2n/3}.$$

Since $2^{2/3} < 3^{1/2}$, we obtain from Lemma 32 for large n

$$C_n(\exists, \exists) \geq m \geq \left(\frac{2 \cdot 3^n}{ne^5 n \ln 3} \right)^{1/2}$$

and the asymptotics (12) for $C_n(\exists, \exists)$ follows.

We proceed in the same way with $D_n(\exists, \exists)$. The upper bound

$$D_n(\exists, \exists) \leq (3^n)^{1/2} + 1$$

follows from Lemma 32 and the relation

$$N = 2|\mathcal{E}| = \sum_{k=1}^{n} \binom{n}{k}(2^{n-k} - 1) = 3^n - 2^{n+1} + 1.$$

The proper lower bound on $D_n(\exists, \exists)$ follows in the same way as the lower bound for $C_n(\exists, \exists)$ with the choice of the ground set of the graph $V = \binom{[n]}{\frac{n}{3}}$. Computations are similar to the case of C_n. ∎

The problem for the type $(\exists, \forall, n)$ in the case of (C) and (D) is not solved. The case (I) has simple solutions for all combinations of pairs from $\{\exists, \forall\}$.

Theorem 21 (Ahlswede, Cai, and Zhang 1996)

$$I_n(\forall,\forall) = I_n(\exists,\forall) = I_n(\forall,\exists) = 2^{n-1}, \tag{14}$$
$$I_n(\exists,\exists) = 2^{n-1} + 2^{n-2} - 1. \tag{15}$$

Proof. Consider the set of families $\{\{A\} : x \in A\}$ for some fixed $x \in [n]$. It has property (I) for $(\forall,\forall)$; therefore

$$I_n(\forall,\exists) \geq I_n(\exists,\forall) \geq I_n(\forall,\forall) \geq 2^{n-1}.$$

Conversely, suppose that for a set of families $(\mathcal{A}_i)_{i=1}^{I_n(\forall,\exists)}$ for some j, $|\mathcal{A}_j| = 1$, that is, $\mathcal{A}_j = \{A_j\}$, then A_j^c cannot occur in any other family, so we may as well count it for $\mathcal{A}_j$. Thus all families can be counted with at least two elements and $I_n(\forall,\exists) \leq 2^{n-1}$.

By the foregoing argument at most 2^{n-1} families can have exactly one element. Therefore,

$$I_n(\exists,\exists) \leq 2^{n-1} + \left\lfloor \frac{2^{n-1}-1}{2} \right\rfloor = 2^{n-1} + 2^{n-2} - 1.$$

On the other hand, the following construction shows that the upper bound is tight.

In the case $n = 2\ell + 1$ we use the elements in $\bigcup_{k=\ell+1}^{n} \binom{[n]}{k}$ as single members of families. The elements in $\bigcup_{k=1}^{m} \binom{[n]}{k}$ can be paired to families, so that in every family $\{A, A'\}$ we have $|A \cup A'| \geq m+1$ and exactly one element is left over. Since $2(m+1) > n$, we have (I) property of type $(\exists,\exists)$ and

$$\left| \bigcup_{k=m+1}^{n} \binom{[n]}{k} \right| + \frac{1}{2}\left(\left| \bigcup_{k=1}^{m} \binom{[n]}{k} \right| - 1 \right) = 2^{n-1} + \frac{1}{2}(2^{n-1} - 2) = 2^{n-1} + 2^{n-2} - 1.$$

In case $n = 2m$, choose singletons from $\bigcup_{k=m+1}^{n} \binom{[n]}{k}$ and from $\binom{[n]}{m}$, but without choosing complementary sets. Pair then the remaining sets such that again

$$|A \cup A'| \geq m+1.$$

∎

Let us mention the following trivial relations:

$$C_n(\forall,\forall) = D_n(\forall,\forall) = n+1.$$

At last we mention an unpublished result by Borden [B86b] (see also [S89]).

Theorem 22 (Borden 1986) *The following relation is valid:*

$$\lim_{n \to \infty} \frac{1}{n} C_n(\forall,\exists) = \frac{\sqrt{5}+1}{2}.$$

Also interesting results arise concerning the different possibilities when one considers the property "incomparable." We indicate it by letter M in the case when the families of sets are disjoint. For nondisjoint families we indicate it by letter N. One can find the proof of the following theorem in [AK94a].

Theorem 23 (Ahlswede and Khachatrian 1994) *The following relations are valid:*

$$M_n(\exists, \forall) \sim 2^{n-1},$$

$$M_n(\forall, \exists) = \begin{cases} 2, \text{ if } n = 2, \\ 2^{n-1} - 1, \text{ if } n \geq 3. \end{cases}$$

$$M_n(\exists, \exists) = \binom{n}{\lfloor \frac{n}{2} \rfloor} + \left\lfloor \frac{2^n - 2 - \binom{n}{\lfloor \frac{n}{2} \rfloor}}{2} \right\rfloor,$$

$$N_n(\exists, \forall) = \binom{k}{\lfloor \frac{k}{2} \rfloor}, \ k = \binom{n}{\lfloor \frac{n}{2} \rfloor},$$

$$N_n(\forall, \exists) \sim 2^{2^n - 2},$$

$$N_n(\exists, \exists) \sim 2^{2^n}.$$

Lecture 12 Properties of Binary Sequences Over Reals

The problems in this lecture are of a quite different nature and have quite different methods of proofs. However, they all deal with binary sequences and linear relations between them over reals. In the first section we start with the most transparent case and discuss a problem that was first stated by Ahlswede and Khachatrian. We consider binary vectors of length n and want to estimate the maximal number of vectors in a family with the property that no vector in this family is a linear combination of other vectors from this family with positive real coefficients. Denote by $f(n)$ the maximal size of such a family for given n. To say it more precisely, $\mathcal{F}$ is positive linear combinations free if for any choice of positive reals c_i and $\mathcal{F}' \subseteq \mathcal{F}$ no $F \in \mathcal{F}$ exists such that

$$F = \sum_{F \neq F_i \in \mathcal{F}' \subseteq \mathcal{F}} c_i F_i.$$

We show that

$$f(n) = (1 + o(1))2^n, \ n \to \infty \tag{1}$$

and find estimations of the term $o(1)$.

In the second section we use the results from Lecture 13 to prove the interesting fact that a hyperplane cuts at most $m\binom{n}{m}$, $m = \lfloor \frac{n}{2} \rfloor + 1$ edges of the unit cube $\mathcal{K}^n = \{0, 1\}^n$, which is imbedded in a natural way into the Euclidean space $\mathbb{R}^n$ and it takes at least n hyperplanes with nonnegative coefficients to cut all edges of the cube. This is the statement of Theorem 25.

At last, in the third section, we deal with the problem of determining the maximal number of n-tuples from $\mathcal{K}^n$ of given Hamming weight with the property that all of them are contained in a k-dimensional subspace of the real space $\mathbb{R}^n$. In Theorem 26 we demonstrate the complete solution of this problem and find an expression for this number. Note also that if one considers k-dimensional spaces not over the reals but over the binary space with componentwise summation *mod* 2, then the same problem becomes considerably more difficult and it is far from a complete solution.

§1 The Size of Positive Linear Combinations Free Sets

In this section we first prove relation (1) and then in Theorem 24 we show more precise estimations of the rest term $o(1)$ in this relation. Now we come to necessary definitions. Consider a partition of $[n]$ into parts $P_1, \ldots, P_m$ each of size $t = \log n - \log\log n$. W.l.o.g. in the proof we can assume that $n = tm$. Let $\mathcal{F}$ consist of all vectors that have at least one element in every part P_i and at least in one part have exactly one element:

$$\mathcal{F} = \left\{ F \in 2^{[n]} : \forall i, \ F \cap P_i \neq \emptyset, \ \exists j : |F \cap P_j| = 1 \right\}. \tag{2}$$

It is true that this family is positive linear combinations free (Exercise 4). We show that this family is sufficiently large. Let $\mathcal{F}_0 \subset 2^{[n]}$ be the family containing all sets that do not intersect at least one of the parts, that is,

$$\mathcal{F}_0 = \left\{ F \in 2^{[n]} : \exists i, \ F \cap P_i = \emptyset \right\}$$

and $\mathcal{F}_1 \subset 2^{[n]}$ be the collection containing all the sets that do intersect every part in at least two elements, that is,

$$\mathcal{F}_1 = \left\{ F \in 2^{[n]} : \forall i, |F \cap P_i| > 1 \right\}.$$

Obviously

$$|\mathcal{F}| = 2^n - |\mathcal{F}_0| - |\mathcal{F}_1| \tag{3}$$

and by the choice of t

$$|\mathcal{F}_0| \leq \frac{n}{t} 2^{n-t} = 2^n O\left(\frac{\log\log n}{\log n} \right),$$

$$|\mathcal{F}_1| \leq (2^t - t - 1)^{n/t} \leq 2^n e^{-n/2^t} = 2^n \frac{1}{\log n}.$$

This with (3) gives (1).

Next we show how to obtain more precise estimates on the rest term $o(1)$ in (1). Here is the main result in this section.

Theorem 24 (Füredi and Ruszinkó 2005) *The following relations are valid:*

$$2^n \left(1 - O\left(\frac{(\log n)^{3/2}}{\sqrt{n}} \right) \right) \leq f(n) \leq 2^n \left(1 - \Omega\left(\frac{1}{\sqrt{n}} \right) \right). \tag{4}$$

Proof. Let us start the proof with the upper bound in (4). Let $\mathcal{F}$ be a positive linear combinations free family and

$$f_k = |\mathcal{F}^k|, \ \mathcal{F}^k = \{ F \in \mathcal{F} : |F| = k \}.$$

For positive integers $p > q$ call a system of sets $A_i \in 2^{[n]}$, $i = 1, \ldots, p$ a $(p, \{0, q\})$-system if the number of sets A_i containing x is either 0 or q for every $x \in [n]$. Notice that a positive linear combinations free family cannot contain a $(p, \{0, q\})$-system $A_1, \ldots, A_p$ together with $A = \bigcup_{i=1}^{p} A_i$, because $A = \frac{1}{q} \sum_{i=1}^{p} A_i$.

Let $\mathcal{A} = \{A_1, \ldots, A_p\}$ be a $(p, \{0, q\})$-system, $A = \bigcup_{i=1}^{p} A_i$, $K(\mathcal{A}) = \{|A| : A \in \mathcal{A}\}$, and $\alpha_k = |\mathcal{A}^k|$. If $|A| = m$ and $f_m \geq c\binom{n}{m}$, then

$$\sum_{k \in K} \frac{\alpha_k f_k}{\binom{n}{k}} \leq p - c. \tag{5}$$

Indeed, consider a permutation π on $[n]$, apply it to $\mathcal{A}$, and consider $\pi(\mathcal{A}) \cap \mathcal{F}$. It consists of at most $p - 1$ elements for every $\pi(A) \in \mathcal{F}^m \subseteq \mathcal{F}$. Therefore,

$$\sum_{\pi \in S_n} |\pi(\mathcal{A}) \cap \mathcal{F}| \leq (p - 1) \sum_{\pi \in S_n : \pi(A) \in \mathcal{F}} 1 + p \sum_{\pi \in S_n : \pi(A) \notin \mathcal{F}} 1$$

$$= (p - 1) \frac{f_m}{\binom{n}{m}} n! + p \left(1 - \frac{f_m}{\binom{n}{m}}\right) n! \leq (p - c) n!.$$

On the other hand, $E \in \mathcal{A}^k$ appears exactly $f_k k! (n - k)!$ times on the left hand side. Hence,

$$\sum_{k \in K(\mathcal{A})} \sum_{E \in \mathcal{A}^k} f_k |E|! (n - |E|)! = \sum_{k \in K(\mathcal{A})} \alpha_k f_k k! (n - k)! \leq (p - c) n!.$$

Now choose $c = 1/2$. If $f_{\lfloor n/2 \rfloor} < 1/2 \binom{n}{\lfloor n/2 \rfloor}$, then by Stirling's formula we obtain the rest term $\Omega\left(\frac{1}{\sqrt{n}}\right)$. Next we suppose that $f_{\lfloor n/2 \rfloor} \geq 1/2 \binom{n}{\lfloor n/2 \rfloor}$. We explicitly construct $(p_i, \{0, q_i\})$-systems $\mathcal{A}_i$ with

$$q_i = p_i - 1 \tag{6}$$

on the underlying set $\lfloor n/2 \rfloor$ and then apply (5) to them. For $\sqrt{n}/2 \leq p_i \leq \sqrt{n}$ let i be a positive integer with

$$p_i \left\lfloor \frac{n}{2} - i \right\rfloor + r_i = (p_i - 1) \left\lfloor \frac{n}{2} \right\rfloor \tag{7}$$

for some $r_i \in 0, \ldots, q_i$. Clearly, for the given range of p_i,

$$\sqrt{n} \leq i \leq 2\sqrt{n}. \tag{8}$$

The vectors in $\{A_1, \ldots, A_{p_i}\} = \mathcal{A}_i$ are defined as follows. A_j intersects the first $[p_i]$ coordinates in q_i positions : $A_j \cap [p_i] = \{j, j+1, \ldots, j+q_i-1\}$ (we take the elements modulo p_i), and for $p_i < x \leq \lfloor n/2 \rfloor$ the position x belongs to the vectors $A_{q_i x + j}$, $j = 1, \ldots, q_i$ (again the indices are taken modulo p_i). Then $\mathcal{A}_i$ consists of vectors of weights $\lfloor n/2 - i \rfloor$ and $\lfloor n/2 - i \rfloor + 1$ only (here we use relations (6) and (7)) and $|A| = \left| \bigcup_{j=1}^{p_i} A_j \right| = \lfloor n/2 \rfloor$. Since $f_{\lfloor n/2 \rfloor} \geq 1/2 \binom{n}{\lfloor n/2 \rfloor}$, it follows from (5) that for every $\sqrt{n} \leq i \leq 2\sqrt{n}$ either

$$\frac{f_{\lfloor n/2-i \rfloor}}{\binom{n}{\lfloor n/2-i \rfloor}} \leq \frac{p_i - 1/2}{p_i} \leq 1 - \frac{1}{2\sqrt{n}}$$

or

$$\frac{f_{\lfloor n/2-i \rfloor+1}}{\binom{n}{\lfloor n/2-i \rfloor+1}} \leq \frac{p_i - 1/2}{p_i} \leq 1 - \frac{1}{2\sqrt{n}}.$$

Let

$$I = \left\{ \frac{n}{2} - 2\sqrt{n} \leq j \leq \frac{n}{2} - \sqrt{n} : f_j \leq \left(1 - \frac{1}{2\sqrt{n}}\right)\binom{n}{i}\right\}.$$

We have $I \cup \{k, k+1\} \neq \emptyset$ for every $n/2 - 2\sqrt{n} \leq k \leq n/2 - \sqrt{n}$. Therefore, $|I| \geq \sqrt{n}/2$ and we have

$$|\mathcal{F}| = \sum_{j=0}^{n} f_j \leq \sum_{j=1}^{n} \binom{n}{i} - \frac{1}{2\sqrt{n}} \sum_{j \in I} \binom{n}{j}$$

$$\leq 2^n - \frac{|I|}{2\sqrt{n}} \binom{n}{n/2 - \sqrt{n}} = 2^n \left(1 - \Omega\left(\frac{1}{\sqrt{n}}\right)\right).$$

This gives the upper bound in (4).

To prove the lower bound we need the following facts. Given $\mathcal{A} \subseteq 2^{[n]}$, let $\mathcal{F}$ contain all the vectors F that intersect every $A \in \mathcal{A}$ and intersect at least one set $A \in \mathcal{A}$ in exactly one point, that is,

$$\mathcal{F} = \{F \in 2^{[n]} : \forall A \in \mathcal{A}, F \cap A \neq \emptyset; \exists A' \in \mathcal{A} : |F \cap A'| = 1\}. \tag{9}$$

Then $\mathcal{F}$ is positive linear combinations free. This fact is to be proved in Exercise 5.

With an arbitrary family $\mathcal{A} \subset 2^{[n]}$ we associate the family $\mathcal{J}(\mathcal{A})$ as follows:

$$\mathcal{J}(\mathcal{A}) = \{J \in 2^{[n]} : \exists A \in \mathcal{A} : J \cap A = \emptyset\}.$$

Now define $\mathcal{N}(\mathcal{A})$ as the family of those vectors from $2^{[n]}$ whose Hamming distance to $\mathcal{A}$ is exactly one, that is,

$$\mathcal{N}(\mathcal{A}) = \{N \in 2^{[n]} : N \notin \mathcal{A}, d_H(N, \mathcal{A}) = 1\}.$$

Note that $\mathcal{A} \cap \mathcal{N}(\mathcal{A}) = \emptyset$. The next fact we need is that for arbitrary $\mathcal{A} \subseteq 2^{[n]}$ the family $\mathcal{N}(\mathcal{J}(\mathcal{A}))$ is positive linear combinations free. To establish this one should use the previous fact (Exercise 6).

In view of the last fact, we need to construct a suitable family $\mathcal{A}$ such that

$$|\mathcal{N}(\mathcal{J}(\mathcal{A}))| > 2^n \left(1 - O\left(\frac{(\log n)^{3/2}}{\sqrt{n}}\right)\right).$$

W.l.o.g. we suppose that $8|n$. Let $B_1 \cup \ldots \cup B_{n/2}$ be a partition of the underlying set into pairs. Let k be an integer with $k \sim \sqrt{n \ln n}$. Also let $\left\{ Y_K : K \in \binom{[n/2]}{k} \right\}$ be a set of i.i.d. random variables with

$$P(Y_K = 1) = \frac{(1000 \log n)^{3/2}}{\sqrt{n}} \binom{n/8}{k}^{-1} = p,$$

$$P(Y_K = 0) = 1 - p.$$

Finally, let $\mathcal{A}$ be a random family defined by

$$\mathcal{A} = \left\{ \bigcup_{i \in K} B_i : Y_K = 1 \right\}.$$

We next show that the expected size of $\mathcal{N}(\mathcal{J}(\mathcal{A}))$ is as large as it was given in Theorem 24. For a fixed vector N denote by N_2 the set of indices of blocks B_i, which are contained in N, $N_2 = \{i : B_i \subset N\}$, and let $n_2(N) = |N_2|$. Similarly, $N_1 = \{i : |B_i \cap N| = 1\}$ and $n_1(N) = |N_1|$. Using simple probabilistic arguments one can derive a formula for the probability $P(N \in \mathcal{N}(\mathcal{J}(\mathcal{A})))$. We see that $N \in \mathcal{N}(\mathcal{J}(\mathcal{A}))$ iff

(i) $\exists K \in \binom{[n/2]}{k}$ such that $K \cap N_2 = \emptyset$, $|K \cap N_1| = 1$, and $Y_K = 1$,
(ii) $\forall K \in \binom{[n/2]}{k}$ such that $K \cap (N_2 \cup N_1) = \emptyset$, $Y_K = 0$.

Since the variables Y_K are independent (see Exercise 6), we obtain

$$P(N \in \mathcal{N}(\mathcal{J}(\mathcal{A}))) \tag{10}$$

$$= (1-p)^{\binom{n/2-n_1-n_2}{k}} \left(1 - (1-p)^{n_1 \binom{n/2-n_1-n_2}{k-1}} \right)$$

$$\geq \left(1 - p \binom{n/2 - n_1 - n_2}{k} \right) \left(1 - \exp\left\{ -p n_1 \binom{n/2 - n_1 - n_2}{k-1} \right\} \right).$$

Here we used the inequalities $1 - xy \leq (1-x)^y$, $x \in [0,1]$, $y \geq 1$ and $(1-x)^y \leq e^{-xy}$, $x \leq 1$, $y \geq 0$. Next we define the collection $\mathcal{T}$ of sets N by the relation

$$\mathcal{T} = \left\{ N \in 2^{[n]} : \left| n_2(N) - \frac{n}{8} \right| < \sqrt{n \log n}; \ \left| n_1(N) - \frac{n}{4} \right| < \sqrt{n \log n} \right\}. \tag{11}$$

The estimation with the help of (1) (Chapter I) shows that

$$|\mathcal{T}| > 2^n \left(1 - O\left(\frac{1}{n} \right) \right). \tag{12}$$

At last, we have for some positive constant c and a set N satisfying (11),

$$p\binom{n/2 - n_1 - n_2}{k} < c\frac{(\log n)^{3/2}}{\sqrt{n}}, \tag{13}$$

$$pn_1\binom{n/2 - n_1 - n_2}{k-1} > 2\log n. \tag{14}$$

Equations (13) and (14) together with (10) imply the inequality

$$P(N \in \mathcal{N}(\mathcal{J}(\mathcal{A}))) > 1 - c\frac{(\log n)^{3/2}}{\sqrt{n}}. \tag{15}$$

Finally (12) and (15) imply that the expected size $\mathbb{E}(\mathcal{N}(\mathcal{J}(\mathcal{A})))$ fulfils the lower bound of the theorem and hence there exists such a family. This completes the proof of the theorem. ∎

§2 The Unit Cube in the Euclidean Space

The problem we consider here concerns the question how hyperplanes cut the edges of a cube in the n-dimensional Euclidean space. The following theorem makes use of the results from Lecture 13, where we modify the LYM inequality to become equality.

We imbed the set $\mathcal{K}^n = \{0,1\}^n$ into the Euclidean space $\mathbb{R}^n$ in such a way that each vertex, which is a binary vector $a^n = (a_1, \ldots, a_n) \in \mathcal{K}^n$, has the same coordinates in $\mathbb{R}^n$. Then we can identify $\mathcal{K}^n$ with the unit cube in $\mathbb{R}^n$. The next theorem states the main result in this section.

Theorem 25 (Ahlswede and Zhang 1990) *Let* $m = \lfloor \frac{n}{2} \rfloor + 1$. *For the n-dimensional unit cube* $\mathcal{K}^n = \{0,1\}^n$ *with* $n2^{n-1}$ *edges:*

(i) a hyperplane cuts at most $m\binom{n}{m}$ *edges and this bound is best possible;*
(ii) it takes at least n hyperplanes with nonnegative coefficients to cut all edges.

Proof. (i) A hyperplane in the n-dimensional space is determined by the vector $\lambda^{n+1} = (\lambda_0, \lambda_1, \ldots, \lambda_n)$ of coefficients of a linear equation $\lambda_0 = \sum_{i=1}^n \lambda_i x_i$.

Let us assume first that $\lambda_i \geq 0$, $i = 1, \ldots, n$. Call a vector $y^n \in \mathcal{K}^n$ minimal if $\sum_{i=1}^n \lambda_i y_i \geq \lambda_0$ and if the replacement of 1 by 0 in any component of y^n results in a vector z^n, with $\sum_{i=1}^n \lambda_i z_i < \lambda_0$. Let $\mathcal{M}(\lambda^{n+1})$ denote the set of minimal vectors. If an edge (x^n, z^n) is cut by the hyperplane such that $\sum_{i=1}^n \lambda_i x_i \geq \lambda_0$ and $\sum_{i=1}^n \lambda_i z_i < \lambda_0$, then x^n and z^n differ in a component in which all vectors from $\mathcal{M}(\lambda^{n+1})$ below x^n have a 1. Recall the correspondence between vectors $x^n \in \{0,1\}^n$ and sets $X \in 2^{[n]}$. Let $\mathcal{A}(\lambda^{n+1})$ be the family of sets corresponding to $\mathcal{M}(\lambda^{n+1})$. By foregoing remarks, the number of edges $\{(x^n, z^n) : z^n < x^n\}$ with vertex x^n fixed, which are cut by the hyperplane, does not exceed $W_{\mathcal{A}(\lambda^{n+1})}(X)$ (for the definition of $W_\mathcal{A}(X)$ see Lecture 13). Denoting by $N(\lambda^{n+1})$ the total number of edges cut by the hyperplane, we thus get

$$N(\lambda^{n+1}) \leq \sum_{X \in 2^{[n]}} W_{\mathcal{A}(\lambda^{n+1})}(X). \tag{16}$$

For hyperplanes with arbitrary coefficients $(\lambda_1, \ldots, \lambda_n)$, a coordinate transformation

$$T(x_i) = \begin{cases} x_i, & if \ \lambda_i \geq 0, \\ 1 - x_i, & if \ \lambda_i < 0 \end{cases}$$

leads to the case of nonnegative coefficients. By (8)

$$\max_{\lambda^{n+1}} N(\lambda^{n+1}) \leq m \binom{n}{m}. \tag{17}$$

The case $\lambda^{n+1} = (m, 1, \ldots, 1)$ shows that this bound is best possible.

(ii) For k hyperplanes with vectors of coefficients $\lambda^{n+1}(i)$, $i = 1, \ldots, k$ we define as before the set $\mathcal{A}(\lambda^{n+1}(j))$ and put $\mathcal{A}_j = \mathcal{A}(\lambda^{n+1}(j))$ in Corollary 4. The number $N(\lambda^{n+1}(1), \ldots, \lambda^{n+1}(k))$ of edges cut by these hyperplanes is bounded by $\sum_{X \in 2^{[n]}} W_{\mathcal{A}_1, \ldots, \mathcal{A}_k}(X)$. Since all edges shall be cut, by (ii) of Corollary 4

$$n2^{n-1} \leq \max_{0 \leq \ell \leq n - N'} \sum_{r=\ell+1}^{\ell+N'} r \binom{n}{r}, \quad \text{where } N' = \min\{n, k\},$$

and, since $n2^{n-1} = \sum_{r=1}^{n} r \binom{n}{r}$, necessarily $k \geq n$. On the other hand, n hyperplanes suffice to cut all edges, which can be seen by the example $\lambda^{n+1}(j) = (j, 1, \ldots, 1)$. ∎

§3 k-Dimensional Sets of Binary Vectors of Given Weight

Another problem concerning the behavior of binary sequences over reals is to find an expression for the maximal number of $\{0, 1\}$-sequences of given Hamming weight, which for given k are contained in a k-dimensional subspace of the real space $\mathbb{R}^n$. In the next theorem we find expressions for this number in all possible cases. In the proof of the theorem we use combinatorial considerations like the LYM inequality, some calculus of binomial coefficients, and tools from Linear Algebra. We give the necessary definitions. Let U_k^n be a k-dimensional subspace of the Euclidean space $\mathbb{R}^n$. Define

$$\Gamma(n, k, \omega) = \max_{U_k^n \subset \mathbb{R}^n} \left| U_k^n \cap \binom{[n]}{\omega} \right|.$$

The following main result of this section solves the problem of finding the value $\Gamma(n, k, \omega)$.

Theorem 26 (Ahlswede, Aydinian, and Khachatrian 2003)

(i) $\Gamma(n,k,\omega) = \Gamma(n,k,n-\omega)$.

(ii) For $\omega \le \frac{n}{2}$ we have

$$\Gamma(n,k,\omega) = \begin{cases} \binom{k}{\omega}, & \text{if } (a)\ 2\omega \le k, \\ \binom{2(k-\omega)}{k-\omega} 2^{2\omega-k}, & \text{if } (b)\ k < 2\omega < 2(k-1), \\ 2^{k-1}, & \text{if } (c)\ k-1 \le \omega. \end{cases}$$

The sets giving the claimed values of $\Gamma(n,k,\omega)$ in the three cases are

(a) $\mathcal{S}_1 = \binom{[k]}{\omega} \times \{0\}^{n-k}$,

(b) $\mathcal{S}_2 = \binom{[2(k-\omega)]}{k-\omega} \times \{10,01\}^{2\omega-k} \times \{0\}^{n-2\omega}$,

(c) $\mathcal{S}_3 = \{10,01\}^{k-1} \times \{1\}^{\omega-k+1} \times \{0\}^{n-k-\omega+1}$.

The corresponding k-dimensional subspaces $V(\mathcal{S}_1), V(\mathcal{S}_2), V(\mathcal{S}_3)$ containing these sets (up to permutation of coordinates) are given by the following basis vectors.

$V(\mathcal{S}_1)$:

$$b_1^n = (1,0,\ldots,0,\ldots,0)$$
$$b_2^n = (0,1,0,\ldots,\ldots,0)$$
$$\vdots$$
$$b_k^n = (0,\ldots,1,0,\ldots,0).$$

$V(\mathcal{S}_2)$:

$$b_1^n = (1,0,\ldots,0,0,\ldots,0,\frac{1}{k-\omega},\ldots,\frac{1}{k-\omega},0,\ldots,0)$$

$$\vdots$$

$$b_{2k-2\omega}^n = (\underbrace{0,\ldots,0,1,0,\ldots,0}_{2k-2\omega},\underbrace{\frac{1}{k-\omega},\ldots,\frac{1}{k-\omega}}_{2\omega-k \quad 2\omega-k},\underbrace{0,\ldots,0}_{n-2\omega})$$

$$b_{2k-2\omega+1}^n = (0,\ldots,0,1,\ldots,0,-1,0,\ldots,0,0,\ldots,0)$$

$$\vdots$$

$$b_k^n = (\underbrace{0,\ldots,0,0,\ldots}_{2k-2\omega},\underbrace{1,0,0,\ldots}_{2\omega-k},\underbrace{-1,0,\ldots,0}_{2\omega-k \qquad n-2\omega}).$$

To obtain $0,1$-vectors we have to consider only linear combinations with coefficients 0 or 1. The linear combinations of the first $2k-2\omega$ vectors must have exactly $k-\omega$ ones in the first $2k-2\omega$ coordinates. Combining each of those vectors with all possible $0,1$-combinations of the remaining basis vectors, we get exactly $\binom{2(k-\omega)}{k-\omega} 2^{2\omega-k}$ vectors of weight ω.

$V(\mathcal{S}_3)$:

$$b_1^n = (1,0,\ldots,-1,0,\ldots,0,0,\ldots,0)$$
$$b_2^n = (0,1,0,\ldots,-1,\ldots,0,0,\ldots,0)$$

$$\vdots$$

$$b_{k-1}^n = (0,\ldots,1,0,\ldots,-1,0,\ldots,0,0,\ldots,0)$$
$$b_k^n = (\underbrace{0,\ldots,0,}_{k-1}\underbrace{1,\ldots,1,}_{k-1}\underbrace{1,\ldots,1,}_{\omega-k+1}0,\ldots,0).$$

Adding all 2^{k-1} possible $0,1$-combinations of the first $k-1$ basis vectors to b_k^n gives $0,1$-vectors of weight ω.

Now we turn to the proof of Theorem 26. We call a nonzero vector $u^n = (u_1,\ldots,u_n) \in \mathbb{R}^n$ nonnegative (resp. positive) if $u_i \geq 0$ (resp. $u_i > 0$) for all $i = 1,\ldots,n$. For the proof we need some auxiliary results. Let us formulate the first one.

Proposition 12 *Assume that a k-dimensional subspace $U_k^n \subset \mathbb{R}^n$ contains a non-negative vector. Then it also contains a nonnegative vector with at least $k-1$ zero coordinates.*

Proof. We apply induction on k and n. The case $k = 1$ is trivial. Assume that the assertion holds for $k' \leq k - 1$ and all n.

Suppose U_k^n is the row space of the $k \times n$ matrix

$$G = \begin{bmatrix} v_1^n \\ \vdots \\ v_k^n \end{bmatrix}$$

and let $u^n \in U_k^n$ be a nonnegative vector. If u^n has zero coordinates, then we are done. Indeed, suppose that $u^n = (u_1,\ldots,u_\ell,0,\ldots,0)$ for $n - k + 1 < \ell < n$ and $u_i > 0$ for $i = 1,\ldots,\ell$. Then G can be transformed into the form shown in Fig. 2, where B is a matrix of rank $rank(B) = s \leq n - \ell < k - 1$, A is a matrix of rank $k - s$, and $\mathbf{0}$ is the all zero matrix.

Now, by the induction hypothesis, the row space of A contains a non-negative vector with at least $k - s - 1$ zero coordinates. Hence, in the row space of G there is a nonnegative vector containing at least $k - s - 1 + n - \ell \geq k - 1$ zeros, proving the proposition in this case.

Suppose now that u^n is a positive vector. Let $v^n \in U_k^n$ with $v^n \neq \alpha u^n, \alpha \in \mathbb{R}$. W.l.o.g. assume $\frac{v_1}{u_1} \geq \frac{v_i}{u_i}$, $i = 2,\ldots,n$. Then one can easily see that $\frac{v_1}{u_1} u^n - v^n \in U_k^n$ is a nonnegative vector with zero in the first coordinate. This completes the proof of the proposition, because we come to the case considered above. $\square$

We say that a matrix M of size $k \times n$ and rank k has a step form if it is of the form shown in Fig. 3 up to permutations of the columns.

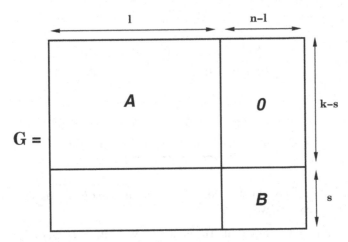

Fig. 2 Proof of Proposition 12

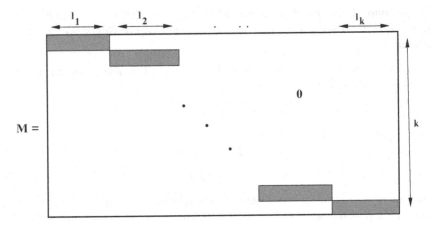

Fig. 3 A matrix with a (positive) step form

Each step (shaded part in Fig. 3) of size $\ell_i \geq 1$, $i = 1, \ldots, k$, $\sum_{i=1}^{k} \ell_i = n$, consists of ℓ_i entries of the ith row, and above the steps M has only zero entries. We say that M has positive step form if all the steps have positive entries.

Using Proposition 12 one can show that the subspace $U_k^n \subset \mathbb{R}^n$ has a generating matrix in a positive step form iff U_k^n contains a positive vector (Exercise 7).

Given an underlying set X, a chain $\mathcal{C} = \{C_1, \ldots C_s\} \subset 2^X$ is called maximal if $s = |X| + 1$ (it agrees with the usual definition of a maximal chain as a chain of maximal possible cardinality).

The second result needed for the proof of the theorem is the following.

Proposition 13 *Let $X = X_1 \cup X_2 \cup \ldots \cup X_s$ be a partition of X with $|X_i| = n_i$ and let $\mathcal{A} \subset \binom{X}{\omega}$ be a family with the following property:*
for any $A, B \in \mathcal{A}$ and $j = 1, \ldots, s$ the relation

$$E = A \cap \left(\bigcup_{i=1}^{j} X_i \right) \neq B \cap \left(\bigcup_{i=1}^{j} X_i \right) = F$$

implies that E and F are incomparable.
Then

$$|\mathcal{A}| \leq \max_{\sum_{i=1}^{s} \omega_i = \omega} \prod_{i=1}^{s} \binom{n_i}{\omega_i}. \tag{18}$$

Proof. Define a product maximal chain (*p*-chain) in X as a sequence $\mathcal{C}^s = (\mathcal{C}_1, \ldots, \mathcal{C}_s)$, where $\mathcal{C}_i \subset 2^{X_i}$ is a maximal chain in X_i. Evidently, the number of *p*-chains is $\prod_{i=1}^{s} n_i!$. Let us represent each element $A \in \mathcal{A}$ as a sequence $A = (A_1, \ldots, A_s)$, where $A_i = A \cap X_i$, $i = 1, \ldots, s$. We write $A \in \mathcal{C}^s$ iff $A_i \in \mathcal{C}_i$, $i = 1, \ldots, s$. The property of $\mathcal{A}$ in the proposition implies that each *p*-chain contains at most one element from $\mathcal{A}$. On the other hand, given $A \in \mathcal{A}$, there are exactly $\prod_{i=1}^{s} |A_i|!(n_i - |A_i|)!$ *p*-chains containing A. Hence

$$\sum_{A \in \mathcal{A}} \prod_{i=1}^{s} |A_i|!(n_i - |A_i|)! \leq \prod_{i=1}^{s} n_i!$$

or

$$\sum_{A \in \mathcal{A}} \frac{1}{\prod_{i=1}^{s} \binom{n_i}{|A_i|}} \leq 1.$$

Next we have

$$\frac{|\mathcal{A}|}{\max_{A \in \mathcal{A}} \prod_{i=1}^{s} \binom{n_i}{|A_i|}} \leq \sum_{A \in \mathcal{A}} \frac{1}{\prod_{i=1}^{s} \binom{n_i}{|A_i|}} \leq 1,$$

which gives the desired result.   □

At last we need the following:

Proposition 14 *Let $n, \omega, s = 1, 2, \ldots$, $s \leq n$, $2\omega \leq n$. Then we have*

$$M \overset{\Delta}{=} \max_{\sum_{i=1}^{s} n_i = n,\, n_i \geq 1; \sum_{i=1}^{s} \omega_i = \omega} \prod_{i=1}^{s} \binom{n_i}{\omega_i}$$

$$= \begin{cases} \binom{n-s+1}{\omega}, & \text{if } 2\omega \leq n - s + 1, \\ \binom{2(n-s+1)-2\omega}{n-s+1-\omega} 2^{2\omega - (n-s+1)}, & \text{if } n - s + 1 < 2\omega < 2(n-s), \\ 2^{n-s}, & \text{if } \omega \geq n - s. \end{cases}$$

Proof. Consider the representation of M in the form

$$M = \prod_{i=1}^{s} \binom{m_i}{k_i}, \tag{19}$$

where $\sum_{i=1}^{s} m_i = n$, $m_i \geq 1$, $\sum_{i=1}^{s} k_i = \omega$, $k_i \geq 0$.

We say that $\binom{\ell}{t}$ is a factor of M iff $\ell = m_i$, $t = k_i$ for some $i = 1,\ldots,s$ in a representation of M in the form (19).

Let now $M = M_1 \binom{2}{1}^{s_1}$ with $s_1 \geq 0$. Then we claim that M_1 does not contain the following factors:

(i) $\binom{m}{k}$ and $\binom{\ell}{t}$ with $m, \ell > 1$ and $k, t > 0$.

(ii) $\binom{m}{k}$ with $m < 2k$,

(iii) $\binom{m}{k}$ with $m > 2k+1$, $s_1 \geq 1$,

(iv) $\binom{m}{k}$ and $\binom{1}{1}$ with $m \neq 1$, $m > 2k$; or $\binom{m}{k}$ and both $\binom{1}{1}$ and $\binom{1}{0}$ with $m \neq 1$, $m = 2k > 2$.

(i) Let $\binom{m}{k}$, $\binom{\ell}{t} \neq \binom{2}{1}$, $m, \ell \neq 1$. Then the following inequalities can be easily verified.

If $m \neq 2k$, $\ell \neq 2t$, then

$$\binom{m}{k}\binom{\ell}{t} < \max\left\{\binom{m+\ell-1}{k+t}\binom{1}{0}, \binom{m+\ell-1}{k+t-1}\binom{1}{1}\right\}.$$

If $m = 2k$ or $\ell = 2t$, then

$$\binom{m}{k}\binom{\ell}{t} < \binom{m+\ell-2}{k+t-1}\binom{2}{1}.$$

Each of these inequalities contradicts the maximality of M if $\binom{m}{k}$ and $\binom{\ell}{t}$ are factors of M_1.

(ii) Suppose M_1 has a factor $\binom{m}{k}$ with $m < 2k$. Then (i) with $2\omega \leq n$ implies the existence of the factor $\binom{1}{0}$, which leads to a contradiction with

$$\binom{m}{k}\binom{1}{0} < \binom{m}{k-1}\binom{1}{1}.$$

(iii) If M_1 has the factor $\binom{m}{k}$ with $m > 2k+1$ and $s_1 \geq 1$, then

$$\binom{m}{k}\binom{2}{1} < \binom{m+1}{k+1}\binom{1}{0}.$$

(iv) Suppose now M_1 possesses the factors $\binom{m}{k}$ and $\binom{1}{1}$ with $m \neq 1$. Then we get a contradiction with

$$\binom{m}{k}\binom{1}{1} < \binom{m-1}{k}\binom{2}{1},$$

when $m > 2k$. If now $m = 2k > 2$, then

$$\binom{m}{k}\binom{1}{1}\binom{1}{0} < \binom{m-2}{k-1}\binom{2}{1}^2$$

leads to a contradiction.

Now we can resume our observations above as follows. M can have only the following form:

$$M = \binom{m_1}{k_1} \binom{2}{1}^{s_1} \binom{1}{1}^{s_2} \binom{1}{0}^{s_3},$$ (20)

where $m_1 + 2s_1 + s_2 + s_3 = n$, $k_1 + s_1 + s_2 = \omega$, $s_1 + s_2 + s_3 + 1 = s$; $s_1, s_2, s_3 \geq 0$, $m_1 \geq 1$, $m_1 \geq 2k_1$.

Finally, we have

(a) $\omega \geq n - s$ implies $s_2 \geq k_1 - 1$. Therefore, in both cases, $s_2 = 0$ or $s_2 > 0$, by (iv) we get $k_1 = 1$, $m_1 = 2$, which means that

$$M = 2^{s_1+1} = 2^{n-s}.$$

(b) $2\omega \leq n - s + 1$ with (iii) implies $s_1 + 2s_2 \leq 1$. Hence $s_2 = 0$ and $s_1 = 0$ or 1, which gives

$$M = \binom{m_1 + s_1}{k_1 + s_1} = \binom{n-s+1}{\omega}.$$

(c) $n - s + 1 < 2\omega \leq 2(n - s)$ implies $s_1 + 2s_2 > 0$, $s_2 < k_1 - 1$, which with (iv) implies $s_2 = 0$. Hence

$$M = 2^{s_1} \binom{2k_1}{k_1},$$

where $s_1 = 2\omega - (n - s + 1)$, $k_1 = n - s + 1 - \omega$.   □

Now we return to the proof of Theorem 26. First we show that $\Gamma(n,k,\omega) = \Gamma(n,k,n-\omega)$. Let $\mathcal{A} \subset \binom{[n]}{\omega}$ with $\dim span(\mathcal{A}) = k$ such that $|\mathcal{A}| = \Gamma(n,k,\omega)$. Suppose that $v_1^n, \ldots, v_k^n$ are linearly independent vectors in $\mathcal{A}$. For every $v^n \in \mathcal{A}$ we have

$$\sum_{i=1}^{k} \alpha_i v_i^n = v^n,$$ (21)

and since $\mathcal{A} \subset \binom{[n]}{\omega}$, we conclude that

$$\sum_{i=1}^{k} \alpha_i = 1.$$ (22)

Consider the complement set $\mathcal{B} = \{1^n - v^n : v^n \in \mathcal{A}\}$ and notice that $\mathcal{B} \subset \binom{[n]}{n-\omega}$, $|\mathcal{B}| = |\mathcal{A}|$. We have

$$\sum_{i=1}^{k} \alpha_i(1^n - v_i^n) = 1^n - v^n,$$

which shows that $\dim span(\mathcal{B}) = k$. Therefore, $\Gamma(n,k,\omega) \leq \Gamma(n,k,n-\omega)$ and by symmetry $\Gamma(n,k,\omega) \geq \Gamma(n,k,n-\omega)$.

Let now U_k^n be an optimal subspace, that is, it contains a maximal number of vectors from $\binom{[n]}{\omega}$. Let further V_{n-k}^n be the space orthogonal to U_k^n with basis a $v_1^n, \ldots, v_{n-k}^n$. Our task is to determine the maximum number of $0, 1$-solutions of the system of $n - k + 1$ equations

$$
\begin{cases}
(v_1^n, x^n) & = 0 \\
\vdots & \\
(v_{n-k}^n, x^n) = 0, & \\
(1^n, x^n) & = \omega
\end{cases}
\tag{23}
$$

for all choices of linearly independent vectors $v_1^n, \ldots, v_{n-k}^n$. In view of Exercise 12 this system of equations can be reduced to the form

$$(a_i^n, x^n) = c_i, \ i = 1, \ldots, n - k + 1,$$

where the matrix $(a_{ij})_{i=1,\ldots,n-k+1}^{j=1,\ldots,n}$ has a positive step form. W.l.o.g. we may assume that this matrix has the form shown in Fig. 3 with steps of size $\ell_i \geq 1$, $i = 1, \ldots, n - k + 1$, and $\sum \ell_i = n$. Let Z denote the set of all $0,1$-solutions of (23). The following property is easily seen. For any solutions $x^n = (x_1, \ldots, x_n)$, $y^n = (y_1, \ldots, y_n)$ and any $t_s = \ell_1 + \ldots, \ell_s$, $s = 1, \ldots, n - k + 1$, if $(x_1, \ldots, x_{t_s}) \neq (y_1, \ldots, y_{t_s})$, there exist $1 \leq i, j \leq t_s$ such that $x_i > y_i$, $x_j < y_j$.

Consider now $(x_1, \ldots, x_{t_s})$ and $(y_1, \ldots, y_{t_s})$ as the characteristic vectors of the sets X and Y, respectively. The property above means that E and H are incomparable. Thus considering the solutions of (23) as a corresponding family of sets $\mathcal{A} \subset \binom{[n]}{\omega}$, where $[n]$ is partitioned into $n - k + 1$ nonempty subsets, we see that $\mathcal{A}$ satisfies the property from Proposition 13. Consequently, we have

$$|Z| \leq |\mathcal{A}| \leq \max_{\sum_{i=1}^{n-k+1} \omega_i = \omega} \prod_{i=1}^{n-k+1} \binom{\ell_i}{\omega_i}.$$

Combining this with Proposition 14 we get the proof of the theorem. ∎

Notes to Chapter IV

Lemma 30 was conjectured in [ACZ96] and proved in the case $k = 2$. The final proof for the general case was done by Alon and Sudakov in [AS95] and we here reproduced their proof. We took most of the material for the second section of Lecture 11 from [ACZ96] and [ACZ94]. Theorem 24 is from [FR05] and [ACZ94]. Theorem 26 is from [AAK03a].

Denote

$$J_t(n,k) = \max_{\mathcal{A} \in \mathcal{I}(n,t)} \{|\mathcal{A}| : \dim span(\mathcal{A}) = k\}.$$

In [AAK03b] it is proved that if $1 \leq t \leq n-k+1$, then $J_t(n,k) = 2^{t-1}$. We formulate their conjecture about the value $J_t(n,k)$ for the case when $t > n-k+1$ in Research Problem 1.

Let us mention one more interesting problem that deals with linear relations of binary vectors over reals: find the probability P_n that a random (with uniform distribution) $n \times n$ matrix M_n with ± 1 entries is singular (with linear combinations over reals). It is conjectured that

$$P_n = P(M_n \text{ is singular }) = (1+o(1))\frac{n^2}{2^{n-1}},$$

that is, P_n is essentially the probability that M_n contains two rows or two columns, which are equal up to the sign. Presently it is proved by Kahn et al. in [KKS95] that $P_n < e^{-\varepsilon n}$ holds for some positive fixed $\varepsilon > 0$. Their proof also has an important corollary: there is a constant $c > 0$ such that if $r < n-c$ and $v_1^n, \ldots, v_r^n$ are chosen (uniformly and independently) at random from $\{\pm 1\}^n$, then

$$P(\dim span(v_1^n, \ldots, v_r^n) = r) = (1+o(1))r\binom{r}{2}2^{-n},$$

$$P(span(v_1^n, \ldots, v_r^n) \cap \{\pm 1\}^n \neq \{v_1^n, \ldots, v_r^n\}) = (1+o(1))4\binom{r}{3}\left(\frac{3}{4}\right)^n.$$

The last relation was first proved by Odlyzko [O88] under the more restrictive assumption that $r < n - 10n/\log n$.

At this moment the best upper bound on $P_n < \left(\frac{3}{4} + o(1)\right)^n$ is obtained by Tao and Vu in [TV05].

Exercises

1. Using (1) (Chapter I) prove the statement associated with formula (5) (Lecture 11).
2. Let $H(Y,p)$ be the induced subgraph of $G(n,p)$ on a given set Y of vertices, $|Y| = n_1$. Let g be a graph-theoretic function given by

$$g(H') = \frac{c_k(H')}{\binom{n_1}{k-2}},$$

where H' is a graph on Y and $c_k(H')$ is the number of k-cliques in H'. Using (3) (Chap. I) and the fact that $g(H)$ is a graph-theoretic function show that

$$P\left(\left|g(H(Y,p)) - \mu(g)\right| > n^{3/4}\sqrt{\binom{n_1}{2}}\right) < 2e^{-n^{3/2}/2}.$$

Derive from this inequality the corresponding inequality for $c_k(H(Y,p))$ and then draw a conclusion from Lemma 31.

3. Prove that

$$I_n(\forall, \forall, k) = \binom{n-1}{k-1}, \quad n \geq 2k.$$

Find the asymptotic behavior of $I_n(\cdot, \cdot, k)$ for other combinations of $\{\exists, \forall\}$.

4. Prove that the family $\mathcal{F}$ in (2) (Lecture 12) is positive linear combinations free.
5. Prove the fact corresponding to the relation (9) (Lecture 12).
6. Prove that the family $\mathcal{N}(\mathcal{J}(\mathcal{F}))$ is positive linear combination free.
   Hint: Indeed, $\mathcal{N}(\mathcal{J}(\mathcal{F})) \cap \mathcal{J}(\mathcal{F}) = \emptyset$, and so the vectors of $\mathcal{N}(\mathcal{J}(\mathcal{F}))$ intersect every vector in $\mathcal{F}$. Take an arbitrary $A \in \mathcal{N}(\mathcal{J}(\mathcal{F}))$. It is a neighbor of some set $A' \in \mathcal{J}(\mathcal{F})$ and there is a set $A^* \in \mathcal{F}$ such that $A^* \cap A' = \emptyset$. Observe that $|A \cap A^*| = 1$. Then use the fact corresponding to the relation (9) (Lecture 12).
7. Using Proposition 12 prove by induction that a subspace $U_k^n \subset \mathbb{R}^n$ has a generator matrix in a positive step form iff U_k^n contains a positive vector.

Research Problems

1. **Conjecture** If $k > 0$ and $t > m = n - k + 1$ we have

$$J_t(n,k) = \begin{cases} \sum_{i=\frac{n+t}{2}-m}^{k-1} \binom{k-1}{i} + \sum_{i=\frac{n+t}{2}}^{k-1} \binom{k-1}{i}, & 2 \mid (n+t), \\ 2\sum_{i=\frac{n+t-1}{2}-m}^{k-2} \binom{k-2}{i} + 2\sum_{i=\frac{n+t-1}{2}}^{k-2} \binom{k-2}{i}, & 2 \nmid (n+t). \end{cases}$$

2. Denote

$$A(n,k) = \max_{\mathcal{A} \subset 2^{[n]}} \{|\mathcal{A}| : \dim span(\mathcal{A}) \leq k, \ \mathcal{A} \ \text{is an antichain}\}.$$

Conjecture The following equality is valid:

$$A(n,k) = \Gamma\left(n, k, \left\lfloor \frac{n}{2} \right\rfloor\right).$$

When $n \geq 2k - 2$ this equality is proved in [AAK03c] (in this case $A(n,k) = 2^{k-1}$).

Chapter V
LYM-Related AZ-Identities, Antichain Splittings and Correlation Inequalities

Lecture 13 LYM-Type Relations

In this lecture we present proofs of several relations that are improvements and generalizations of the so-called LYM (Lubell-Yamamoto-Meshalkin) inequality. Initially this inequality set a relation for the distribution of cardinalities of members in an antichain of subsets. A family $\mathcal{A} \subset 2^{[n]}$ is a chain if any two sets $A_1, A_2 \in \mathcal{A}$ are comparable, that is, $A_1 \subset A_2$ or $A_2 \subset A_1$. Obviously the maximal length of a chain is $n+1$. A family $\mathcal{A} \subset 2^{[n]}$ is an antichain if any two sets $A_1, A_2 \in \mathcal{A}$ are incomparable, that is, $A_1 \not\subset A_2$ and $A_2 \not\subset A_1$. Obviously the family of k-sets $\mathcal{A} \subset \binom{[n]}{k}$ is an antichain and Sperner's Lemma says that the antichain $|\mathcal{A}| = \binom{n}{\lfloor \frac{n}{2} \rfloor}$ has maximal size (and it also settles the uniqueness property discussed below). The upper bound $|\mathcal{A}| \leq \binom{n}{\lfloor \frac{n}{2} \rfloor}$ is not obvious and follows from the LYM inequality. If $f_0, f_1, \ldots, f_n$ is a sequence such that $f_i = \left| \mathcal{A} \cap \binom{[n]}{i} \right|$ for some antichain $\mathcal{A}$, then the LYM inequality says that

$$\sum_{i=0}^{n} \frac{f_i}{\binom{n}{i}} \leq 1. \tag{1}$$

Hence

$$|\mathcal{A}| = \sum_{i=0}^{n} f_i \leq \binom{n}{\lfloor \frac{n}{2} \rfloor} \sum_{i=0}^{n} \frac{f_i}{\binom{n}{i}} \leq \binom{n}{\lfloor \frac{n}{2} \rfloor}.$$

The main statement of Sperner's Lemma follows. This lecture has only one section. Here we investigate the possibility of improving the LYM inequality (1). Actually taking into account other terms we derive the so-called AZ (Ahlswede–Zhang) identity. Also we consider the uniqueness of the antichain $\mathcal{A}$ for which equality $|\mathcal{A}| = \binom{n}{\lfloor \frac{n}{2} \rfloor}$ is achieved and related problems. In Theorems 28 and 30 we demonstrate further generalizations of the AZ-identity. The proof of Theorem 29 is similar to the proof of the AZ-identity and we ask to do it in Exercise 1.

Let us give some definitions. For every $X \in 2^{[n]}$ and every $\mathcal{A} \subset 2^{[n]}$ we define

$$X_{\mathcal{A}} = \bigcap_{X \supset A \in \mathcal{A}} A, \; W_{\mathcal{A}}(X) = |X_{\mathcal{A}}|.$$

Using the functions

$$W_{\mathcal{A}}^{(i)} = \sum_{X \in \binom{[n]}{i}} W_{\mathcal{A}}(X)$$

we can write

$$\sum_{X} \frac{W_{\mathcal{A}}(X)}{|X| \binom{n}{|X|}} = \sum_{i=1}^{n} \frac{W_{\mathcal{A}}^{(i)}}{i \binom{n}{i}}.$$

§1 The AZ-Identity and Related Results

We start with the proof of the following:

Theorem 27 (Ahlswede and Zhang 1990) *For every family $\mathcal{A}$ of nonempty subsets of $[n]$*

$$\sum_{i=1}^{n} \frac{W_{\mathcal{A}}^{(i)}}{i \binom{n}{i}} = 1. \tag{2}$$

Identity (2) is called the AZ-identity.

When $\mathcal{A}$ is an antichain, identity (2) becomes

$$\sum_{X \in \mathcal{A}} \frac{1}{\binom{n}{|X|}} + \sum_{X \in \mathcal{U}(\mathcal{A}) \setminus \mathcal{A}} \frac{W_{\mathcal{A}}(X)}{|X| \binom{n}{|X|}} = 1, \tag{3}$$

where $\mathcal{U}(\mathcal{A}) = \{Y \in 2^{[n]} : \exists A \in \mathcal{A}, A \subseteq Y\}$ is the upset generated by $\mathcal{A}$. Hence (2) is a strengthening of the LYM inequality, whose deficiency can be measured by the second summand.

Proof. Note first that only minimal elements in $\mathcal{A}$ determine $X_{\mathcal{A}}$ and therefore matter. Hence we can assume that $\mathcal{A}$ is an antichain.

The LYM inequality is obtained as follows. All saturated chains (saturated means that the chain is not contained in a larger chain), which pass through members of $\mathcal{A}$, are counted:

$$\sum_{A \in \mathcal{A}} |A|!(n - |A|)!.$$

No chain is counted twice, because $\mathcal{A}$ is an antichain. Since there are $n!$ saturated chains in total, clearly

$$\sum_{A \in \mathcal{A}} |A|!(n - |A|)! \leq n!.$$

Hence

$$\sum_{A \in \mathcal{A}} \frac{1}{\binom{n}{|A|}} \leq 1.$$

This is the LYM inequality.

Next observe that we can also count the saturated chains not passing through $\mathcal{A}$. Denote $\mathcal{U} = \mathcal{U}(\mathcal{A})$ and count the saturated chains according to their exit in $\mathcal{U}$. For this we view $2^{[n]}$ as a directed graph with an edge between vertices B, C exactly if $B \supset C$ and $|B \setminus C| = 1$.

Since $\emptyset \notin \mathcal{A}$, clearly $\emptyset \notin \mathcal{U}$. Therefore, every saturated chain starting in $[n] \in \mathcal{U}$ has a last set, say exit set, in $\mathcal{U}$. For every $U \in \mathcal{U}$ we call $E = (U, V)$ an exit edge, if $V \in 2^{[n]} - \mathcal{U}$, and we denote the set of exit edges by $\mathcal{E}_{\mathcal{A}}(U)$. The number of saturated chains leaving $\mathcal{U}$ in U is then

$$(n - |U|)! |\mathcal{E}_{\mathcal{A}}(U)| (|U| - 1)!.$$

Therefore,

$$\sum_{U \in \mathcal{U}} (n - |U|)! |\mathcal{E}_{\mathcal{A}}(U)| (|U| - 1)! = n! \tag{4}$$

and since $\mathcal{E}_{\mathcal{A}}(X) = \emptyset$ for $X \in 2^{[n]} - \mathcal{U}$, also

$$\sum_{X \in 2^{[n]}} \frac{|\mathcal{E}_{\mathcal{A}}(X)|}{|X| \binom{n}{|X|}} = 1. \tag{5}$$

∎

Notice the validity of the equality

$$|\mathcal{E}_{\mathcal{A}}(X)| = W_{\mathcal{A}}(X). \tag{6}$$

For any antichain $\mathcal{A} \subset 2^{[n]}$ set $\mathcal{A}^i = \mathcal{A} \cap \binom{[n]}{i}$. Then the LYM-inequality can be rewritten in the form

$$\sum_{i=1}^{n} \frac{|\mathcal{A}^i|}{\binom{n}{i}} \leq 1.$$

Hence

$$\sum_{i=1}^{n} \frac{i |\mathcal{A}^i|}{i \binom{n}{i}} \leq 1$$

and

$$\sum_{i=1}^{n} i |\mathcal{A}^i| \leq \max_i i \binom{n}{i} = m \binom{n}{m}, \tag{7}$$

where in this lecture $m = \lfloor \frac{n}{2} \rfloor + 1$. Obviously bound (7) is best possible. Theorem 27 implies

$$\sum_{X \in 2^{[n]}} W_{\mathcal{A}}(X) \leq m \binom{n}{m}, \quad \mathcal{A} \subset 2^{[n]}. \tag{8}$$

One can also extend the previous theorem to the case of several families.

Corollary 4 *For N families $\mathcal{A}_1, \ldots, \mathcal{A}_N$ of nonempty subsets of $2^{[n]}$ and $X \in 2^{[n]}$ define*

$$W_{\mathcal{A}_1, \ldots, \mathcal{A}_N}(X) = \left| \bigcup_{j=1}^{N} X_{\mathcal{A}_j} \right|$$

and

$$W^{(i)}_{\mathcal{A}_1, \ldots, \mathcal{A}_N} = \sum_{X \in \binom{[n]}{i}} W_{\mathcal{A}_1, \ldots, \mathcal{A}_N}(X).$$

Then

(i)

$$\sum_{i=1}^{n} \frac{W^{(i)}_{\mathcal{A}_1, \ldots, \mathcal{A}_N}}{i \binom{n}{i}} \leq N$$

   and

(ii)

$$\sum_{X \in 2^{[n]}} W_{\mathcal{A}_1, \ldots, \mathcal{A}_N} \leq \max_{0 \leq \ell \leq n - N'} \sum_{r=\ell+1}^{\ell+N'} r \binom{n}{r}, \quad \text{where } N' = \min\{n, N\}.$$

Proof. Since by the definitions

$$W^{(i)}_{\mathcal{A}_1, \ldots, \mathcal{A}_N} \leq \sum_{j=1}^{N} W^{(i)}_{\mathcal{A}_j},$$

(i) follows from Theorem 27.

From its definition we have

$$W^{(i)}_{\mathcal{A}_1, \ldots, \mathcal{A}_N} \leq i \binom{n}{i}$$

and therefore the numbers $\alpha_i = W^{(i)}_{\mathcal{A}_1, \ldots, \mathcal{A}_N} / i \binom{n}{i}$ satisfy $0 \leq \alpha_i \leq 1$ and $\sum_{i=1}^{n} \alpha_i \leq N'$. Hence

$$\sum_{i=1}^{n} \alpha_i \binom{n}{i} \leq \max_{0 \leq \beta_i \leq 1, \, \sum_{i=1}^{n} \beta_i \leq N'} \sum_{i=1}^{n} \beta_i \binom{n}{i}$$

$$= \max_{0 \leq \ell \leq n - N'} \sum_{r=\ell+1}^{\ell+N'} r \binom{n}{r}$$

and since

$$\sum_{X \in 2^{[n]}} W_{\mathcal{A}_1, \ldots, \mathcal{A}_N}(X) = \sum_{i=1}^{n} W^{(i)}_{\mathcal{A}_1, \ldots, \mathcal{A}_N} = \sum_{i=1}^{n} \alpha_i \binom{n}{i},$$

relation (ii) follows.   □

Some consequences of these results are discussed in Lecture 12. Here we demonstrate simple uniqueness proofs that can be given using Theorem 27.

We use the elementary facts that $\max_{1\le\ell\le n}\ell\binom{n}{\ell} = m\binom{n}{m}$, and if $\ell \ne m, m-1$ and n is even or $\ell \ne m$ and n is odd, then

$$\ell\binom{n}{\ell} < m\binom{n}{m}. \tag{9}$$

For an antichain the identity from the theorem says

$$1 = \sum_{X\in\mathcal{A}} \frac{1}{\binom{n}{|X|}} + \sum_{A\notin\mathcal{A}} \frac{W_{\mathcal{A}}(X)}{|X|\binom{n}{|X|}}. \tag{10}$$

If $\mathcal{A}$ is maximal, then $|\mathcal{A}| = \binom{n}{\lfloor\frac{n}{2}\rfloor}$, and therefore

$$\mathcal{A} \subset \binom{[n]}{\lfloor\frac{n}{2}\rfloor} \cup \binom{[n]}{\lceil\frac{n}{2}\rceil}, \tag{11}$$

$$\sum_{X\notin\mathcal{A}} \frac{W_{\mathcal{A}}(X)}{|X|\binom{n}{|X|}} = 0. \tag{12}$$

Obviously from (12) follows that $\mathcal{A} = \binom{[n]}{k}$ if $n = 2k$. But in the case $n = 2k+1$ the uniqueness of the maximal antichain is not so obvious and it does not follow from the LYM inequality. All what we can say using the LYM inequality is that $\mathcal{A} \subset \binom{[n]}{k} \cup \binom{[n]}{m}$. Next we show that for $n = 2k+1$ the assumption $\mathcal{A} \ne \binom{[n]}{k}, \binom{[n]}{m}$ violates (12). From this it follows that for the maximal antichain in this case there are only two possibilities $\mathcal{A} = \binom{[n]}{k}$ or $\mathcal{A} = \binom{[n]}{m}$. To prove this note that $W_{\mathcal{A}}(X) = 0$ for $X \in \binom{[n]}{m} \setminus \mathcal{A}$ implies that in the graph defined on $\binom{[n]}{k} \cup \binom{[n]}{m}$ by containment X has no connections with $\binom{[n]}{k} \setminus \mathcal{A}$. Since there are no connections between $\mathcal{A}_m = \binom{[n]}{m} \cap \mathcal{A}$ and $\mathcal{A}_k = \binom{[n]}{k} \cap \mathcal{A}$, we have two connected components $\mathcal{A}_m \cup (\binom{[n]}{k} \setminus \mathcal{A})$ and $\mathcal{A}_k \cup (\binom{[n]}{m} \setminus \mathcal{A})$. However, $\binom{[n]}{k} \cup \binom{[n]}{m}$ is easily shown to be connected.

Next we consider the uniqueness of the family $\mathcal{A}$ for which there is equality in (8). This is a more difficult problem than the previous one and it is solved completely by Corollary 5 below. Since for every family $\mathcal{A}$ with the set of minimal elements $\mathcal{M}$ $W_{\mathcal{A}}(X) = W_{\mathcal{M}}(X)$, and since every antichain occurs as a set of minimal elements, it suffices to characterize those antichains $\mathcal{A}$ with

$$\sum_{X\in 2^{[n]}} W_{\mathcal{A}}(X) = m\binom{n}{m}. \tag{13}$$

Corollary 5 *Equality holds in (13) for $\mathcal{A} = \binom{[n]}{m}$, if $n = 2k+1$ and for every antichain $\mathcal{A} \subset \binom{[n]}{m} \cup \binom{[n]}{k}$, which is saturated in $\binom{[n]}{m}$, if $n = 2k$.*

Proof. It follows from (13) and identity (2) that $W_{\mathcal{A}}(X) = 0$ if $|X|\binom{n}{|X|} < m\binom{n}{m}$. For $n = 2k+1$ we have therefore $W_{\mathcal{A}}(X) = 0$, if $|X| \ne m$, and thus $\mathcal{A} = \binom{[n]}{m}$.

On the other hand, for $n = 2k$, $k\binom{n}{k} = m\binom{n}{m}$ and we can conclude that $\mathcal{A} \subset \binom{[n]}{m} \cup$ $\binom{[n]}{k}$. $\mathcal{A}$ is saturated in $\binom{[n]}{m}$, because otherwise there would be an antichain $\mathcal{A}' = \mathcal{A} \cup \{X\}$ with $X \in \binom{[n]}{m} \setminus \mathcal{A}$ and $\sum_{Y \in 2^{[n]}} W_{\mathcal{A}'}(Y) = \sum_{Y \in 2^{[n]}} W_{\mathcal{A}}(Y) + |X|$ contradicting the optimality of $\mathcal{A}$.

It remains to show that equality (13) holds for these antichains. Let $\mathcal{A}_+^k$ be the family of elements of $\binom{[n]}{m}$, which are connected with an element of $\mathcal{A}^k$, and let $d(\mathcal{A}^k, X)$ count the number of connections of $\mathcal{A}^k$ with X. Then we have

$$\sum_X W_{\mathcal{A}}(X) = k|\mathcal{A}^k| + (k+1)\left(\binom{2k}{k+1} - |\mathcal{A}_+^k|\right)$$

$$+ \sum_{X \in \mathcal{A}_+^k} ((k+1) - d(\mathcal{A}^k, X)) = k|\mathcal{A}^k| + (k+1)\binom{2k}{k+1} - (n-k)|\mathcal{A}^k|$$

$$= (k+1)\binom{2k}{k+1}. \tag{14}$$

The notion of antichains $\mathcal{A} \subset \binom{[n]}{m} \cup \binom{[n]}{k}$ which are saturated in $\binom{[n]}{m}$ but not in $\binom{[n]}{k}$ is meaningful. As an illustration, let $(n, m, k) = (4, 3, 2)$ and

$$\mathcal{A}^m = \{\{1,2,3\}\}, \quad \mathcal{A}^k = \{\{1,4\}, \{2,4\}\}.$$

The antichain $\mathcal{A} = \mathcal{A}^m \cup \mathcal{A}^k$ cannot be extended by $\{1,2,4\}$, $\{1,3,4\}$, or $\{2,3,4\}$; however, it can be extended by $\{3,4\}$.

Here we solve the problem of uniqueness of a maximal antichain satisfying equality in relation (7). Since for an antichain $\mathcal{A}$ the equality

$$\sum_{A \in \mathcal{A}} |A| = m\binom{n}{m} \tag{15}$$

can hold only if $\mathcal{A}$ is contained in the class characterized by Corollary 5, and so (14) implies necessarily $\sum_{X \in \mathcal{A}_+^k} (k+1) = \sum_{X \in \mathcal{A}_+^k} d(\mathcal{A}^k, X)$ or equivalently that $\mathcal{A}^k \cup \mathcal{A}_+^k$ is a connected component of $\binom{[n]}{k} \cup \binom{[n]}{m}$. This is possible only if $\mathcal{A}^k = \emptyset$ or $\mathcal{A} = \binom{[n]}{k}$. Hence equality holds in (15) exactly for $\mathcal{A} = \binom{[n]}{m}$, if $n = 2k+1$, and for $\mathcal{A} \in \left\{\binom{[n]}{m}, \binom{[n]}{k}\right\}$, if $n = 2k$.   □

Next we give several interesting generalizations of Theorem 27. We give a proof of Theorem 28. Theorems 29 and 30 are asked to be proved in Exercise 1.

The special case (3) of identity (2) can be generalized as follows.

Theorem 28 (Ahlswede and Zhang 1990) *Suppose* $A_1 \subseteq B_1, \ldots, A_N \subseteq B_N$ *are subsets of* $[n]$ *such that* $A_i \not\subseteq B_j$ *for* $i \neq j$, *then*

$$\sum_{i=1}^N \binom{n - |B_i - A_i|}{|A_i|}^{-1} + \sum_{X \in \mathcal{O}} \frac{W_{\mathcal{A}}(X)}{|X|}\binom{n}{|X|}^{-1} = 1, \tag{16}$$

where

$$\mathcal{O} = \{Y : \nexists i : A_i \subset Y \subset B_i\} \tag{17}$$

and $W_{\mathcal{A}}(X) = \left| \bigcap_{A_i \subset X} A_i \right|$.

Notice that the special case $A_i = B_i$ for $i \in [N]$ corresponds to the identity from Theorem 27. Also from this theorem follows the Bollobás inequality

$$\sum_{i=1}^{N} \binom{n - |B_i - A_i|}{|A_i|}^{-1} \leq 1. \tag{18}$$

Instead of $X \in \mathcal{O}$ in (16) we can write $X \notin \mathcal{D}(\{B_i\}_{i=1}^{N})$, where $\mathcal{D}(\{B_i\}_{i=1}^{N}) = \{Y : \exists B_i, Y \subseteq B_i\}$ is the downset generated by $\{B_i\}_{i=1}^{N}$.

Proof. To prove this theorem we calculate first the number of maximal chains $[n] = S_n \supset S_{n-1} \supset \cdots \supset S_1 \supset S_0 = \emptyset$, which intersect an interval $[A, B] = \{C : A \subseteq C \subseteq B\}$. We can construct the chain by subtracting one by one elements from $[n]$ to obtain $S_{n-1}, S_{n-2}, \ldots, S_0$. A maximal chain meets $[A, B]$ iff all elements from $[n] - B$ are subtracted before any element from A is subtracted. We can classify the maximal chains by the first $n - |B|$ elements that are subtracted from S_n in the positions $[n] - B + A$. As we have already mentioned, only one possibility leads to the chains which meet $[A, B]$. The order in which these elements are subtracted are equally likely. Hence the number of maximal antichains that meet $[A, B]$ is equal to the whole number of chains $n!$ divided by the number of possibilities of the first subtraction of $n - |B|$ elements from $[n] - B + A$, that is, $\binom{n - |B - A|}{|A|}$. Hence we have an exact formula for the number of antichains that meet $[A, B]$:

$$n! \binom{n - |B - A|}{|A|}^{-1}. \tag{19}$$

Next we follow the same ideas as in the proof of Theorem 27. For the upset $\mathcal{U}(\mathcal{A})$ the number of saturated chains leaving $\mathcal{U}$ in U is

$$(n - |U|)! W_{\mathcal{A}}(U)(|U| - 1)!.$$

Since intervals $[A_i, B_i] = \{X : A_i \subseteq X \subseteq B_i\}$ are disjoint, we have

$$\sum_{i=1}^{N} \sum_{U \in [A_i, B_i]} (n - |U|)! |A_i| (|U| - 1)! + \sum_{U \in \mathcal{O}} (n - |U|)! W_{\mathcal{A}}(U)(|U| - 1)! = n!.$$

From (19) we get

$$\sum_{U \in [A_i, B_i]} (n - |U|)! |A_i| (|U| - 1)! = n! \binom{n - |B_i - A_i|}{|A_i|}^{-1}$$

and the theorem follows.   ∎

One can follow the approach in the proof of Theorem 27, which applies to any poset, to derive the following combinatorial identity.

Theorem 29 (Ahlswede and Zhang 1990) *In* $\{0,1,\ldots,m\} \times \{0,1,\ldots,n\}$ *write* $(i,j) \le (i',j')$ *iff* $i \le i'$ *and* $j \le j'$. *Let* $I = \{(i_t,j_t) : 1 \le t \le T\}$ *be an antichain. Then*

$$\sum_{t=1}^{T} \left[\sum_{\ell=j_t-1}^{j_{t+1}-1} \frac{\binom{m}{i_t} i_t \binom{n}{\ell}}{(i_t+\ell)\binom{m+n}{i_t+\ell}} + \sum_{k=i_t-1}^{i_{t-1}-1} \frac{\binom{m}{k} j_t \binom{n}{j_t}}{(j_t+k)\binom{m+n}{k+j_t}} + \frac{\binom{m}{i_t}\binom{n}{j_t}}{\binom{m+n}{i_t+j_t}} \right] = 1. \qquad (20)$$

This is asked to be proved in Exercise 1. From (20) follows the inequality

$$\sum_{(i,j)\in I} \binom{m}{i}\binom{n}{j}\binom{m+n}{i+j}^{-1} \le 1.$$

A further generalization of Theorem 28 is the following:

Theorem 30 (Ahlswede and Cai 1993) *Suppose that for a family* $\mathcal{B} = \{B_1,\ldots, B_N\}$ *of subsets of* $[n]$ *and a system* $\mathcal{S} = \{\mathcal{A}_1,\ldots,\mathcal{A}_N\}$ *of subsets of* $2^{[n]}$, *where* $\mathcal{A}_i = \{A_i^t : t \in T_i\}$ *for a finite index set* T_i , *we have the properties*

(i) $A_i^t \subset B_i$ *for* $t \in T_i$ *and* $i = 1,\ldots,N$
(ii) $A_i^t \not\subset B_j$ *for* $t \in T_i$ *and* $i \ne j$.

Then with $\mathcal{A} = \bigcup_{i=1}^{N} \mathcal{A}_i$

$$\sum_{i=1}^{N} \sum_{k=1}^{|T_i|} (-1)^{k-1} \sum_{S \subset T_i, |S|=k} \binom{n - |B_i - \bigcup_{t\in S} A_i^t|}{|\bigcup_{t\in S} A_i^t|}^{-1} + \sum_{X \notin \mathcal{D}(\mathcal{B})} \frac{W_{\mathcal{A}}(X)}{|X|\binom{n}{|X|}} = 1. \qquad (21)$$

Theorem 28 corresponds to the special case $|T_i| = 1$. Exercise 1 asks to prove this theorem. For additional information about Sperner systems and LYM-type inequalities see [B73b].

Lecture 14 The Splitting Property

Here we describe an interesting property of subsets of a poset, which is called the splitting property. We give the definition of the splitting property later and now we recall some necessary notations. Suppose $\mathcal{P} = (\mathcal{R},<)$ is a poset and $\mathcal{A}$ is a subset of $\mathcal{R}$. The downset $\mathcal{D}(\mathcal{A})$ of the subset $\mathcal{A}$ is defined as

$$\mathcal{D}(\mathcal{A}) = \{X \in \mathcal{R} : \exists A \in \mathcal{A}, X \le A\}.$$

The upset of $\mathcal{A}$ is
$$\mathcal{U}(\mathcal{A}) = \{X \in \mathcal{R} : \exists A \in \mathcal{A}, A \le X\}.$$

We also define the sets

$$\mathcal{D}^*(\mathcal{A}) = \{X \in \mathcal{R} : \exists A \in \mathcal{A}, X < A\},$$
$$\mathcal{U}^*(\mathcal{A}) = \{X \in \mathcal{R} : \exists A \in \mathcal{A}, A < X\}.$$

Let $\mathcal{P} = (\mathcal{R}, <) = (2^{[n]}, \subset)$ and let $\mathcal{S} = \binom{[n]}{k}$. Actually when we say poset we mean a family of sets ordered by inclusion operation. For $\mathcal{A}, \mathcal{B} \subset \mathcal{R}$ we write $\mathcal{A} > | < \mathcal{B}$ iff for all $A \in \mathcal{A}$, $B \in \mathcal{B}$ the elements A and B are incomparable. We also write $A > | < B$ instead of $\{A\} > | < \{B\}$ and $A > | < \mathcal{B}$ instead of $\{A\} > | < \mathcal{B}$. Remind that an antichain $\mathcal{S}$ is a poset, which contains only incomparable elements. An antichain $\mathcal{S} \subset \mathcal{R}$ is said to be saturated if

$$\mathcal{D}(\mathcal{S}) \cup \mathcal{U}(\mathcal{S}) = \mathcal{R}. \tag{1}$$

Let $\mathcal{P} = (\mathcal{R}, <) = (2^{[n]}, \subset)$ and let $\mathcal{S} = \binom{[n]}{k}$. Denote by $\mathcal{S}_1 \subset \mathcal{S}$ the family of all k-element subsets of $[n]$ containing a fixed element of $[n]$ and let $\mathcal{S}_2 = \mathcal{S} \setminus \mathcal{S}_1$. Then it is easy to check that

$$\mathcal{D}(\mathcal{S}_1) \cup \mathcal{U}(\mathcal{S}_2) = 2^{[n]}.$$

This property of the set § is called splitting. In fact, a similar splitting can be achieved for every saturated antichain of $2^{[n]}$. We are going to establish a general condition on posets and antichains under which the latter ones have a splitting property.

Let $\mathcal{P}$ be a finite poset. A subset $\mathcal{A} \subseteq \mathcal{R}$ is called dense in poset $\mathcal{P}$ if any nonempty open interval $< X, Y >= \{z \in \mathcal{R} : X < Z < Y\}$ that intersects $\mathcal{A}$ contains at least two elements of $\mathcal{A}$. If the dense $\mathcal{A} = \mathcal{R}$ we call $\mathcal{P}$ dense.

We say that a saturated antichain $\mathcal{S}$ satisfies the splitting property if there exists a partition of $\mathcal{S}$ into disjoint subsets $\mathcal{S}_1$ and $\mathcal{S}_2$ such that

$$\mathcal{D}(\mathcal{S}_1) \cup \mathcal{U}(\mathcal{S}_2) = \mathcal{R} \tag{2}$$

holds. The poset $\mathcal{P}$ has the splitting property if every saturated antichain $\mathcal{S}$ in $\mathcal{P}$ satisfies the splitting condition expressed in (2). The splitting property of poset $\mathcal{P} = (\mathcal{R}, <)$ means that we can divide every saturated antichain into two parts such that the upset of the one part and the downset of the other part together give the whole set $\mathcal{R}$.

§1 Antichains that Satisfy the Splitting Condition

The main results of this section are contained in two theorems. The first Theorem 31 says that all saturated dense antichains satisfy the splitting condition and the second Theorem 32 states that in dense posets every saturated antichain satisfies the splitting condition. At the end we show that these theorems are independent: there exists a dense poset and no dense saturated antichain in it.

It is easy to find a poset $\mathcal{P}$ and a saturated antichain $\mathcal{S}$ in $\mathcal{P}$ such that $\mathcal{S}$ has no splitting. Indeed, if $\mathcal{P}$ has a (nonminimal and nonmaximal) element s, which is comparable with any other element of $\mathcal{R}$, then $\mathcal{S} = \{S\}$ is a saturated antichain and does not satisfy the splitting property. The next theorem establishes a sufficient condition on an antichain to have the splitting property.

Theorem 31 (Ahlswede, Erdös, and Graham 1995) *Let S be a saturated, dense antichain in the poset $\mathcal{P}$. Then S satisfies the splitting condition.*

Proof. Let S be a dense and saturated antichain in the poset $\mathcal{P}$ and let $<_{ord}$ be an arbitrary linear ordering on S. For every element $X \in \mathcal{D}^*(S)$ let $f(X)$ be the greatest element $S \in S$ (with respect to $<_{ord}$) such that $X < S$. Set

$$S_1 = \{f(X): \ X \in \mathcal{D}^*(S)\},$$
$$S_2 = S \setminus S_1.$$

We claim that S_1 and S_2 satisfy the splitting condition (2). Assume the contrary. Then there exists a $Y \in \mathcal{U}^*(S)$ such that $Y \notin \mathcal{U}(S_2)$. Let $f(Y)$ be the smallest element $S \in S$ (with respect to $<_{ord}$) such that $S < Y$. We know that $f(Y) \notin S_2$; therefore, there exists an element $X \in \mathcal{D}^*(S)$ satisfying $f(X) = f(Y) = S$. Then the open interval $< X, Y >$ contains the element S; hence there exists an $s' \in < X, Y > \cap (S \setminus \{S\})$. The linear ordering gives us the order between S and S'. If, say $S <_{ord} S'$, then, due to the definitions, $f(X)$ cannot be S, a contradiction. The other case, $S' <_{ord} S$, leads to a contradiction with $f(Y) = S$. ∎

Next we give a sufficient condition on a poset guaranteeing that every saturated antichain in the poset has the splitting property.

Theorem 32 (Ahlswede, Erdös, and Graham 1995) *Let $\mathcal{P}$ be a dense poset. Then every saturated antichain S satisfies the splitting condition.*

The proof of this theorem consists in reducing it to the previous Theorem 31.
Proof. We construct a subposet $\mathcal{P}'$ in $\mathcal{P}$ such that a saturated antichain S has the splitting property in $\mathcal{P}'$ if and only if it has that property in $\mathcal{P}$. Let $\mathcal{D}$ be the set of maximal elements of $\mathcal{D}^*(S)$ in $\mathcal{P}$ and let $\mathcal{U}$ be the set of minimal elements of $\mathcal{U}^*(S)$ in $\mathcal{P}$. Consider the subposet $\mathcal{P}'$ with $P' = \mathcal{D} \cup S \cup \mathcal{U}$ and $<_{\mathcal{P}'}$ being the induced order. Our definition of $\mathcal{P}'$ necessitates that a good splitting of S in $\mathcal{P}'$ supplies a good splitting in $\mathcal{P}$. Obviously, S is saturated in $\mathcal{P}'$. Furthermore, the antichain S is dense in $\mathcal{P}'$, otherwise either $\mathcal{P}$ would not be dense or S would not be saturated. Applying Theorem 31 completes the proof. ∎

Notice that Theorem 31 and Theorem 32 are independent. This follows from the fact that there exists a dense poset $\mathcal{P}$ and a saturated antichain S in $\mathcal{P}$ such that S is not dense there. To show this, let $\mathcal{R} = \{A, B, C, D, E, F, G\}$ and let the covering relations be $A < C < F$, $A < D < F$, $B < C < G$, $B < D < G$, and $B < E < G$. This poset is dense, but the saturated antichain $\{A, E\}$ is not dense in $\mathcal{P}$. (Observe that $S_1 = \{E\}, S_2 = \{A\}$ is a good splitting.)

Lecture 15 Correlation Inequalities

In the development of probabilistic correlation inequalities perhaps the most basic discovery originated with the work [FKG71], producing the so-called FKG inequality. There the problem of characterizing probability distributions on a finite set, such that the correlation between two random variables X, Y is nonnegative, was stated:

$$\mathbb{E}(XY) - \mathbb{E}(X)\mathbb{E}(Y) \geq 0. \tag{1}$$

As shown below it implies earlier and later probabilistic and combinatorial inequalities like the Harris inequality [H60], the Kleitman inequality [K66a], the Seymour inequality [S73], and the Holley inequality [H74]. Somewhat earlier the seemingly very special Marica/Schönheim inequality

$$|A \setminus A| \geq |A| \tag{2}$$

for $A \subset 2^{[n]}$ [MS69] opened another direction, because it makes no monotonicity assumptions as all previously mentioned inequalities do. It inspired Daykin, Kleitman, and West [DKW79] to derive for all $A, B \subset \Gamma$, a distributive lattice,

$$|A||B| \leq |A \vee B||\Gamma|, \tag{3}$$

which finally led Daykin [D77] to the significant improvement

$$|A||B| \leq |A \vee B||A \wedge B|. \tag{4}$$

These combinatorial inequalities do not reflect correlation between random variables, but in a certain sense correlation between operations on set systems (intersection and union, "meets" and "joins" or sums and products, which we introduce later).

While trying to find his own proof for (4) after having heard about it from Daykin, Ahlswede introduced "weights" for the purpose of induction. This led to a new method of proof and a new product theorem [AD79b] involving the concept of $\mathcal{M}$-expansiveness. The original hope was to also prove this way that

$$|A||B| \leq |A \Delta B||A \wedge B|, \tag{5}$$

but this did not work, because $(\Delta, \wedge)$ is not $\mathcal{M}$-expansive! However, a closely related method worked and led in particular to Theorem 43 in §5. Furthermore, by another method, Ahlswede and Daykin obtained the second AD inequality in Theorem 36 of §2, which implies (5). The most important special case of the Product Theorem for the operations $(\vee, \wedge)$ was published separately in [AD78], because of its apparent appeal to probabilists and combinatorialists: it includes all probabilistic and combinatorial correlation inequalities mentioned above and is often even sharper. It became known as the AD inequality. Chebyshev's inequality from Analysis is an example of an inequality, which is not implied by FKG but is implied by AD. In §1 we first address this inequality and its generalization in [RS93]. Then we consider instead of $(\vee, \wedge)$ other Boolean operations in §2, arithmetical operations in §3, and in §4 implications for order extensions. Then we present the general theory in §5 and conclude in §6 with number theoretical correlation inequalities.

For work on infinite spaces see [P74], [K77], [B80], and [GR99].

§1 AD Inequality

Let us start with some definitions. We consider a finite distributive lattice $(\Gamma, <)$. Remind that distributivity means that

$$a \wedge (b \vee c) = (a \wedge b) \vee (a \wedge c) \text{ for all } a, b, c \in \Gamma$$

or equivalently

$$a \vee (b \wedge c) = (a \vee b) \wedge (a \vee c) \text{ for all } a, b, c \in \Gamma.$$

Remind also that any finite distributive lattice is order-isomorphic to a restriction of $(2^{[n]}, \subset)$, which is the family of subsets of $[n]$ for some n with the order generated by the inclusion. In this isomorphism $\wedge = \cap$ and $\vee = \cup$. For two sets $A, B \subset \Gamma$ we set

$$A \wedge B = \{a \wedge b : a \in A, b \in B\},$$
$$A \vee B = \{a \vee b : a \in A, b \in B\}.$$

In this lecture we introduce several conditions that yield correlation inequalities.

Now we turn to the formulation of the results and their proofs. For an arbitrary function $f : \Gamma \to [0, \infty)$ we set $f(A) = \sum_{a \in A} f(a)$. In [D77] Daykin proved the remarkable fact that the lattice Γ is distributive iff (4) holds for all $A, B \subset \Gamma$.

Despite the significance of this theorem, its direct part and many other relations follow from the following result:

Theorem 33 (Ahlswede and Daykin 1978) *Let* f_1, f_2, f_3, f_4 *be functions* $\Gamma \to [0, \infty)$ *on a finite distributive lattice* Γ.

If

$$f_1(a) f_2(b) \leq f_3(a \vee b) f_4(a \wedge b) \text{ for all } a, b \in \Gamma, \tag{6}$$

then

$$f_1(A) f_2(B) \leq f_3(A \vee B) f_4(A \wedge B) \tag{7}$$

for all $A, B \subset \Gamma$.

We consider further consequences of this theorem in Section 6 and in the Appendix.

Proof. The present-day proof of this theorem is quite simple and elegant. It is due to Bollmann, who noticed that it suffices to establish (7) for $A = B = \Gamma$ ([B80]), that is,

$$f_1(\Gamma) f_2(\Gamma) \leq f_3(\Gamma) f_4(\Gamma). \tag{8}$$

Indeed if we set $I_1 = 1_A f_1, I_2 = 1_B f_2, I_3 = 1_{A \vee B} f_3, I_4 = 1_{A \wedge B} f_4$, then these functions also satisfy (6) and from (8) follows (7).

Imbed now Γ into $2^{[n]}$ for some n and use induction. For $n = 1$ from (6) it follows that

$$f_1(0) f_2(0) \leq f_3(0) f_4(0),$$
$$f_1(0) f_2(1) \leq f_3(1) f_4(0),$$

$$f_1(1)f_2(0) \leq f_3(1)f_4(0),$$
$$f_1(1)f_2(1) \leq f_3(1)f_4(1).$$

Hence it suffices to prove that

$$f_1(0)f_2(1) + f_1(1)f_2(0) \leq f_3(1)f_4(0) + f_3(0)f_4(1) \qquad (9)$$

or denoting by

$$x = f_1(0)f_2(1), \ y = f_1(1)f_2(0), \ z = f_3(1)f_4(0), \ q = f_3(0)f_4(1)$$

we have to prove $x + y \leq z + q$. But $x, y \leq z$ and $xy \leq zq$, hence

$$z + q - (x + y) = \frac{1}{z}((z - x)(z - y) + (zq - xy)) \geq 0.$$

Next when $n > 1$ we consider the functions of one dimensional variables

$$\varphi_1(c) = f_1(a^{n-1}, c), \ \varphi_2(c) = f_2(b^{n-1}, c),$$
$$\varphi_3(c) = f_3(a^{n-1} \vee b^{n-1}, c), \ \varphi_4(c) = f_4(a^{n-1} \wedge b^{n-1}, c), c \in \{0, 1\}.$$

It is easy to see that these functions satisfy (6) and hence our previous proof satisfy (9). Hence the functions $\tilde{\varphi}_i(a^{n-1}) = f_i(a^{n-1}, 0) + f_i(a^{n-1}, 1)$ satisfy (6) and this completes the induction step. ∎

Consequences of Theorem 33

1. Daykin inequality [D77].

$$|A||B| \leq |A \vee B||A \wedge B| \ for \ all \ A, B \subset \Gamma.$$

Indeed if we set $f_1 = f_2 = f_3 = f_4 = 1$ we get the inequality.
It implies several known inequalities. We list here some under (a)–(d).

(a) Marica, Schönheim inequality [MS69] (see also [L96]).

$$|A| \leq |A \setminus A| \ for \ all \ A \subset \Gamma.$$

Indeed

$$|A||B| = |A||\bar{B}| \leq |A \vee \bar{B}||A \wedge \bar{B}|$$
$$= \overline{|A \vee \bar{B}|}|A \wedge \bar{B}| = |\bar{A} \wedge B||A \wedge \bar{B}| = |A \setminus B||B \setminus A|.$$

Choose $A = B$.

(b) Daykin, Kleitman, West inequality [DKW79].

$$|A||B| \leq |A \vee B||\Gamma| \ for \ all \ A, B \subset \Gamma.$$

Indeed $|A \vee B| \leq |\Gamma|$.

(c) Kleitman inequality [K66a].

$$|U \cap D||\Gamma| \leq |U||D| \text{ for all upsets } U \text{ and all downsets } D \text{ of } \Gamma.$$

Indeed put $A = U \cap D$ and notice that $\Gamma \vee A \subset U$, $\Gamma \wedge A \subset D$. Then

$$|A||\Gamma| \leq |\Gamma \vee A||\Gamma \wedge A| \leq |U||D|.$$

(d) Seymour inequality [S73].

$$|U_1||U_2| \leq |U_1 \cap U_2||\Gamma| \text{ for upsets } U_1, U_2 \subset \Gamma.$$

Indeed notice that $U_1 \cap U_2 = U_1 \vee U_2$.

2. Chebyshev inequality.

Let $\alpha_0, \alpha_1, \ldots, \alpha_n; \beta_0, \beta_1, \ldots, \beta_n \in \mathbb{R}_+$.

Put $\gamma_k = \max\{\alpha_i \beta_k, \alpha_k \beta_i : 0 \leq i \leq k\}$, $0 \leq k \leq n$, then

$$\sum_{i=0}^{n} \alpha_i \sum_{j=0}^{n} \beta_j \leq (n+1) \sum_{k=0}^{n} \gamma_k.$$

In particular, if $0 \leq \alpha_1 \leq \cdots \leq \alpha_n$ and $0 \leq \beta_1 \leq \cdots \leq \beta_n$, then $\gamma_k = \alpha_k \beta_k$ and we get the Chebyshev's inequality

$$\sum_{i=0}^{n} \alpha_i \sum_{j=0}^{n} \beta_j \leq (n+1) \sum_{k=0}^{n} \alpha_k \beta_k.$$

Indeed let $f_1, f_2 : \{0,1\}^n \to \mathbb{R}_+$ and define

$$f_3(a) = \max_{c=a \vee b} f_1(a) f_2(b) \text{ for } c \in \{0,1\}^n.$$

then

$$f_1(a) f_2(b) \leq f_3(a \vee b) \cdot 1, \ f_4 = 1$$

and hence

$$f_1(A) f_2(B) \leq f_3(A \vee B)|A \wedge B|.$$

Choose now $f_1, f_2 : \{0,1\}^n \to \mathbb{R}_+$ such that

$$f_1(\{1,2,\ldots,k\}) = \alpha_k, \ f_2(\{1,2,\ldots,k\}) = \beta_k, \ 0 \leq k \leq n$$

and $f_1 = f_2 = 0$ otherwise.

For $A = B = \{\{1,2,\ldots,k\} : 0 \leq k \leq n\}$, $A \vee B = A \wedge B = A$, $|A| = n+1$, and therefore $f_1(A) = \sum_{k=0}^{n} \alpha_k$, $f_2(B) = \sum_{k=0}^{n} \beta_k$, and

$$\sum_{i=0}^{n} \alpha_i \sum_{j=0}^{n} \beta_j \leq (n+1) \sum_{k=0}^{n} \gamma_k.$$

3. Holley inequality [H74].

If $\alpha, \beta : \{0,1\}^n \to \mathbb{R}_+$ and

$$\alpha(a)\beta(b) \le \alpha(a \wedge b)\beta(a \vee b) \text{ for all } a, b \in \{0,1\}^n \tag{10}$$

then

$$\alpha(\Gamma)\beta(U) \le \alpha(U)\beta(\Gamma) \text{ for an upset } U \text{ and } \Gamma = \{0,1\}^n. \tag{11}$$

Indeed choose $f_1 = f_3 = \alpha$, $f_2 = f_4 = \beta$, $A = \Gamma$, $B = U$. Then $\Gamma \vee U = U$, $\Gamma \wedge U = \Gamma$.

More generally, for a monotone function $f : \{0,1\}^n \to \mathbb{R}_+$ rather than just the characteristic function of an up-set, Holley's inequality says that under hypothesis (10) one has

$$\alpha(\Gamma) \sum_{p \in \Gamma} f(p)\beta(p) \le \sum_{p \in \Gamma} f(p)\alpha(p)\beta(\Gamma).$$

This follows immediately from (11) by writing f as $f = \sum_{i=0}^n \lambda_i 1_{U_i}$ with $\lambda_i \ge 0$ suitable. For probability distributions α and β the factors $\alpha(\Gamma) = \beta(\Gamma) = 1$ drop out.

4. FKG inequality [FKG71].

Suppose that for $\mu : \{0,1\}^n \to \mathbb{R}_+$

$$\mu(a)\mu(b) \le \mu(a \vee b)\mu(a \wedge b), \text{ for all } a, b, \in \Gamma = \{0,1\}^n,$$

then for two up-functions $f, g : \Gamma \to \mathbb{R}$

$$\sum_{x \in \Gamma} \mu(x)f(x) \cdot \sum_{x \in \Gamma} \mu(x)g(x) \le \sum_{x \in \Gamma} \mu(x)f(x)g(x) \cdot \sum_{x \in \Gamma} \mu(x). \tag{12}$$

Hint: one can assume that f, g are nonnegative, because (12) is invariant under addition of constants to these functions.

In particular, if U, V are upsets, $f = 1_U$ and $g = 1_V$, then (12) says that

$$\mu(U)\mu(V) \le \mu(U \vee V)\mu(\Gamma). \tag{13}$$

Indeed, since $U \cap V = U \vee V$, the last inequality follows from (12).

Notice also that we actually get the sharper estimate

$$\mu(U)\mu(V) \le \mu(U \vee V)\mu(U \wedge V)$$

by setting $\mu = f_1 = f_2 = f_3 = f_4$. The derivation of (12) from (13) is standard, and one just writes f and g as

$$f(p) = \sum_{i=0}^n \lambda_i 1_{U_i}(p), \quad g(p) = \sum_{j=0}^n v_j 1_{V_j}(p)$$

and calculates the expression (12).

The next result is a generalization of Theorem 33. For arbitrary $a_1, a_2, \ldots, a_m \in \Gamma$ and given set $I \subset \{1, 2, \ldots, m\}$ we define $a^I = \bigwedge_{i \in I} a_i$ and

$$a^{[\ell]} = \bigvee_{I:|I|=\ell} a^I, \ell = 1, 2, \ldots, m.$$

Theorem 34 (Rinott and Saks 1993) *Let $f_1, f_2, \ldots, f_m; g_1, g_2, \ldots, g_m$ be (nonnegative) real functions $2^{[n]} \to [0, \infty)$, such that for every $a_1, a_2, \ldots, a_m \in 2^{[n]}$*

$$\prod_{j=1}^{m} f_j(a_j) \le \prod_{j=1}^{m} g_j(a^{[j]}). \tag{14}$$

Then for a product measure μ on $2^{[n]}$ $(\mu(a) = \prod_{i=1}^{m} \mu_i(a_i))$ the following inequality is valid:

$$\prod_{j=1}^{m} \sum_{a \in 2^{[n]}} f_j(a)\mu(a) \le \prod_{j=1}^{m} \sum_{a \in 2^{[n]}} g_j(a)\mu(a). \tag{15}$$

Proof. The proof of this theorem is like the proof of Theorem 33 by induction and the most involved part here is the validity of the theorem in case $n = 1$. The induction step is the same as in the proof of the first AD inequality. Define

$$\varphi_i(a) = f_i(a_1^{n-1}, a), \quad \psi_i(a) = g_i((a_1^{n-1})^{[i]}, a).$$

Then φ_i, ψ_i satisfy the hypothesis of the theorem in one variable a. Hence if the theorem is valid in the case $n = 1$, then the functions

$$\tilde{\varphi}_i(a_1^{n-1}) = f_i(a_1^{n-1}, 0)\mu_n(0) + f_i(a_1^{n-1}, 1)\mu_n(1),$$
$$\tilde{\psi}_i(a_1^{n-1}) = g_i((a_1^{n-1})^{[i]}, 0)\mu_n(0) + g_i((a_1^{n-1})^{[i]}, 1)\mu_n(1)$$

again satisfy the hypothesis of the theorem. This completes the induction step. Hence it remains to prove the theorem in case $n = 1$. We need the following:

Lemma 33 (Rinott and Saks 1993) *Let $A = \{A_{i,j}\}, B = \{B_{i,j}\}$ be $m \times m$ matrices with nonnegative entries such that for some $r \in \{1, 2, \ldots, m\}$ A consists of r identical rows followed by $m - r$ (possibly another) identical rows. B has the same structure as well. If for any $1 \le i_1 \le i_2 \le \cdots \le i_m \le m$ and any permutation π of $\{1, 2, \ldots, m\}$*

$$A_{i_{\pi(1)}, 1} A_{i_{\pi(2)}, 2} \cdots A_{i_{\pi(m)}, m} \le B_{i_1, 1} B_{i_2, 2} \cdots B_{i_m, m}, \tag{16}$$

then

$$\mathbf{Per}(A) \le \mathbf{Per}(B),$$

where

$$\mathbf{Per}(A) = \sum_{\pi} A_{\pi(1), 1} A_{\pi(2), 2} \cdots A_{\pi(m), m}$$

is the permanent of matrix A.

Assume for a while that this lemma is valid. Then to prove the theorem for the case $n = 1$ we set $A_{i,j} = f_j(a_i)$, $B_{i,j} = g_j(a_i)$; $i,j = 1,2,\ldots,m$. This implies

$$\mathbf{Per}(A) = \sum_\pi \prod_{i=1}^m f_i(a_{\pi(i)}),$$

$$\mathbf{Per}(B) = \sum_\pi \prod_{i=1}^m g_i(a_{\pi(i)}).$$

Hence the theorem will follow from the inequality

$$\mathbf{Per}(A) \le \mathbf{Per}(B). \tag{17}$$

We suppose that $a_1 \ge a_2 \ge \cdots \ge a_m$. Otherwise we rearrange the order of columns. As $a_i \in \{0,1\}$, there exists r such that $a_1,\ldots,a_r = 1, a_{r+1},\ldots,a_m = 0$. Then the first r rows of matrix A are $(f_1(1), f_2(1),\ldots,f_m(1))$ and the last $m - r$ rows are $(f_1(0), f_2(0),\ldots,f_m(0))$. We have the same configuration for matrix B. Then

$$\prod_{j=1}^m A_{i_{\pi(j)},j} = \prod_{j=1}^m f_j(a_{i_{\pi(j)}}) \le \prod_{j=1}^m g_j(a_{i_j}) = \prod_{j=1}^m B_{i_j,j}. \tag{18}$$

Here the inequality is a consequence of inequality (14). By (18) we get equality (17) and the theorem is proved, after Lemma 33 is proved.

Proof of the lemma. For a vector $x^n = (x_1, x_2,\ldots,x_n) \in \mathbb{R}^n$ denote its nonincreasing rearrangement by $\hat{x}^n = (\hat{x}_1 \ldots, \hat{x}_n)$. For pairs of vectors $x^n, y^n \in \mathbb{R}^n$ consider the order $\prec_w$ as follows: $x^n \prec_w y^n$ iff for all $j \in \{1, 2,\ldots,n\}$, $\sum_{i=1}^j \hat{x}_i \le \sum_{i=1}^j \hat{y}_i$. The sign w in the order means that we consider the weak order, which differs from the (strong) order when the additional condition $\sum_{i=1}^n \hat{x}_j = \sum_{i=1}^n \hat{y}_j$ is imposed. Different properties of structures with such order can be found in [MO79]. We use only a result from [T49].

Theorem 35 (Tomić 1949) *If a real function $f : \mathbb{R}^n \to \mathbb{R}$ is convex and increasing, then from $x^n \prec_w y^n$ it follows*

$$\sum_{i=1}^n f(x_i) \le \sum_{i=1}^n f(y_i).$$

This is recommended to prove in Exercise 4. For every given permutation π define $\alpha_\pi = \prod_{i=1}^m A_{\pi(i),i}$, $\beta_\pi = \prod_{i=1}^m B_{\pi(i),i}$. Then we have to prove that $\mathbf{Per}(A) = \sum_\pi \alpha_\pi \le \sum_\pi \beta_\pi = \mathbf{Per}(B)$.

Applying the theorem with $f = \exp(x)$ we see that it is enough to prove that $(\ln \alpha_{\pi_1},\ldots,\ln \alpha_{\pi_{m!}}) \prec_w (\ln \beta_{\pi_1},\ldots,\ln \beta_{\pi_{m!}})$. To prove this we show that for any set of permutations V there exists a set of permutations W such that $|V| = |W|$ and

$$\prod_{\pi \in V} \alpha_\pi \le \prod_{\pi \in W} \beta_\pi. \tag{19}$$

Let $(s_1, \ldots, s_m)$, $(t_1, \ldots, t_m)$ be two distinct rows of A. Then

$$\prod_{\pi \in V} \alpha_\pi = \prod_{j=1}^{m} s_j^{k_j} t_j^{h_j}, \tag{20}$$

with $k_j, h_j \geq 0$, $k_j + h_j = |V|$, $j = 1, 2, \ldots, m$. Suppose $u = (u_1, \ldots, u_m)$, $v = (v_1, \ldots, v_m)$ are two distinct rows of the matrix B, u is preceding v and π' is a permutation such that $k_{\pi'(1)} \geq \cdots \geq k_{\pi'(m)}$. Then the expression in (20) does not exceed the value

$$\prod_{j=1}^{m} u_j^{k_{\pi'(j)}} v_j^{h_{\pi'(j)}}. \tag{21}$$

Inequality (16) takes the form

$$s_{\pi(1)} \cdots s_{\pi(j)} t_{\pi(j+1)} \cdots t_{\pi(m)} \leq u_1 \ldots u_j v_{j+1} \ldots v_m, j = 0, 1, \ldots, m, \tag{22}$$

where π is an arbitrary permutation.

Setting $S_j = s_{\pi'(1)} \cdots s_{\pi'(j)} t_{\pi'(j+1)} \cdots t_{\pi'(m)}$, $U_j = u_1 \ldots u_j v_{j+1} \ldots v_m$ we have $S_j \leq U_j$. If $k_{\pi'(0)} = |V|$, $k_{\pi'(m+1)} = 0$, then

$$\prod_{\pi \in V} \alpha_\pi = \prod_{j=1}^{m} S_j^{k_{\pi'(j)} - k_{\pi'(j+1)}} \leq \prod_{j=1}^{m} U_j^{k_{\pi'(j)} - k_{\pi'(j+1)}} = \prod_{j=1}^{m} u_j^{k_{\pi'(j)}} v_j^{h_{\pi'(j)}}.$$

As

$$\prod_{j=1}^{m} u_j^{k_{\pi'(j)}} v_j^{h_{\pi'(j)}} = \prod_{\pi \in W} \beta_\pi$$

when $W = \{\pi \pi'; \pi \in V\}$, we obtain inequality (19). Thus (19) is established and therefore also Lemma 33.

Hence the proof of Theorem 34 is complete. ∎

§2 Other Boolean Operations

We start with the second AD inequality that uses Boolean operations symmetric difference (Δ) and minimum ($\wedge$).

Theorem 36 (Ahlswede and Daykin 1979) *Let* $f_i : 2^{[n]} \to [0, \infty)$, $i = 1, 2, 3, 4$ *be given such that* f_4 *is monotonically nondecreasing* ($f_4(a) \leq f_4(b)$ *if* $a \subseteq b$) *and for all* $a, b \in 2^{[n]}$

$$f_1(a) f_2(b) \leq f_3(a \Delta b) f_4(a \wedge b).$$

Then

$$f_1(A) f_2(B) \leq f_3(A \Delta B) f_4(A \wedge B), \tag{23}$$

where A, B *are arbitrary subsets of* $2^{[n]}$ *and* $A \Delta B = \{a \Delta b : a \in A, b \in B\}$.

Proof. To prove this theorem consider the following sets:

$$C_{st} = \{(a,b) : a,b \in 2^{[n]}, a\Delta b = s, a \wedge b = t\},$$
$$C_{st}^+ = \{a : \text{there exists } b \text{ such that } (a,b) \in C_{st}\},$$
$$D_{st} = \{(s,d) : d \in C_{st}^+\},$$
$$E_{st} = \{A \times B\} \cap C_{st},$$
$$F_{st} = \{(A\Delta B) \times (A \wedge B)\} \cap D_{st}$$
$$= \{(s,d) : (s,d) \in (A\Delta B) \times (A \wedge B), d \in C_{st}^+\}$$

It is easy to see that $\{C_{st} : s,t \in 2^{[n]}\}$ is a partition of $2^{[n]} \times 2^{[n]}$, also $C_{st}^+ \cap C_{st'}^+ = \emptyset$ if $t \neq t'$, $D_{st} \cap D_{s't'} = \emptyset$ if $(s,t) \neq (s',t')$ and $|D_{st}| = |C_{st}^+| = |C_{st}|$. Hence $\{D_{st} : s,t \in 2^{[n]}\}$ is also a partition of $2^{[n]} \times 2^{[n]}$. Next if $(a,b),(a',b') \in C_{st}$, then $(a \vee a', b \wedge b') \in C_{st}$. Hence C_{st}^+ is a sublattice of $2^{[n]}$. The key relation in the proof of the theorem is the inequality

$$|E_{st}| \leq |F_{st}|.$$

The set E_{st} has the form $E = \{(e_1, \bar{e}_1), \ldots, (e_k, \bar{e}_k)\}$, where each $\bar{e}_m$ is complement to e_m in the sublattice C_{st}^+. Hence

$$|F_{st}| \geq |\{(s,d) : d = e_i \wedge \bar{e}_j, 1 \leq i,j \leq k\}|$$
$$= |\{e_i - e_j : 1 \leq i,j \leq k\}| \geq |E_{st}|.$$

The last inequality here is the Marica/Schönheim inequality.

Now we have

$$f_1(A)f_2(B) = \sum_{(a,b)\in A\times B} f_1(a)f_2(b)$$
$$\leq \sum_{s,t\in 2^{[n]}} \sum_{(a,b)\in E_{st}} f_3(a\Delta b)f_4(a \wedge b) = \sum_{s,t\in 2^{[n]}} \sum_{(a,b)\in E_{st}} f_3(s)f_4(t)$$
$$\leq \sum_{s,t\in 2^{[n]}} \sum_{(c,d)\in F_{st}} f_3(c)f_4(d) = \sum_{(c,d)\in(A\Delta B)\times(A\wedge B)} f_3(c)f_4(d)$$
$$= f_3(A\Delta B)f_4(A \wedge B).$$

This completes the proof of the theorem. ∎

Next we prove theorems, which are generalizations of the Marica/Schönheim inequality.

Theorem 37 (Ahlswede and Daykin 1979) *If $S_1, \ldots, S_m$ are distinct finite sets, and $T_1, \ldots, T_n$ are sets such that each S_i has some T_j as a subset, then there are at least m distinct differences $S_i \setminus T_j$.*

If $m = n$ and $S_i = T_i$ for all $i = 1, \ldots, n$, then from this theorem we obtain the Marica/Schönheim inequality.

Proof. We may assume that all the sets S_i, T_j are subsets of $\{1, \ldots, r\}$ and use induction on r. The case $r = 1$ is trivial. Put $\mathcal{S} = \{S_1, \ldots, S_m\}$, $\mathcal{T} = \{T_1, \ldots, T_n\}$, $\mathcal{S}_1 = \{S \setminus r : S \setminus r \in \mathcal{S}, S \cup r \in \mathcal{S}\}$, $\mathcal{S}_2 = \{S \setminus r : S \in \mathcal{S}\}$, $\mathcal{T}_1 = \{T : r \notin T \in \mathcal{T}\}$, and $\mathcal{T}_2 = \{T \setminus r : T \in \mathcal{T}\}$. Then $m = |\mathcal{S}_1| + |\mathcal{S}_2|$. Also $\mathcal{S}_1, \mathcal{T}_1$ and $\mathcal{S}_2, \mathcal{T}_2$ satisfy the hypothesis on $\{1, 2, \ldots, r-1\}$, and so $|\mathcal{S}_1| \leq |\mathcal{S}_1 \setminus \mathcal{T}_1|$ and $|\mathcal{S}_2| \leq |\mathcal{S}_2 \setminus \mathcal{T}_2|$. If $E \in \mathcal{S}_1 \setminus \mathcal{T}_1$, then $E = S \setminus T$ for some $S \in \mathcal{S}_1$, $T \in \mathcal{T}_1$. Thus $S \setminus r$, $S \cup r \in \mathcal{S}$, and $r \notin T \in \mathcal{T}$, and so $E \setminus r, E \cup r \in \mathcal{S} \setminus \mathcal{T}$. On the other hand, if $E \in \mathcal{S}_2 \setminus \mathcal{T}_2$, then clearly either $E \setminus r$ or $E \cup r$ is in $\mathcal{S} \setminus \mathcal{T}$. Hence

$$|\mathcal{S}_1 \setminus \mathcal{T}_1| + |\mathcal{S}_2 \setminus \mathcal{T}_2| \leq |\mathcal{S} \setminus \mathcal{T}|$$

and the result follows.   ■

Theorem 38 (Daykin and Lovász 1976) *Any nontrivial Boolean function takes at least m distinct values when evaluated over m distinct sets.*

This was first proved by Daykin and Lovàsz in [DL76]. Here we give in Theorem 39 a generalization of this theorem from [AD79a]. In a special case this will yield Theorem 38. Let c be a fixed positive integer. If S is a set, then S^c denotes the set of all tuples of length c with elements in S. Given a map $f : S^c \to S$, put for $A_1, \ldots, A_c \subset S$

$$f(A_1, \ldots, A_c) = \{f(a_1, \ldots, a_c) : a_i \in A_i, i = 1, \ldots, c\}.$$

Call f expansive if

$$|A| \leq |f(A, \ldots, A)| \text{ for all } A \subset S.$$

Call f c-expansive if

$$|A_1| \leq |f(A_1, \ldots, A_c)| \text{ for all } A_1, \ldots, A_c \subset S \text{ with } |A_1| = \cdots = |A_c|.$$

Notice that when $|S| = 2$, expansive is the same as c-expansive and simply means nonconstant Boolean functions.

If S, T are sets and $f : S^c \to S$ while $g : T^c \to T$, we define the direct product h of f and g to be the map $h : (S \times T)^c \to S \times T$ satisfying

$$h((s_1, t_1), \ldots, (s_c, t_c)) = (f(s_1, \ldots, s_c), g(t_1, \ldots, t_c)), \ s_i \in S, \ t_j \in T.$$

The direct product of expansive maps is not necessarily expansive. For example, let $c = 2$, $S = \{0, 1, 2\}$, $f(a, b) = \max\{0, a - b\}$. Now take the direct product of f with itself and $A = (S \times S) \setminus \{(0, 0), (2, 2)\}$.

Theorem 39 (Ahlswede and Daykin 1979) *In the above notation, if f is expansive and g is c-expansive, then h is expansive.*

Proof. If $B \subset S \times T$ and m is a positive integer, let B_m be the set of all $s \in S$ such that $(s, t) \in B$ for at least m different $t \in T$. Let $A \subset S \times T$ be given and $x \in f(A_m, \ldots, A_m)$. Thus there are $s_1, \ldots, s_c \in A_m$ with $x = f(s_1, \ldots, s_c)$. For $1 \leq i \leq c$ there are distinct $t_{i1}, \ldots, t_{im} \in T$ with $(s_i, t_{ij}) \in A$ for $1 \leq j \leq m$. By condition on g we have

$$m \leq |\{g(t_{1j_1}, \ldots, t_{cj_c}) : 1 \leq j_1, \ldots, j_c \leq m\}|$$

and this means that $x \in (h(A, \ldots, A))_m$. Finally

$$|A| = \sum_m |A_m| \le \sum_m |f(A_m, \ldots, A_m)| \le \sum_m |(h(A, \ldots, A))_m| = |h(A, \ldots, A)|$$

and the proof is complete. ∎

Now let $|S| = 2$ and $f_1, \ldots, f_n : S^c \to S$. Further, let Λ_c, Λ_1 be the sets of all matrices of order $n \times c$, $n \times 1$, respectively, with elements in S. Define $e : \Lambda_c \to \Lambda_1$ by

$$e((a_{ij})) = \begin{pmatrix} f_1(a_{11}, \ldots, a_{1c}) \\ \vdots \\ f_n(a_{n1}, \ldots, a_{nc}) \end{pmatrix},$$

where $(a_{ij}) \in \Lambda_c$. By induction on n we get from Theorem 39 that if $f_1, \ldots, f_n$ are nonconstant, then e is expansive. The case when $f_1 = \ldots = f_n$ is Theorem 38.

§3 Arithmetical Operations

One remarkable problem that we are going to consider here is the problem of estimating from below the product

$$|A + A||A \cdot A|,$$

where A is an arbitrary finite set of reals and $A + A = \{a + a' : a, a' \in A\}$, $A \cdot A = \{aa' : a, a' \in A\}$. The following inequality is a conjecture of Erdös and Szemerédi [ES83]. For arbitrary $\varepsilon > 0$, some $f(\varepsilon)$ and all A, $|A| > f(\varepsilon)$,

$$\max\{|A + A|, |A \cdot A|\} \ge f(\varepsilon)|A|^{2-\varepsilon}. \tag{24}$$

The following result for arbitrary A was obtained by Elekes [E97a]:

$$|A + A||A \cdot A| \ge c|A|^{5/2}, \tag{25}$$

for some constant c. This inequality reflects the correlation between the sum and the product of real numbers. From here it follows that

$$\max\{|A + A|, |A \cdot A|\} \ge c|A|^{5/4}.$$

We prove this inequality. For this we use the Euler relation

$$n - e + f = 2, \tag{26}$$

where n is the number of vertices, e is the number of edges, and f is the number of faces of a planar connected graph. We use the simple consequence from this formula that if a (nonempty) graph is simple, then

$$e \le 3n - 6. \tag{27}$$

To prove this inequality we can assume that the graph is connected. Let f_i be the number of faces bounded by exactly i edges (an edge that on both sides borders the same region has to be counted twice). As the graph is simple, we have that $f_1 = f_2 = 0$. Then

$$f = f_3 + f_4 + \cdots,$$
$$2e = 3f_3 + 4f_4 + \cdots$$

and hence $2e - 3f \geq 0$ or $3n - 6 = 3e - 3f \geq e$. In the case of an arbitrary (non-planar) graph G we define the crossing number $\mathbf{cr}(G)$ as the minimal number of points, where the edges of the graph intersect. Here the minimum is taken over all imbeddings of the graph into the plane, in which the crossing points are not allowed to coincide with vertices of the graph or with each other.

Theorem 40 (Chazelle, Sharir, and Welzl 1998)
First (Ajtai, Chvátal, Newborn, and Szemerédi 1982; Leighton 1983) proved it independently with the constant $\frac{1}{100}$.

If $e \geq 4n$ and graph G is simple, then

$$\mathbf{cr}(G) \geq \frac{1}{64} \frac{e^3}{n^2}. \tag{28}$$

To prove this note that

$$\mathbf{cr}(G) \geq e - 3n + 6. \tag{29}$$

Proof. Indeed, if we consider the crossing points as additional vertices of the graph, then we obtain a simple planar graph on $n + \mathbf{cr}(G)$ vertices and $e + 2\mathbf{cr}(G)$ edges. Hence by inequality (29) we have $e + 2\mathbf{cr}(G) \leq 3(n + \mathbf{cr}(G)) - 6$, which leads to inequality (28).

Next, for a given imbedding of the graph G into the plane we consider a random graph, which is obtained from G by choosing the vertices of G independently with equal probability $p = 4n/e$. The edges of the random graph are induced by the edges of the initial graph, and the crossing number is defined analogously to the whole graph G. Then if $\tilde{X}$, $\tilde{e}$, $\tilde{n}$ are the crossing number, the number of edges, and the number of vertices of the random graph, respectively, then because of $\tilde{X} - \tilde{e} + 3\tilde{n} \geq 0$, we have

$$\mathbb{E}(\tilde{X} - \tilde{e} + 3\tilde{n}) \geq 0.$$

Also $\mathbb{E}(\tilde{n}) = np$, $\mathbb{E}(\tilde{e}) = ep^2$, and $\mathbb{E}(\tilde{X}) = \mathbf{cr}(G)p^4$, thus

$$\mathbf{cr}(G)p^4 - ep^2 + 3np \geq 0.$$

Substitution of $p = 4n/e$ in the last inequality gives relation (28). $\blacksquare$

The next result is a direct consequence of Theorem 40.

Corollary 6 *Let $2 \leq m \leq \sqrt{n}$. For n points in the Euclidean plane, the number ℓ of lines containing at least m of the points is at most cn^2/m^3 (for some constant c).*

Proof. We construct a graph G, which satisfies the conditions of Theorem 40, by taking only the lines passing through at least e points. Note that G has at least $\ell(m-1)$ edges. Hence we have either

$$\ell^2 \geq \frac{e^3}{64n^2} \geq \frac{c(\ell(m-1))^3}{n^2}$$

or $\ell(m-1) < 4n$. In the first case we have the statement, but also in the second case, as $\ell < 4n/(m-1) < cn^2/m^3$.   □

Now we return to the proof of (25). Define $|A|^2$ linear functions

$$f_{a,b}(x) = a(x-b), \; a,b \in A.$$

If we imagine the points from the set $A+A$ imbedded into the x-axis and the points from the set $A \cdot A$ imbedded into the y-axis, then the number of pairs (c,d), $c \in A+A$, $d \in A \cdot A$ is $n = |A+A||A \cdot A|$. Each line $f_{a,b}$ intersects at least $e = |A|$ points and the number of lines $\ell = |A|^2$. If $|A| = 1$, then (25) is trivial. Otherwise, we have $2 \leq |A| \leq \sqrt{|A+A||A \cdot A|}$ and using Corollary 6 we obtain

$$|A|^2 \leq c\frac{(|A+A||A \cdot A|)^2}{|A|^3},$$

from which (25) follows.

§4 Implications for Order Extensions and Random Permutations

In [FDS88] the following interesting fact was stated, which is connected to correlation inequalities. Let π be a permutation of $[n]$. The set of fixed points of π is

$$F(\pi) = \{i \in [n] : \pi(i) = i\}.$$

Consider the uniform distribution on the set of all permutations:

$$P(\pi) = \frac{1}{n!}.$$

Let $P(a)$ be the probability that $F(\pi) = a$. Then for arbitrary upsets $A, B \subset 2^{[n]}$

$$P(A \cap B) \geq P(A)P(B). \tag{30}$$

From here the correlation inequality for a pair of nondecreasing functions $f, g :$ $2^{[n]} \to \mathbb{R}$ follows. However, the proof of this relation does not follow directly from Theorem 33, because condition (6) is not true in this case. Although this condition is valid for almost all $a, b \in 2^{[n]}$, it fails in the case $|a \cup b| = n - 1$. In that case

$P(a \cup b) = 0$, because there is no permutation with exactly $n - 1$ fixed points. The proof of (30) is rather involved and still uses some considerations from the proof of Theorem 33.

The statement of Theorem 33 can be written in another form as it was done in Theorem 34. It is easy to see that the product probability measure $\mu = \mu_1 \times \ldots \times \mu_n$ on $2^{[n]}$ satisfies the relation

$$\mu(a)\mu(b) = \mu(a \cup b)\,\mu(a \cap b).$$

Thus functions $f_i\mu$ satisfy the hypothesis of Theorem 33 and hence by setting $A = B = \Gamma$ we obtain

$$\mathbb{E}(f_1)\mathbb{E}(f_2) \le \mathbb{E}(f_3)\mathbb{E}(f_4).$$

Here is a consequence of relation (4): if both A, B are upsets or downsets, then

$$2^n |A \cap B| \ge |A||B|.$$

If $A \subset 2^{[n]}$ is an upset and $B \subset 2^{[n]}$ is a downset, then

$$2^n |A \cap B| \le |A||B|.$$

Another application of the correlation inequalities introduced above is obtaining new correlation inequalities of the following kind. Let $(X, <)$ be a finite partially ordered set. We impose the uniform distribution P on the set of all linear extensions of $(X, <)$. Let the set X be decomposed into two nonintersecting sets $A = \{a_i\}$ and $B = \{b_i\}$. Next we consider two cases: in the first case we have the poset $\mathcal{P}_0 = \{a_1 < a_2 < \ldots < a_m\} \cap \{b_1 < b_2 < \ldots < b_n\}$. Let each of $\mathcal{P}_1, \mathcal{P}_2, \mathcal{P}_3$ be a set of given relations $\cap\{a_i < b_j\}$. Then

$$P(\mathcal{P}_1 \cap \mathcal{P}_3 | \mathcal{P}_2 \cap \mathcal{P}_0) \ge P(\mathcal{P}_1 | \mathcal{P}_2 \cap \mathcal{P}_0) P(\mathcal{P}_3 | \mathcal{P}_2 \cap \mathcal{P}_0). \tag{31}$$

In the second case we have posets $\mathcal{P}_1, \mathcal{P}_2$ like previously and $\mathcal{Q}_0 = \{a_{i_1} < a_{j_1}, \ldots, a_{i_r} < a_{j_r}\} \cap \{b_{k_1} < b_{\ell_1}, \ldots, b_{k_s} < b_{\ell_s}\}$, that is, in the poset $\mathcal{Q}_0$ the order relations are permitted only inside A or B. This condition is weaker than in the previous example, because previously we required linear order on A and B. Then

$$P(\mathcal{P}_1 \cap \mathcal{P}_2 | \mathcal{Q}_0) \ge P(\mathcal{P}_1 | \mathcal{Q}_0) P(\mathcal{P}_2 | \mathcal{Q}_0). \tag{32}$$

The condition of the linear order on A and B cannot be weakened as in the second example: it is easy to produce an example when there is a nonlinear (strictly partial) order on A and B and (31) is not valid (see [S80]). Proofs of these two inequalities use the FKG inequality. Inequality (31) was first proved in [GYY80], see also [S80]. Inequality (32) was proved in [S80].

The next example of correlation between orderings is as follows. Let $\mathcal{P}_0 = (X, <)$ be an arbitrary poset and $x, y, z \in X$, then

$$P(\{x < y\} \cap \{x < z\} | \mathcal{P}_0) \ge P(x < y | \mathcal{P}_0) P(x < z | \mathcal{P}_0). \tag{33}$$

This inequality was first proved in [S82]. Moreover, if x, y, z are mutually incomparable in $\mathcal{P}_0$, then strict inequality holds in (33). This was proved in [F84]. Actually, when x, y, z are not mutually incomparable, relation (33) is obvious.

Note that inequality (33) is valid for any poset $\mathcal{P}_0$. It is a natural question for which posets $\mathcal{P}_1, \mathcal{P}_2$ the inequality

$$P(\mathcal{P}_1 \cap \mathcal{P}_2 | \mathcal{P}_0) \geq P(\mathcal{P}_1 | \mathcal{P}_0) P(\mathcal{P}_2 | \mathcal{P}_0) \qquad (34)$$

holds for an arbitrary poset $\mathcal{P}_0$. Let us call such pairs of posets universally positive correlated. The answer is the following. For a given poset $\mathcal{P}$ let $\Delta(\mathcal{P})$ be the set of covering pairs of $\mathcal{P}$, that is, the set of pairs (x, y), $x, y \in \mathcal{P}$ such that $x < y$ without existence of $z \in \mathcal{P}$ with $x < z < y$. Then $\mathcal{P}_1, \mathcal{P}_2$ are universally positive correlated iff the order relations on $\mathcal{P}_1$ and $\mathcal{P}_2$ are consistent and for every pair $(x, y) \in (\Delta(\mathcal{P}_1 \cup \mathcal{P}_2) - \Delta(\mathcal{P}_2))$ and every pair $(u, v) \in (\Delta(\mathcal{P}_1 \cup \mathcal{P}_2) - \Delta(\mathcal{P}_1))$, either $x = u$ or $y = v$. This fact was proved in [W83], see also [B85]. It is a generalization of (33). In [B85] also necessary and sufficient conditions for the posets $\mathcal{P}_1, \mathcal{P}_2$ to be universally negative correlated, that is, for the validity of the opposite inequality of (34) for all posets $\mathcal{P}_0$ were found. Actually, the conditions for the validity of relation (34) are quite restrictive. Related work can be found in [B86c], [BB90], [F86], [F91], [F92], [G82], [J-DSV84], [KR80], [KS81], and [W86].

§5 General Correlation Inequalities: Methods for Proving Them

We begin with a general product theorem, the most important contribution of [AD79b]. It is based on the following two concepts.

For finite set S and binary operations

$$\varphi_S, \psi_S : S \times S \to S$$

the quadruple $(\alpha, \beta, \gamma, \delta)$ with

$$\alpha, \beta, \gamma, \delta : S \to \mathbb{R}_+$$

is compatible with (φ_S, ψ_S) if

$$\alpha(a)\beta(b) \leq \gamma(\varphi_S(a, b))\delta(\psi_S(a, b)) \qquad (35)$$

for all $a, b \in S$.

The pair (φ_S, ψ_S) is $\mathcal{M}$-expansive, if

$$\alpha(A)\beta(B) \leq \gamma(\varphi_S(A, B))\delta(\psi_S(A, B)), \qquad (36)$$

for all compatible quadruples $(\alpha, \beta, \gamma, \delta)$ and all $A, B \subseteq S$, where

$$(\alpha(A), \beta(B), \gamma(\varphi_S(A,B)), \delta(\psi_S(A,B)))$$

$$= \left(\sum_{a \in A} \alpha(a), \sum_{b \in B} \beta(b), \sum_{c \in \varphi_S(A,B)} \gamma(c), \sum_{c \in \psi_S(A,B)} \delta(c) \right)$$

and

$$\varphi_S(A,B) = \{\varphi_S(a,b),\ a \in A,\ b \in b\},$$
$$\psi_S(A,B) = \{\psi_S(a,b),\ a \in A,\ b \in b\}.$$

Theorem 41 (Ahlswede and Daykin 1979) *The direct product*

$$(\varphi, \psi) = ((\varphi_S, \varphi_T), (\psi_S, \psi_T))$$

of an $\mathcal{M}$-expansive pair of maps (φ_S, ψ_S) with an $\mathcal{M}$-expansive pair of maps (φ_S, ψ_S) is $\mathcal{M}$-expansive.

Remark. The use of Theorem 41 lies in the fact that one can apply it iteratively. This leaves us of course with the task to decide whether a component pair is $\mathcal{M}$-expansive. For the Boolean pair of operations $(\vee, \wedge)$ this is a little exercise and thus the familiar AD inequality follows from Theorem 41. Otherwise it is essentially unexplored, which pairs (φ_S, ψ_S) are $\mathcal{M}$-expansive.

The result shows that AD inequalities are based on Cartesian products and not necessarily on lattice structure as is often said in the literature.
Proof of Theorem 41. We are given $\mathcal{M}$-expansive pairs of maps (φ_S, ψ_S), (φ_T, ψ_T) and their direct product (φ, ψ) on $S \times T$. Let $\alpha, \beta, \gamma, \delta : S \times T \to \mathbb{R}$ satisfy

$$\alpha(a)\beta(b) \leq \gamma(\varphi(a,b))\delta(\psi(a,b)),\ a, b \in S \times T. \tag{37}$$

Let $A, B \subset S \times T$. We must show that

$$\alpha(A)\beta(B) \leq \gamma(\varphi(A,B))\delta(\psi(A,B)). \tag{38}$$

We define now marginal weights that depend on A, B. Define $\alpha_S, \beta_S, \gamma_S, \delta_S : S \to \mathbb{R}$ by

$$\alpha_S(s) = \sum_{t \in T,\ (s,t) \in A} \alpha(s,t),$$

$$\beta_S(s) = \sum_{t \in T,\ (s,t) \in B} \beta(s,t),$$

$$\gamma_S(s) = \sum_{t \in T,\ (s,t) \in \varphi(A,B)} \gamma(s,t),$$

$$\delta_S(s) = \sum_{t \in T,\ (s,t) \in \psi(A,B)} \delta(s,t).$$

Then

$$\alpha(A) = \sum_{(s,t)\in A} \alpha(s,t) = \sum_{s\in S}\sum_{t\in T,\ (s,t)\in A} \alpha(s,t) = \sum_{s\in S}\alpha_S(s) = \alpha_S(S).$$

Similarly $\beta(B) = \beta_S(S)$.

$$\begin{aligned}
\gamma(\varphi(A,B)) &= \sum_{(s,t)\in\varphi(A,B)} \gamma(s,t) \\
&= \sum_{s\in\varphi_S(S,S)}\sum_{t\in T,\ (s,t)\in\varphi(A,B)} \gamma(s,t) \\
&= \sum_{s\in\varphi_S(S,S)} \gamma_S(s) = \gamma_S(\varphi_S(S,S)).
\end{aligned}$$

Similarly $\delta(\psi(A,B)) = \delta_S(\psi_S(S,S))$.

Assume for a moment that

$$\alpha_S(s_1)\beta_S(s_2) \le \gamma_S(\varphi_S(s_1,s_2))\delta_S(\psi_S(s_1,s_2)), \quad s_1,s_2 \in S. \tag{39}$$

Since (φ_S,ψ_S) is $\mathcal{M}$-expansive this implies

$$\begin{aligned}
\alpha(A)\beta(B) = \alpha_S(S)\beta_S(S) &\le \gamma_S(\varphi_S(S,S))\delta_S(\psi_S(S,S)) \\
&= \gamma(\varphi(A,B))\delta(\psi(A,B)),
\end{aligned}$$

which is (38) as required. Thus it remains to prove (39). Let s_1,s_2 be fixed arbitrarily, and put $s_3 = \varphi_S(s_1,s_2)$ and $s_4 = \psi_S(s_1,s_2)$. Define $\alpha_T,\beta_T,\gamma_T,\delta_T : T \to \mathbb{R}$ by

$$\alpha_T(t) = \begin{cases} \alpha(s_1,t), & \text{if } (s_1,t)\in A \\ 0, & \text{otherwise} \end{cases}$$

$$\beta_T(t) = \begin{cases} \beta_S(s_2,t), & \text{if } (s_2,t)\in B \\ 0, & \text{otherwise} \end{cases}$$

$$\gamma_T(t) = \begin{cases} \gamma(s_3,t), & \text{if } (s_3,t)\in\varphi(A,B) \\ 0, & \text{otherwise} \end{cases}$$

$$\delta_T(t) = \begin{cases} \delta(s_4,t), & \text{if } (s_4,t)\in\psi(A,B) \\ 0, & \text{otherwise} \end{cases}.$$

Then

$$\alpha_S(s_1) = \sum_{t\in T,\ (s_1,t)\in A} \alpha(s_1,t) = \sum_{t\in T}\alpha_T(t).$$

Similarly

$$\beta_S(s_2) = \beta_T(T).$$

Next

$$\gamma_S(s_3) = \sum_{t\in T,\ (s_3,t)\in\varphi(A,B)} \gamma(s_3,t)$$

$$= \sum_{t \in \varphi_T(T,T),\ (s_3,t)\varphi(A,B)} \gamma(s_3,t) = \sum_{t \in \varphi_T(T,T)} \gamma_T(t)$$

$$= \gamma_T(\varphi_T(T,T)).$$

Similarly $\delta_S(s_4) = \delta_T(\psi_T(T,T))$.

We show now

$$\alpha_T(t_1)\beta_T(t_2) \leq \gamma_T(\varphi_T(t_1,t_2))\delta_T(\psi_T(t_1,t_2)),\ t_1,t_2 \in T. \tag{40}$$

Then by hypothesis on (φ_T,ψ_T) we have

$$\alpha_T(T)\beta_T(T) \leq \gamma_T(\varphi_T(T,T))\delta_T(\psi_T(T,T)), \tag{41}$$

in other words (39) follows.

The LHS of (40) is zero unless $(s_1,t_1) \in A$ and $(s_2,t_2) \in B$, in which case it is $\alpha(s_1,t_1)\beta(s_2,t_2)$. Furthermore, in this case

$$\varphi((s_1,t_1),(s_2,t_2)) = (\varphi_S(s_1,s_2),\varphi_T(t_1,t_2)) = (s_3,\varphi_T(t_1,t_2)) \in \varphi(A,B)$$
$$\psi((s_1,t_1),(s_2,t_2)) = (\psi_S(s_1,s_2),\psi_T(t_1,t_2)) = (s_4,\psi_T(t_1,t_2)) \in \psi(A,B),$$

and so the RHS of (40) is $\gamma(\varphi((s_1,t_1),(s_2,t_2)))\delta(\psi((s_1,t_1),(s_2,t_2)))$ and (40) follows from (37). ∎

We mentioned at the beginning of this lecture that $(\Delta,\wedge)$ is not $\mathcal{M}$-expansive (see also Exercise 7). An immediate idea was to enforce something like $\mathcal{M}$-expansiveness by adding a suitable constant factor. It has the obvious drawback that with weights one goes beyond probability distributions. But still it is mathematically meaningful. All this work shows that there is no reason to limit ourself to lattices, because Cartesian product theorems revealed a basic truth about correlation. Moreover, the machinery of induction, once started, can be set to work with any number of factors on both sides of the inequality. This has been carried out in [D80]. At that time (see also the proof of Theorem 41 above from [AD79b]) we did not have the following simple, but very helpful observation, stated in the following Lemma 34, which together with Lemma 35 gives us a simpler proof of Theorem 41.

The pair (φ_S,ψ_S) is called $\mathcal{M}_S$-expansive if (36) is valid for $A = B = S$.

Lemma 34 $\mathcal{M}$-*expansiveness and* $\mathcal{M}_S$-*expansiveness are equivalent for all pairs* $\varphi,\psi: S \times S \to S$.

Proof. For every $A,B \subseteq S$, the quadruple of indicator functions

$$\left(1_A,1_B,1_{\varphi(A,B)},1_{\psi(A,B)}\right)$$

is obviously compatible with (φ,ψ). Furthermore, for two compatible quadruples $(\alpha,\beta,\gamma,\delta)$ and $(\alpha',\beta',\gamma',\delta')$ obviously also the quadruple

$$(\alpha\alpha',\beta\beta',\gamma\gamma',\delta\delta')$$

is compatible. In particular, if $(\alpha, \beta, \gamma, \delta)$ is compatible, then always

$$(\alpha 1_A, \beta 1_B, \gamma 1_{\varphi(A,B)}, \delta 1_{\psi(A,B)})$$

is compatible. Since

$$\sum_{s \in S} \alpha(s) 1_A(s) = \alpha(A), \text{ etc.,}$$

the $\mathcal{M}_S$-expansiveness implies the formally stronger $\mathcal{M}$-expansiveness. $\qquad \square$

Lemma 35 *Let (φ_S, ψ_S) be an $\mathcal{M}_S$-expansive pair of maps $S \times S \to S$, let ϕ_T, ψ_T be maps $T \times T \to T$, and let the quadruple $(\alpha, \beta, \gamma, \delta)$ be compatible with*

$$(\varphi, \psi) = ((\varphi_S, \varphi_T), (\psi_S, \psi_T)).$$

Then the quadruple $(\alpha_T, \beta_T, \gamma_T, \delta_T)$ such that

$$(\alpha_T(t), \beta_T(t), \gamma_T(t), \delta_T(t))$$
$$= \left(\sum_{s \in S} \alpha(s,t), \sum_{s \in S} \beta(s,t), \gamma(\varphi_S(S,S),t), \delta(\psi_S(S,S),t) \right) \qquad (42)$$

is compatible with (φ_T, ψ_T).

Proof. Given $t_1, t_2 \in T$, we assume that

$$\alpha(s_1,t_1)\beta(s_2,t_2) \leq$$

$$\gamma(\varphi_S(s_1,s_2), \varphi_T(t_1,t_2))\delta(\psi_S(s_1,s_2), \psi_T(t_1,t_2)).$$

Therefore, the quadruple

$$(\alpha(\cdot,t_1), \beta(\cdot,t_2), \gamma(\cdot, \varphi_T(t_1,t_2)), \delta(\cdot, \psi_T(t_1,t_2)))$$

is compatible with (φ_S, ψ_S). Since (φ_S, ψ_S) is $\mathcal{M}_S$-expansive, we conclude that

$$\alpha(S,t_1)\beta(S,t_2) \leq \gamma(\varphi_S(S,S), \varphi_T(t_1,t_2))\delta(\psi_S(S,S), \psi_T(t_1,t_2)).$$

Hence, using (42) we obtain

$$\alpha_T(t_1)\beta_T(t_2) \leq \gamma_T(\varphi_T(t_1,t_2))\delta_T(\psi_T(t_1,t_2)).$$

$\qquad \square$

Second proof of Theorem 41. Suppose that the pair (φ_T, ψ_T) satisfies conditions of Lemma 35 and that this pair is $\mathcal{M}_T$-expansive. Then

$$\alpha(S,T)\beta(S,T) \leq \gamma(\varphi_S(S,S), \varphi_T(T,T))\delta(\psi_S(S,S), \psi_T(T,T)),$$

that is, we state $\mathcal{M}_{S \times T}$-expansiveness of (φ, ψ). Hence Lemma 34 gives the claimed $\mathcal{M}$-expansiveness. $\qquad \blacksquare$

We say that pair (φ, ψ), $\varphi, \psi: S \times S \to S$ is expansive if

$$|A||B| \leq |\varphi(A,B)||\psi(A,B)|, \, A, B \subset S.$$

Theorem 42 (Ahlswede and Daykin 1979) *If (φ_S, ψ_S) is $\mathcal{M}$-expansive and (φ_T, ψ_T) is expansive, then the direct product $((\varphi_S, \varphi_T), (\psi_S, \psi_T))$ is expansive.*

Proof. We follow the first proof of Theorem 41 with $\alpha = \beta = \gamma = \delta = 1$, except that we cannot use the $\mathcal{M}$-expansiveness of (φ_T, ψ_T) to go from (40) to (41). Instead we put

$$T_1 = \{t: \alpha_T(t) = 1\}, \, T_2 = \{t: \beta_T(t) = 1\},$$
$$T_3 = \{t: \gamma_T(t) = 1\}, \, T_4 = \{t: \delta_T(t) = 1\},$$

then

$$\alpha_T(T) = |T_1|, \, \beta_T(T) = |T_2|, \, \gamma_T(T) = |T_3|, \, \delta_T(T) = |T_4|.$$

Since (40) holds again this implies for $t_1 \in T_1$ and $t_2 \in T_2$ that $\varphi_T(t_1, t_2) \in T_3$ and $\psi(t_1, t_2) \in T_4$.

Therefore, $\varphi_T(T_1, T_2) \subset T_3$ and $\psi_T(T_1, T_2) \subset T_4$. Since (φ_T, ψ_T) is expansive

$$\alpha_T(T)\beta_T(T) = |T_1||T_2| \leq |\varphi_T(T_1, T_2)||\psi_T(T_1, T_2)| \leq |T_3||T_4| = \gamma_T(T)\delta_T(T)$$

so (41) follows as required.

At the present state of knowledge, most important is to classify pairs (or tuples) of maps that are $\mathcal{M}$-expansive, because then we can apply Theorems 41 and 42. A striking example of this approach is the proof of Rinott and Saks.

We use now a straightforward generalization of Lemmas 34 and 35 to give a very simple proof of a generalization of Theorem 41. Let m, n be positive integers and let S be a finite set. For m-ary operations $\varphi_1, \ldots, \varphi_n: S^m \to S$ we introduce the following concepts.

The maps $\alpha_1, \ldots, \alpha_m; \beta_1, \ldots, \beta_n: S \to \mathbb{R}_+$ are compatible with $\varphi_1, \ldots, \varphi_n$ if

$$\alpha_1(s_1) \ldots \alpha_m(s_m) \leq \beta_1(\varphi_1(s_1, \ldots, s_m)) \ldots \beta_n(\varphi_n(s_1, \ldots, s_m)) \qquad (43)$$

for all $s_1, \ldots, s_m \in S$.

As before we give two different concepts of expansiveness and show that they are equivalent.

Let $\underline{\sigma}_S = \underline{\sigma}_S(S, \varphi_1, \ldots, \varphi_n)$ be the smallest real number with

$$\alpha_1(S) \ldots \alpha_m(S) \leq \underline{\sigma}_S \beta_1(\varphi_1(S, \ldots, S)) \ldots \beta_n(\varphi_n(S, \ldots, S)) \qquad (44)$$

for all compatible weights $\alpha_1, \ldots, \alpha_m; \beta_1, \ldots, \beta_n$.

Let $\underline{\sigma} = \underline{\sigma}(S, \varphi_1, \ldots, '\varphi_n)$ be the smallest real number with

$$\alpha_1(A_1) \ldots \alpha_m(A_m) \leq \underline{\sigma} \beta_1(\varphi_1(A_1, \ldots, A_m)) \ldots \beta_n(\varphi_n(A_1, \ldots, A_m)) \qquad (45)$$

for all compatible weights $\alpha_1, \ldots, \alpha_m$; $\beta_1, \ldots, \beta_n$; and for all $A_1, \ldots, A_m \subset S$. Here we used the notation $\varphi(A_1, \ldots, A_m) = \{\varphi(a_1, \ldots, a_m) : a_i \in A_i\}$ and $\alpha(A) = \sum_{a \in A} \alpha(a)$.

$$(\varphi_1, \ldots, \varphi_n) \text{ is } \sigma \times \mathcal{M} - \text{ expansive, if } \sigma \geq \underline{\sigma}. \tag{46}$$

$$(\varphi_1, \ldots, \varphi_n) \text{ is } \sigma \times \mathcal{M}_S - \text{ expansive, if } \sigma \geq \underline{\sigma}_S. \tag{47}$$

∎

We derive now the generalization of Lemma 34.

Lemma 36 *For* $\varphi_1, \ldots, \varphi_n : S^m \to S$ $\sigma \times \mathcal{M}_S$-*expansiveness and* $\sigma \times \mathcal{M}-$*expansiveness are equivalent and* $\underline{\sigma} = \underline{\sigma}_S$.

Proof. For every $A_1, \ldots, A_m \subset S$ the tuple of indicator functions

$$\left(1_{A_1}, \ldots 1_{A_m}; 1_{\varphi_1(A_1, \ldots, A_m)}, \ldots, 1_{\varphi_m(A_1, \ldots, A_m)} \right)$$

is obviously compatible with $(\varphi_1, \ldots, \varphi_n)$.

For two compatible tuples $(\alpha_1, \ldots, \alpha_m; \beta_1, \ldots, \beta_n)$ and $(\alpha_1', \ldots, \alpha_m'; \beta_1', \ldots, \beta_n')$ obviously also the tuple

$$(\alpha_1 \alpha_1', \ldots, \alpha_m \alpha_m'; \beta_1 \beta_1', \ldots, \beta_n \beta_n')$$

is compatible. In particular,

$$\left(\alpha_1 1_{A_1}, \ldots, \alpha_m 1_{A_m}; \beta_1 1_{\varphi_1(A_1, \ldots, A_m)}, \ldots, \beta_n 1_{\varphi_n(A_1, \ldots, A_m)} \right)$$

is compatible. Since

$$\sum_{s \in S} \alpha(s) 1_A(s) = \alpha(A),$$

the $\sigma \times \mathcal{M}_S$-expansiveness implies the $\sigma \times \mathcal{M}-$expansiveness since obviously $\underline{\sigma}_S \leq \underline{\sigma}$, actually $\underline{\sigma}_S = \underline{\sigma}$. □

Next we generalize Lemma 35.

Lemma 37 *Let* $\varphi_1, \ldots, \varphi_n$ *be a* $\sigma \times \mathcal{M}_S$-*expansive tuple of maps* $S \times S \to S$ *and let* $\psi_1, \ldots, \psi_n$ *be maps* $T \times T \to T$ *and let* $(\alpha_1, \ldots, \alpha_m; \beta_1, \ldots, \beta_n)$ *be compatible with* $(\varphi_1, \psi_1), \ldots, (\varphi_n, \psi_n)$. *Then the tuple*

$$(\alpha_{1T}, \ldots, \alpha_{mT}; \beta_{1T}, \ldots, \beta_{nT}),$$

defined by

$$(\alpha_{1T}(t), \ldots, \alpha_{mT}(t); \beta_{1T}(t), \ldots, \beta_{nT}(t)) \tag{48}$$

$$= \left(\sum_{s \in S} \alpha_1(s,t), \ldots, \sum_{s \in S} \alpha_m(s,t); \sigma\beta_1(\varphi_1(S, \ldots, S), t), \ldots, \beta_n(\varphi_n(S, \ldots, S), t) \right),$$

is compatible with $(\psi_1, \ldots, \psi_n)$.

Proof. Given $t_1, \ldots, t_m \in T$ we assume that

$$\alpha_1(s_1, t_1) \ldots \alpha_m(s_m, t_m)$$

$$\leq \beta_1(\varphi_1(s_1, \ldots, s_m), \psi_1(t_1, \ldots, t_m)) \ldots \beta_n(\varphi_n(s_1, \ldots, s_m), \psi_n(t_1, \ldots, t_m)).$$

Therefore, the tuple

$$\alpha_1(\cdot, t_1), \ldots, \alpha_m(\cdot, t_m); \beta_1(\cdot, \psi_1(t_1, \ldots, t_m)), \ldots, \beta_n(\cdot, \psi_n(t_1, \ldots, t_m))$$

is compatible with $(\varphi_1, \ldots, \varphi_n)$. Since it is $\sigma \times \mathcal{M}_S$-expansive, we conclude that

$$\alpha_1(s, t_1) \ldots \alpha_n(s, t_m)$$

$$\leq \sigma\beta_1(\varphi_1(S, \ldots, S), \psi_1(t_1, \ldots, t_m)) \ldots \beta_n(\varphi_n(S, \ldots, S), \psi_n(t_1, \ldots, t_m))$$

Hence, using (48) we obtain

$$\alpha_{1T}(t_1) \ldots \alpha_{mT}(t_m) \leq \beta_{1T}(\varphi_1(t_1, \ldots, t_m)) \ldots \beta_{nT}(\varphi_m(t_1, \ldots, t_m)).$$

$\square$

Theorem 43 (Ahlswede and Daykin 1979) *The direct product*

$$(\varphi, \psi) = ((\varphi_1, \psi_1), \ldots, (\varphi_n, \psi_n))$$

of an $\sigma \times \mathcal{M}$-expansive tuple of maps $(\varphi_1, \ldots, \varphi_n)$ with a $\tau \times \mathcal{M}$-expansive tuple of maps $(\psi_1, \ldots, \psi_n)$ is $\sigma\tau \times \mathcal{M}$-expansive.

Proof. Suppose that $(\varphi_1, \ldots, \varphi_n)$ and $(\psi_1, \ldots, \psi_n)$ satisfy the conditions of Lemma 37 and that $(\psi_1, \ldots, \psi_n)$ is $\tau \times \mathcal{M}_T$-expansive. Then

$$\alpha_1(S, T) \ldots \alpha_n(S, T)$$

$$\leq \sigma\tau\beta_1(\varphi_1(S, \ldots, S), \psi_1(T, \ldots, T)) \ldots \beta_n(\psi_n(S, \ldots, S), \psi_n(T, \ldots, T)),$$

that is, we have $\sigma\tau \times \mathcal{M}_{S \times T}$-expansiveness of $(\varphi_1, \ldots, \varphi_n; \psi_1, \ldots, \psi_n)$. Hence Lemma 36 gives the claimed $\sigma\tau \times \mathcal{M}$-expansiveness. ∎

§6 Number Theoretical Correlation Inequalities

Finally we come to Number Theory and learn that there is a series of inequalities, which can be traced back with their beginning to Dirichlet [DD63]. They concern asymptotic densities d and for sets $A, B \subset \mathbb{N}$ the set of least common multiples $[A, B] = \{[a, b] : a \in A, b \in B\}$ (or set of largest common divisors $(A, B) = \{(a, b) : a \in A, b \in B\}$). Actually they all deal with sets of multiples $M(A) = A \times \mathbb{N}$, $M(B)$, $M([A, B])$, $M((A, B))$.

In [AK97a] for finite sets A, B the following equivalent inequalities are obtained:

$$dM(A)dM(B) \leq dM((A,B))dM([A,B]). \tag{49}$$

$$dN(A) \cdot dN(B) + dN(A,B) \cdot (1 - dN[A,B]) \leq dN(A \cup B). \tag{50}$$

The inequality (49) is by the factor $dM((A,B))$ sharper than Behrends's well-known inequality. This in turn is a generalization of an earlier inequality in [R37] and in [H37], which settled a conjecture of Hasse concerning an identity due to Dirichlet [DD63]. In [AK95a] the inequality

$$dM(A)dM(B) \geq dM((A \cdot B)), \tag{51}$$

where $A \cdot B = \{ab : a \in A, b \in B\}$ was also obtained for finite sets A, B. It does not seem to have predecessors. Observing the similarity of (49) to the AD inequality led to the main inequality from [AK97a]

$$\underline{D}A\underline{D}B \leq \underline{D}[A,B]\underline{D}(A,B), \tag{52}$$

where A, B are arbitrary sets of positive integers and $\underline{D}$ denotes the lower Dirichlet density. It is much more general than the previous inequality (49) for multiples of sets. This is more than an analogy: AD implies this number theoretical correlation inequality. For reasons of scaling it is important to work with Dirichlet density. Now AD has not only combinatorial and probabilistic correlation inequalities (as shown in §1) as consequences, but also the known correlation inequality in number theory. Emphasizing their relation we refer to AD and (52) inequalities as twins.

Similarly Behrend's inequality can also be derived from the FKG inequality, but it actually preceded it. We refer to them also as twins. Dealing with multiples of sets of numbers corresponds to dealing with upsets or monotone functions.

In the same sense one can ask for the twin of inequality (51). It will be shown below that it is the van den Berg/Kesten inequality.

A twin brother of the inequality by Rinott and Saks (see Theorem 34 in §1) is mentioned in the Problems. Finally, we close this section with a new inequality for the operations $a/(a,b)$ and $b/(a,b)$.

There are several interesting relations between sets of multiples and corresponding sets of non-multiples yielding inequalities. They are also delegated to the collection of problems.

As we can write

$$M(A) = \{m \in \mathbb{N} : a|m \text{ for some } a \in A\}$$

we define the set of non-multiples of A

$$N(A) = \{m \in \mathbb{N} : a \nmid m \text{ for all } a \in A\}.$$

For two numbers $u, v \in \mathbb{N}$ we write $u|v$ iff u divides v; (u, v) stands for the largest common divisor of u and v and $[u, v]$ denotes the smallest common multiple of u

and v. This definition will not be in conflict with the definition of the interval $[i, j] = \{i, i+1, \ldots, j\}$, $i < j$. In case $(u, v) = 1$, u and v are said to be relatively prime and are also called coprimes. If $u \leq v$ are integers, then $[u, v] = \{u, u+1, \ldots, v\}$ and for any $A \subset \mathbb{N}$ we set $A_n = A \cap [n]$.

The asymptotic density dA of $A \subset \mathbb{N}$ is defined

$$dA = \lim_{n \to \infty} \frac{|A_n|}{n}$$

if the limit exists.

We are now prepared to present the classical results. For $Q = \{q_1, \ldots, q_t\} \subset \mathbb{N}$, let us consider the set of integers in $[q]$, where

$$q = \prod_{i=1}^{t} q_i,$$

not divisible by any number in Q, that is, $N_q(Q)$. Dirichlet observed that

$$|N_q(Q)| = q \prod_{i=1}^{t} \left(1 - \frac{1}{q_i}\right)$$

if Q contains only relative primes (note that if Q contains only primes, then $|N_q(Q)|$ is known as Euler's function); the equivalent statement is

$$dN(Q) = \prod_{i=1}^{t} \left(1 - \frac{1}{q_i}\right). \tag{53}$$

For arbitrary Q, using inclusion–exclusion method, we have

$$|N_q(Q)| = q \left(1 - \sum_{i=1}^{t} \frac{1}{q_i} + \sum_{i<j} \frac{1}{[q_i, q_j]} - \cdots\right)$$

and

$$dN(Q) = 1 - \sum_{i=1}^{t} \frac{1}{q_i} + \sum_{i<j} \frac{1}{[q_i, q_j]} - \cdots. \tag{54}$$

Since

$$\prod_{i=1}^{t} \left(1 - \frac{1}{q_i}\right) = 1 - \sum_{i=1}^{t} \frac{1}{q_i} + \sum_{i<j} \frac{1}{q_i q_j} - \cdots,$$

the expressions in (53) and (54) do not coincide. Hasse conjectured that

$$1 - \sum_{i=1}^{t} \frac{1}{q_i} + \sum_{i<j} \frac{1}{[q_i, q_j]} - \cdots \leq 1 - \sum_{i=1}^{t} \frac{1}{q_i} + \sum_{i<j} \frac{1}{q_i q_j} - \cdots,$$

or equivalently

$$\prod_{i=1}^{t} dN(\{q_i\}) \leq dN(Q). \tag{55}$$

This conjecture was independently proved by Rohrbach and Heilbronn in 1937. Furthermore, they proved that the equality in (54) holds iff Q contains only relative primes, that is, Dirichlet's case is unique.

In [B48], see also [HR66], the following equivalent inequalities are proved:

$$dM([A,B]) \geq dM(A)dM(B),$$
$$dN(A \cup B) \geq dN(A)dN(B).$$

In [AK95a] the following equivalent inequalities are proved, which are improvements of the previous inequalities:

$$dM([A,B])dM((A,B)) \geq dM(A)dM(B), \tag{56}$$
$$dN(A \cup B) \geq dN(A)dN(B) + dN((A,B))(1 - dN([A,B])).$$

Behrend's inequalities can be represented as direct generalization of (55). The next step in estimating densities are inequalities (56). Their proof is based on the following observation: for any finite sets $A, B \subset \mathbb{N}$ and any number $m \in \mathbb{N}$

$$d(M(A) \cap M(B)) \leq md(M(mA) \cap M(B)),$$
$$dM(mA \cap B) \leq \frac{1}{m}dM(A \cup B) + \frac{m-1}{m}dM(B).$$

With every $A \subset \mathbb{N}$ we associate the Dirichlet series

$$D(A,s) = \sum_{n \in A} \frac{1}{n^s},$$

where s is complex number. The number

$$DA = \lim_{s \to 1^+} (s-1)D(A,s)$$

is called the Dirichlet density of A, if the limit exists. Furthermore,

$$\underline{D}A = \liminf_{s \to 1^+}(s-1)D(A,s)$$

and

$$\overline{D}A = \limsup_{s \to 1^+}(s-1)D(A,s)$$

are lower and upper Dirichlet densities, respectively. The proofs of the next results about densities and further references can be found in [HR66].

Theorem 44 *For all $A \subset \mathbb{N}$*

$$0 \leq \underline{d}A \leq \underline{D}A \leq \overline{D}A \leq \overline{d}A.$$

It is known that DA does not need to exist. While the existence of dA implies the existence of DA, the converse is not true. However, under quite general conditions also the converse implication holds as it follows from Ikehara's Tauberian theorem. Good references here are also the book [O56] and the book [N00].

Theorem 45 *If $D(A,s)$ is convergent in $s = \sigma + \tau i$ for the half plane $\sigma > 1$, if the corresponding analytic function has for $\sigma \geq 1$ at most one singularity at $s = 1$, and if this singularity is a simple pole with residuum ρ, then dA exists and*

$$dA = \rho = \lim_{s \to 1^+} (s-1) D(A,s).$$

We come now to a third notion of density. The quantities

$$\underline{\delta}A = \liminf_{n \to \infty} \frac{1}{\log n} \sum_{a \in A_n} \frac{1}{a}$$

and

$$\overline{\delta}A = \limsup_{n \to \infty} \frac{1}{\log n} \sum_{a \in A_n} \frac{1}{a}$$

are, respectively, the lower and upper logarithmic densities of A, and if the two are equal we say that A possesses logarithmic density δA, given by

$$\delta A = \lim_{n \to \infty} \frac{1}{\log n} \sum_{a \in A_n} \frac{1}{a}.$$

It is easy to derive with Abel summation the following relations:

$$1 \leq \underline{d}A \leq \underline{\delta}A \leq \overline{\delta}A \leq \overline{d}A.$$

Comparing this with Theorem 45 we see that logarithmic density has similar properties as Dirichlet density. In fact, much more is true. The following result is attributed to Dirichlet.

Theorem 46 *The existence of DA implies the existence of δA, and conversely. Furthermore, $DA = \delta A$, if the densities exist.*

Remark Inspection of the proof shows that actually always $\underline{DA} = \underline{\delta}A$ and $\overline{DA} = \overline{\delta}A$.

A famous example of Besicovitch shows that even sets of multiples need not have an asymptotic density. However, it was shown by Davenport and Erdös ([DE36]) that they do have logarithmic density.

Theorem 47 (Davenport and Erdös 1936) *For an arbitrary $A \subset \mathbb{N}$ $DM(A)$ always exists.*

Concerning asymptotic densities these authors proved

Theorem 48 (Davenport and Erdös 1936) *For $A = \{a_1, a_2, \ldots\}$ a necessary and sufficient condition for $dM(A)$ to exist is*

$$\lim_{\varepsilon \to 0^+} \limsup_{n \to \infty} \frac{1}{n} \sum_{n^{1-\varepsilon} < a_i \le n} \left| \left(M(\{a_i\}) \setminus M(\{a_1, \ldots, a_{i-1}\}) \right) \cap [n] \right| = 0.$$

Furthermore, if $|A_n| = O(n/\log n)$, then $dM(A)$ exists.

We present now the twin for AD.

Theorem 49 (Ahlswede and Khachatrian 1997) *For arbitrary sets of positive integers A, B the following inequality is valid:*

$$\underline{D}A\underline{D}B \le \underline{D}[A, B]\underline{D}(A, B). \tag{57}$$

In particular,

$$DM(A)DM(B) \le DM[A, B]DM(A, B) \tag{58}$$

holds.

Proof. Let $\{p_1, \ldots, p_m\}$ be the set of all primes in $\{1, \ldots, n\}$. The set of natural numbers generates a lattice under operations $a \vee b = [a, b]$, $a \wedge b = (a, b)$. We can embed the sublattice $L' = \{\prod_{i=1}^m p_i^{\pi_i}, \pi_i = 0, 1, \ldots, n\}$ into lattice $2^{[mn]}$ in the following way: if

$$a' = \prod_{i=1}^m p_i^{\pi_i(a')},$$

then for $\varphi : L' \to 2^{[mn]}$

$$\varphi(a') = a = (a_{11}, \ldots, a_{n1}, a_{12}, \ldots, a_{n2}, \ldots, a_{1m}, \ldots, a_{nm}),$$

where

$$a_{ji} = \begin{cases} 1, & \text{if } \pi_i(a') \ge j, \\ 0, & \text{otherwise.} \end{cases}$$

Then in (7) for the lattice $L = 2^{[mn]}$ we put $f_1 = f_3 = \alpha$, $f_2 = f_4 = \beta$, where

$$\alpha(a) = (a')^{-s}, \ \beta(a) = (a')^{-t}, a \in L$$

and $\alpha(a) = \beta(a) = 0$, if $a \notin L$. As $[a, b](a, b) = ab$ for $1 < s < t$, we have

$$\frac{1}{a^s} \frac{1}{b^t} \le \frac{1}{[a, b]^s} \frac{1}{(a, b)^t}.$$

Next, since

$$[a, b] = \prod_{i=1}^m p_i^{\max\{\pi_i(a), \pi_i(b)\}}, (a, b) = \prod_{i=1}^m p_i^{\min\{\pi_i(a), \pi_i(b)\}},$$

the last inequality is condition (6) and hence from Theorem 33 we have

$$\sum_{a \in A_n} \frac{1}{a^s} \sum_{b \in B_n} \frac{1}{b^t} \le \sum_{c \in [A_n, B_n]} \frac{1}{c^s} \sum_{d \in (A_n, B_n)} \frac{1}{d^t},$$

or

$$D(A_n, s)D(B_n, t) \leq D([A_n, B_n], s)D((A_n, B_n), t). \tag{59}$$

Taking $\liminf_{n \to \infty}$ on both sides of the last inequality and then consequently $\lim_{t \to 1^+}$, $\lim_{s \to 1^+}$, we obtain inequality (57) and (58) follows with Theorem 47, which implies (49).   ∎

Next we formulate another theorem, which deals with multiples of sets in $\mathbb{N}$ and generalizes (51) to infinite sets.

Theorem 50 (Ahlswede and Khachatrian 1997) *The following relation is valid:*

$$D(M(A))D(M(B)) \geq D(M(A \cdot B)). \tag{60}$$

We found as its twin the van den Berg/Kesten inequality, which we proved first. It concerns the property NBU (new better than used) for a random variable, which was investigated for multivariate distributions in [MS82].

Theorem 51 (van den Berg and Kesten 1985) *Assume that a random variable X taking values in $\mathbb{R}_+$ satisfies for all $x_1, x_2 \geq 0$ the following relation:*

$$P(X > x_1 + x_2) \leq P(X > x_1)P(X > x_2). \tag{61}$$

Consider $\mathbb{R}^n$ as a lattice with $a^n \geq b^n$, $a^n, b^n \in \mathbb{R}^n$ iff $a_i \geq b_i$ for all $i = 1, \ldots, n$. Let $X_1, X_2, \ldots, X_n$ be independent random variables taking values in $\mathbb{R}_+$, which satisfy (61). Let also $X = (X_1, \ldots, X_n)$. Then for arbitrary upsets $A, B \subset \mathbb{R}^n$ the following inequality is valid:

$$P(X \in A + B) \leq P(X \in A)P(X \in B). \tag{62}$$

Proof. Let $X_1, X_2, \ldots, X_n$ be independent random variables taking values in $\mathbb{R}_+$ and satisfying (61). Assume that X_1, X_2 are identically distributed and $A \in \mathbb{R}^n$ is an upset. Then

$$P((X_1, X_2, \ldots, X_n) \in A) \geq P((X_1, X_3, X_4, \ldots, X_n) \in A^*), \tag{63}$$

where $A^* = \{(x_1 + x_2, x_3, x_4, \ldots, x_n) : (x_1, x_2, \ldots, x_n) \in A\} \subset \mathbb{R}^{n-1}$.

Indeed the conditional probability of the event on the LHS of (63) for each x_1, x_2 : $(x_1, x_2, \ldots, x_n) \in A$ when $X_3 = x_3, \ldots, X_n = x_n$ are given is not less than $P(X_1 \geq x_1, X_2 \geq x_2) = P(X_1 \geq x_1)P(X_2 \geq x_2)$. Thus this conditional probability is not less than

$$\sup_{(x_1, x_2):(x_1, x_2, \ldots, x_n) \in A} P(X_1 \geq x_1)P(X_2 \geq x_2).$$

The corresponding conditional probability on the RHS of (63) is

$$P(X_1 \in \{x_1 + x_2 : (x_1, x_2, \ldots, x_n) \in A\}) = \sup_{(x_1, x_2):(x_1, x_2, \ldots, x_n) \in A} P(X_1 \geq x_1 + x_2).$$

Taking into account property (61), one can see that the last expression is less than or equal to

$$\sup_{(x_1, x_2):(x_1, x_2, \ldots, x_n) \in A} P(X_1 \geq x_1)P(X_2 \geq x_2).$$

This proves (63).

To complete the proof of Theorem 51 let now $m = 2n$ be even. If in the assumption all distributions $X_1, X_2, \ldots, X_m$ can be divided into n pairs of equal distributions: $(v_1, v_{n+1}), \ldots, (v_n, v_{2n})$, then applying (63) n times we obtain that for every upsets $A, B \in \mathbb{R}^n$ we have

$$v(A + B) \leq (v \times v)(A \times B), \tag{64}$$

where $v = v_1 \times v_2 \times \ldots \times v_n$. From relation (64), taking into account the independence of v_i, we obtain (62). ∎

Proof of Theorem 50. For the product of chains $L(\ell, m) = \{0, 1, 2, \ldots, \ell\}^m$, we define the probability distribution $v^m = \prod_{i=1}^m v_{i,3}$ where

$$v_i(\ell_i) = p_i^{-s\ell_i} \cdot \left(\sum_{j=0}^{\ell} p_i^{-sj} \right)^{-1}$$

and $p_1, \ldots, p_m$ are the first m prime numbers. Let $N(\ell, m) = \{\prod_{i=1}^m p_i^{\ell_i} : (\ell_1, \ldots, \ell_m) \in L(\ell, m)\}$, then

$$\lim_{\ell, m \to \infty} D\left(M(A) \cap N(\ell, m), s\right) = D(M(A), s),$$

$$\lim_{\ell, m \to \infty} D\left(M(A \cdot B) \cap N(\ell, m), s\right) = D(M(A \times B), s).$$

Let us show that

$$v_i(\{\lambda_1, \ldots, \ell\}) v_i(\{\lambda_2, \ldots, \ell\}) \geq v_i(\{\lambda_1 + \lambda_2, \ldots, \ell\}). \tag{65}$$

Note that this inequality is equivalent to inequality (61) in the new notation.

The last inequality is true when $\lambda_1 + \lambda_2 > \ell$. Next for $\lambda_1 + \lambda_2 \leq \ell$ inequality (65) can be rewritten as

$$\sum_{j=\lambda_1}^{\ell} \frac{1}{p^j} \sum_{j=\lambda_2}^{\ell} \frac{1}{p^j} \geq \sum_{j=\lambda_1 + \lambda_2}^{\ell} \frac{1}{p^j} \sum_{j=0}^{\ell} \frac{1}{p^j} \tag{66}$$

which, in turn, is equivalent to

$$(1 + p + \ldots + p^{\ell - \lambda_1})(1 + p + \ldots + p^{\ell - \lambda_2}) \geq$$
$$(1 + p + \ldots + p^{\ell - \lambda_1 - \lambda_2})(1 + p + \ldots + p^{\ell})$$

or

$$(p^{\ell - \lambda_1 + 1} - 1)(p^{\ell - \lambda_2 + 1} - 1) \geq (p^{\ell - \lambda_1 - \lambda_2 + 1} - 1)(p^{\ell + 1} - 1).$$

This inequality, in turn, is equivalent to

$$p^{\ell - \lambda_1 - \lambda_2 + 1} + p^{\ell + 1} \geq p^{\ell - \lambda_1 + 1} + p^{\ell - \lambda_2 + 1}$$

or

$$1 + p^{\lambda_1 + \lambda_2} \geq p^{\lambda_1} + p^{\lambda_2},$$

which is true. Thus inequality (65) is proved.

Now $U = M(A) \cap N(\ell,m)$ and $V = M(B) \cap N(\ell,m)$ are upsets. Denoting $W = M(A \cdot B) \cap N(\ell,m)$, from Proposition 51 we obtain the inequality

$$\sum_{u \in U} \frac{1}{u^s} \Big(\sum_{n \in N(\ell,m)} \frac{1}{n^s} \Big)^{-1} \cdot \sum_{v \in V} \frac{1}{v^s} \Big(\sum_{n \in N(\ell,m)} \frac{1}{n^s} \Big)^{-1} \geq \sum_{w \in W} \frac{1}{w^s} \Big(\sum_{n \in N(\ell,m)} \frac{1}{n^s} \Big)^{-1}.$$

This yields

$$\sum_{u \in U} \frac{1}{u^s} \sum_{v \in V} \frac{1}{v^s} \geq \sum_{w \in W} \frac{1}{w^s} \sum_{n \in N(\ell,m)} \frac{1}{n^s}$$

and

$$D(M(A),s)D(M(B),s) \geq D(M(A \cdot B),s)\zeta(s). \tag{67}$$

As the Dirichlet density exists for the set of multiples and $\lim_{s \to 1^+}(s-1)\zeta(s) = 1$, the last inequality implies the statement of the theorem. ∎

It is time now to emphasize that Number Theory also gave birth to combinatorial (or operational) correlation inequalities with the following conjectures of Graham [G70].

Given a finite set $A \subset \mathbb{N}$ of integers and let $G = G(A) = \{a/(a,b) : a,b \in A\}$ then

(i) $|A| \leq \max\{c : c \in G\}$.
(ii) $|A| \leq |G|$.

Clearly (ii) implies (i). However, (ii) was disproved by Levin and Szemeredi. They gave as counterexample the set $A = \{2,3,4,6,9,12,18\}$ of nontrivial divisors of 36. Since $G(A) = \{1,2,3,4,6,9\}$ $7 = |A| \not\leq |G| = 6$.

However, (ii) is true if A contains only squarefree integers. This is a consequence of a combinatorial conjecture by Schönheim [S69], which became the Marica/Schönheim inequality [MS69]. We state as its

Corollary 7 *If a finite $A \subset \mathbb{N}$ contains only squarefree integers, then*

$$|G(A)| \geq |A| \text{ and a fortiori } \max\{c : c \in G(A)\} \geq |A|.$$

Much later, conjecture (i) was proved for all finite $A \subset \mathbb{N}$ in [BS96].

However, the max-operation does not fit into the frame of operations in which correlation inequalities could be established (set operations like $[A,B]$ or (A,B)).

On the other hand, counterexamples to (ii) seem to destroy hopes for correlation inequalities here too.

However, we have learned previously that for infinite sets $B \subset \mathbb{N}$ with Dirichlet densities as measurements the world looks different. We define now for arbitrary sets $B \subset \mathbb{N}$ $B_n = B \cap [n]$. The corollary implies that for an infinite set A of squarefree numbers

$$|A_n| \leq |G(A_n)| \leq |G_n|$$

and therefore for Dirichlet densities

$$\overline{D}A \leq \overline{D}G, \ \underline{D}A \leq \underline{D}G.$$

Since $\underline{d}B \leq \underline{D}B \leq \overline{D}B \leq \overline{d}B$ this implies for existing asymptotic density dA that

$$dA \leq \underline{D}G. \tag{68}$$

But now comes a new discovery for the general not necessarily squarefree case. From one set A with one operation we go to two sets A, B with two operations $\frac{a}{(a,b)}$ and $\frac{b}{(a,b)}$ for a and b. Naturally we define

$$G(A,B) = \{a/(a,b) : a \in A, b \in B\} \text{ and}$$

$$G(B,A) = \{b/(a,b) : b \in B, a \in A\}.$$

Theorem 49 suggests to look for an analogous inequality for $G(A,B)$ and $G(B,A)$. We show first that the pair of operations $(\backslash, \backslash)$ is M_S-expansive for $S = \{0,1\}$. An argument like in the proof of Theorem 33 works.

For $\alpha, \beta, \gamma, \delta : \{0,1\} \to \mathbb{R}_+$ compatible with $(\backslash, \backslash)$ we have

$$\alpha(0)\beta(0) \leq \gamma(0)\delta(0)$$
$$\alpha(0)\beta(1) \leq \gamma(0)\delta(1)$$
$$\alpha(1)\beta(0) \leq \gamma(1)\delta(0)$$
$$\alpha(1)\beta(1) \leq \gamma(0)\delta(0) \tag{69}$$

It suffices to show that

$$\alpha(0)\beta(0) + \alpha(1)\beta(1) \leq \gamma(0)\delta(0) + \gamma(1)\delta(1).$$

With the abbreviations $\alpha(0)\beta(0) = x$, $\alpha(1)\beta(1) = y$, $\gamma(0)\delta(0) = z$, and $\gamma(1)\delta(1) = q$ to be shown is

$$x + y \leq z + q. \tag{70}$$

Clearly by (69) $x, y \leq z$ and $xy \leq zq$. Hence

$$z + q - (x+y) = \frac{1}{z}((z-x)(z-y)) + (zq - xy) \geq 0$$

and (70) is established.

By Theorem 43 we have an AD-type inequality with weights for $(\backslash, \backslash)$ and instead of weights

$\alpha(a) = \frac{1}{a^s}$, $\beta(b) = \frac{1}{b^t}$, $\gamma([a,b]) = \alpha([a,b])$, $\delta((a,b)) = \beta((a,b))$

we use now the same α and β but $\gamma(\frac{a}{(a,b)}) = \left(\frac{(a,b)}{a}\right)^s$, $\delta(\frac{b}{(a,b)}) = \left(\frac{(a,b)}{b}\right)^t$.

$(\alpha, \beta, \gamma, \delta)$ is compatible, because obviously (here even for all $s, t > 0$)

$$\frac{1}{a^s}\frac{1}{b^t} \leq \frac{(a,b)^s}{a^s} \cdot \frac{(a,b)^t}{b^t} \tag{71}$$

and we get the following:

Theorem 52 (Ahlswede, proved in 2007) *For all $A, B \subset \mathbb{N}$*

$$\underline{D}A\underline{D}B \leq \underline{D}G(A,B)\underline{D}G(B,A)$$

and for $A = B$

$$\underline{D}A \leq \underline{D}G(A).$$

Thus Graham's story has also another happy end.

Remark We know that there is no inequality

$$|A||B| \leq |G(A,B)||G(B,A)| \tag{72}$$

as the example $A = B = \{2,3,4,6,9,12,18\}$ discussed above shows. This is an amazing difference to the behavior of the sets $[A,B]$ and (A,B) (see Exercise 9).

Notes to Chapter V

Theorem 27 was proved by Ahlswede and Zhang [AZ90a]. Theorems 28 and 29 were proved by the same authors in [AZ90b]. Theorem 30 was proved by Ahlswede and Cai [AC93a]. In these papers, conditions when the second term on the LHS of the identities vanishes were also formulated.

Note one interesting identity, which is a generalization of the AZ-identity obtained by Thu in [T07]. Let m be an integer and let $\mathcal{A}$ be a family of nonempty subsets of $2^{[n]}$. If $|A| + m > 0$ for each $A \in \mathcal{A}$, then

$$\sum_{X \in \mathcal{U}(\mathcal{A})} \frac{W_{\mathcal{A}}(X) + m}{|X| + m} \binom{n+m}{|X|+m}^{-1} = 1. \tag{73}$$

The case $m = 0$ gives the original AZ-identity.
The case $m = -1$ gives for antichains $\mathcal{A}$ with $|A| > 1$ for each $A \in \mathcal{A}$

$$\sum_{A \in \mathcal{A}} \frac{1}{\binom{n-1}{|A|-1}} + \sum_{X \in \mathcal{U}(\mathcal{A})\setminus\mathcal{A}} \frac{W_{\mathcal{A}}(X) - 1}{(|X|-1)\binom{n-1}{|X|-1}} = 1. \tag{74}$$

This implies for an antichain $\mathcal{A}$, which is a star,

$$\sum_{A \in \mathcal{A}} \frac{1}{\binom{n-1}{|A|-1}} \leq 1.$$

There are further related results in [T07] and especially in [T08]. Again methods of [T93] are used.

The motivation comes from a theorem of Bollobás (18 in Lecture 13), which can also be found in the book [B88]. In that theorem on intersecting antichains of subsets of $[n]$ it is assumed that all subsets have cardinalities not exceeding $\frac{n}{2}$. This assumption is termed by Thu "half-way." It means that the poset $\mathcal{P}_{\lfloor \frac{n}{2} \rfloor} = \bigcup_{k \leq \frac{n}{2}} \binom{[n]}{k}$ is considered. Actually in [AZ90a] already upsets in general posets are considered to derive abstract AZ-identities.

The main result in [T08] is the analogue of (73).

$$\sum_{X \in \mathcal{U}(A), |X| \leq \lfloor \frac{n}{2} \rfloor} \frac{W_A(X) + m}{|X| + m \binom{n+m}{|X|+m}} = \frac{|\mathcal{U}_A \cap \mathcal{P}_{\lfloor \frac{n}{2} \rfloor}|}{\binom{n+m}{\lfloor \frac{n}{2} \rfloor + m}} \tag{75}$$

for non-empty families $A \subset 2^{[n]}$ with $\emptyset \notin A$ and $|A| + m > 0$ for all $A \in A$.

Among the consequences there are dual identities of (75) in the spirit of [DT94] and extensions to two families of sets in the spirit of [AZ90a].

We say that the family $\mathcal{D} \subset 2^{[n]}$ is k-Sperner if the length of the maximal chain $\mathcal{C} \subset \mathcal{D}$ is k. Generalizing Sperner's Theorem Erdös [E45] proved that the size of a k-Sperner family is not larger than that of the sum of the k largest binomial coefficients and this bound is attained iff the family $\mathcal{D}$ consists of the corresponding levels (full families of the sets of equal cardinalities). We take the following proof of Erdös' result and the uniqueness of an optimal k-Sperner family from [P05]. It seems that a uniqueness proof was not written down before this.

Theorem 53 (Erdös 1945) *If $\mathcal{D} \subset 2^{[n]}$ is a k-Sperner family, then*

$$|\mathcal{D}| \leq \sum_{i=\lfloor \frac{n-k+1}{2} \rfloor}^{\lfloor \frac{n+k-1}{2} \rfloor} \binom{n}{i}$$

and equality holds iff $\mathcal{D}$ consists of the levels from $\lfloor \frac{n-k+1}{2} \rfloor$ to $\lfloor \frac{n+k-1}{2} \rfloor$ or from $\lceil \frac{n-k+1}{2} \rceil$ to $\lceil \frac{n+k-1}{2} \rceil$.

Proof. If A is an antichain, then the second sum from (3) (Lecture 13) equals 0 iff A is a level. To see this observe that if A is not a level, then there are sets $A_1 \in A$, $A_2 \notin A$, with $|A_1| = |A_2|$, $|A_2 \setminus A_2| = 1$, and saturated (not extendable) chains containing A_2 and $A_2 \cup A_1$ do not meet A. Consider the partition of $\mathcal{D}$ into antichains in the following way:

$$A_1 = \min \mathcal{D}, \quad A_{j+1} = \min \left(\mathcal{D} \setminus \bigcup_{i=1}^{j} A_i \right),$$

where $\min(\mathcal{F})$ is the family of minimal sets in $\mathcal{F}$. If $\mathcal{D}$ has no $k+1$-chains, then $A_{k+1} = \emptyset$. Adding identities (3) (Lecture 13) for all A_i, $i = 1, \ldots, k$, we get the following identity

$$\sum_{X \in \mathcal{D}} \frac{1}{\binom{n}{|X|}} + \sum_{i=1}^{k} \sum_{X \in \mathcal{U}(\mathcal{A}_i) \setminus \mathcal{A}_i} \frac{W_{\mathcal{A}}(X)}{|X| \binom{n}{|X|}} = k.$$

Since the double sum equals 0 iff all $\mathcal{A}_i$'s are levels, the uniqueness in Erdös' result follows.   □

The notion of the splitting property was first introduced in [AEG95]. The material for Lecture 14 is taken from that work. The splitting property for infinite posets is investigated in [AK00a].

Theorem 36 was proved in [AD79b]. Both, Lemma 33 and Theorem 34 were proved in [RS93]. Theorem 34 was also proved in [AK96] by using another method.

Theorem 37 is from [AD79a].

Theorem 40 with worse constant $\frac{1}{100}$ was first proved in [ACNS82] and [L83]. The proof that we reproduced here is from [AZ98]. The proof of Corollary 6 is from [S97]. The proof of inequality (25) we took from [E97a]. Further results about the correlation between sums and products of integers can be found in [C99], [ENR00], [F98], [N97], and [NT99].

Theorems 49 and 50 are from [AK97a], see also [AK95a].

Theorem 52 was proved by Ahlswede 2007, previously unpublished.

Exercises

1. Follow the proof of Theorem 28 and prove Theorem 29:

   (a) To $I = \{(i_t, j_t) : 1 \leq t \leq T\}$ assign the upset

   $$U = \{(i, j) : (i, j) \geq (i_t, j_t) \text{ for some } t\}.$$

   (b) Count the saturated chains according to their exit from U. We can assume $i_1 \geq \ldots \geq i_T$. Then necessarily $j_1 \leq \ldots \leq j_T$. Therefore, exits from U occur in three kinds of elements:

     i. (i_t, j) with $j_t < j \leq j_{t+1} - 1$,

     ii. (i, j_t) with $i_t < i \leq i_{t-1} - 1$,

     iii. (i_t, j_t).

   (c) Using an analogy with Theorem 27 prove Theorem 30. First using the inclusion–exclusion principle prove the following fact, which generalizes (19): for $B \subset [n]$ and $\mathcal{A} \subset 2^{[n]}$ with $A \subset B$ for all $A \in \mathcal{A}$ exactly

   $$n! \sum_{k=1}^{|\mathcal{A}|} (-1)^{k-1} \sum_{\mathcal{A}' \subset \mathcal{A}, |\mathcal{A}'|=k} \binom{n - |B - \bigcup_{A \in \mathcal{A}'} A|}{|\bigcup_{A \in \mathcal{A}'} A|}^{-1}$$

   maximal chains in $2^{[n]}$ meet $\{X : A \subset X \subset B \text{ for some } A \in \mathcal{A}\}$.

   (d) (A bipartite extension of Sperner's Theorem [AL06]). Let $U \subset [u]$, $V = [n] \setminus U$, $u \leq n$. Prove, that if every two sets $A \neq B \in \mathcal{A}$ satisfy that either

$A \cap U$, $B \cap U$ are incomparable or $A \cup V$, $B \cup V$ are incomparable, then $|A| \leq \binom{u}{\lfloor u/2 \rfloor}\binom{n-u}{\lfloor (n-u)/2 \rfloor}$ and this bound is tight.

2. Use the AZ-identity for posets (see [AZ90a]) and calculate it for

   (a) regular trees of depth k and improve thus Kraft's inequality for prefix codes.
   (b) subspaces over $GF(q)$.
   (c) cascade graphs. Show that Rényi's inequality [R69] in his uniform flow theorem can be replaced by an identity.

3. Prove that if the probability measure μ on $[n]$ depends only on the cardinality of the set $A \subset [n]$, then the condition

$$\mu(A)\mu(B) \leq \mu(A \cup B)\mu(A \cap B)$$

is equivalent to the condition

$$\mu^2([k]) \leq \mu([k-1])\mu([k+1]).$$

4. Prove Theorem 35. Define the partial order $\prec$ on $D = \{x^n : x_1 \geq x_2 \geq \ldots \geq x_n\} \subset \mathbb{R}^n$ by the relations $x^n \prec y^n$ iff $\tilde{x}_k = \sum_{i=1}^{k} x_i \leq \tilde{y}_k = \sum_{i=1}^{k} y_i$, $k = 1, 2, \ldots, n-1$, and $\sum_{i=1}^{n} x_i = \sum_{i=1}^{n} y_i$. A function $\varphi : \mathbb{R}^n \to \mathbb{R}$ is Schur convex (on D) iff $x^n \prec y^n \Rightarrow \varphi(x^n) \leq \varphi(y^n)$.

   (a) Show that $x^n \prec_w y^n$ implies $\varphi(x^n) \leq \varphi(y^n)$ iff for $z^n \in D$, $\varphi(z_1, z_2, \ldots, z_n) = \varphi(\tilde{z}_1, \tilde{z}_2 - \tilde{z}_1, \ldots, \tilde{z}_n - \tilde{z}_{n-1})$ is increasing in $\tilde{z}_i, i = 1, 2, \ldots, n$.
   (b) Show that $x^n \prec y^n$ implies $\varphi(x^n) \leq \varphi(y^n)$ iff for $z^n \in D$, $\varphi(z_1, z_2, \ldots, z_n) = \varphi(\tilde{z}_1, \tilde{z}_2 - \tilde{z}_1, \ldots, \tilde{z}_n - \tilde{z}_{n-1})$ is increasing in $\tilde{z}_i, i = 1, 2, \ldots, n-1$.
   (c) Show that $x^n \prec_w y^n$ implies $\varphi(x^n) \leq \varphi(y^n)$ iff $x^n \prec y^n$ implies $\varphi(x^n) \leq \varphi(y^n)$ (it is Schur convex) and in addition $\varphi(z_1, z_2, \ldots, z_n + \varepsilon)$ is increasing in $\varepsilon \in [0, z_{n-1} - z_n]$ for all $z^n \in D$.
   (d) Define the T-transform by the $n \times n$ matrix

$$T = \lambda I + (1 - \lambda)Q,$$

   where $\lambda \in [0, 1]$, I is a unit matrix, and Q is a permutation matrix, which interchanges two coordinates. Then $x^n \prec y^n$ iff $x^n = \tilde{T}y^n$, where $\tilde{T}$ is a combination of a finite number of successive T-transforms, that is,

$$x^n \tilde{T} = (x_1, \ldots, x_{j-1}, \lambda x_j + (1 - \lambda)x_k, x_{j+1}, \ldots, x_{k-1}, \lambda x_k + \\ + (1 - \lambda)x_j, x_{k+1}, \ldots, x_n).$$

   Let j be the largest index such that $x_j < y_j$ and k be the smallest index greater than j, such that $x_k > y_k$. By choice of j, k

$$y_j > x_j \geq x_k > y_k.$$

Setting $\delta = \min\{y_j - x_j, x_k - y_k\}$, $\lambda = 1 - \delta/(y_j - y_k)$ we obtain

$$\hat{y}^n = (y_1, \ldots, y_j - \delta, y_{j+1}, \ldots, y_{k-1}, y_k + \delta, y_{k+1}, \ldots, y_n) =$$
$$= \lambda y^n + (1 - \lambda)(y_1, \ldots, y_{j-1}, y_k, y_{j+1}, \ldots, y_{k-1}, y_j, y_{k+1}, \ldots, y_n).$$

Note that $\hat{y}^n \prec y^n$ and at least one of the following equalities is valid: $\hat{y}_j = x_j$, $\hat{y}_k = x_k$.

(e) From the previous item it follows that if the function $\varphi(z_1, z_2, \ldots, z_n)$ is Schur convex in any two components when the remaining $n - 2$ components are fixed, then it is Schur convex. Conclude from this that if $g : \mathbb{R} \to \mathbb{R}$ is convex, then

$$\varphi(z^n) = \sum_{i=1}^{n} g(z_i)$$

is Schur convex (if $x^n \prec y^n$ and x^n differs from y^n in two components $1, 2$, then $x_1 = \alpha y_1 + (1 - \alpha)y_2$, $x_2 = (1 - \alpha)y_1 + \alpha y_2$).

5. Show that

$$\min\{\kappa : (\varphi_1, \ldots, \varphi_n; \psi_1, \ldots, \psi_n) \text{ is } \kappa \times \mathcal{M}_{S \times T} - \text{expansive}\} = \underline{\sigma}\underline{\tau}.$$

6. Show that then for the operations

$$(\vee, \triangle) : \{0, 1\} \times \{0, 1\} \to \{0, 1\} = S$$

we have

$$\underline{\sigma}(S, \vee, \triangle) = \frac{1}{2}(1 + \sqrt{2}).$$

7. Show that $(\triangle, \wedge)$ is not $\mathcal{M}$-expansive
8. If $d(A)$ exists does then $D(G(A))$ exist?
9. Show that
$$|A||B| \le |[A, B]||(A, B)|.$$

10. The set theoretic operation $a \triangle b$ has a number theoretical correspondence $\frac{[a,b]}{(a,b)} = \frac{a}{(a,b)} \frac{b}{(a,b)}$. Establish inequalities for this operation.

Research Problems

1. **Conjecture** Let

$$a^{(i)} = a^{[i]} \setminus a^{[i+1]}, \ i = 1, 2, \ldots, m - 1,$$
$$a^{(m)} = a^{[m]}, \ a_i \in A_i.$$

The following generalization of the second AD inequality is valid:

$$\prod_{i=1}^{m}|A_i| \leq \prod_{i=1}^{m}|A^{(i)}|.$$

2. **Conjecture** (Generalization of Lemma 33 suggested by Rinott and Saks in [RS93]) Prove that if A, B are matrices with nonnegative entries satisfying (16) (Lecture 15), then **Per**$(A) \leq$ **Per**(B). This would include Muirhead's inequality (see [M1903], [MO79]).

3. One can easily derive by the approach in Lecture 15, Section 6 twins of the generalized AD inequality of Rinott and Saks. To avoid notation we formulate only the case $r = 3$.

The inequalities

$$\underline{D}A \cdot \underline{D}B \cdot \underline{D}C \leq \underline{D}[A,B,C] \cdot \underline{D}[(A,B),(A,C),(B,C)] \cdot \underline{D}(A,B,C)$$

and

$$\underline{D}A \cdot \underline{D}B \cdot \underline{D}C \leq (\underline{D}[A,B] \cdot \underline{D}(A,B) \cdot \underline{D}[A,C] \cdot \underline{D}(A,C) \cdot \underline{D}[B,C] \cdot \underline{D}(B,C))^2$$

are valid for arbitrary $A, B, C \subset \mathbb{N}$.

Conjecture The first upper bound is always at least as good as the second upper bound. If this is true, this gives an inequality for these bounds.

Chapter VI
Basic Problems from Combinatorial Number Theory

Lecture 16 Solutions of Problems of P. Erdös

§1 Definitions, Formulation of Problems, and Conjectures

Let $\mathbb{N}$ denote the set of natural numbers and $\mathbb{P} = \{2, 3, 5, \ldots\}$ the set of all primes. For two numbers $u, v \in \mathbb{N}$ we write $u|v$ iff u divides v, (u, v) stands for the largest common divisor of u and v and $[u, v]$ denotes the smallest common multiple of u and v. The numbers u and v are called coprime iff $(u, v) = 1$.

We are interested in the sets

$$\mathbb{N}_s = \left\{ u \in \mathbb{N} : \left(u, \prod_{i=1}^{s-1} p_i \right) = 1 \right\}$$

and

$$\mathbb{N}_s(n) = \mathbb{N}_s \cap [n].$$

Let $f(n, k, s)$ be the largest integer r for which an

$$A \subset \mathbb{N}_s(n), \quad |A| = r$$

exists with no $k + 1$ numbers in A being pairwise coprime. Certainly the set

$$\mathbb{E}(n, k, s) = \{ u \in \mathbb{N}_s(n) : u = p_{s+i} v \text{ for some } i = 0, \ldots, k - 1 \}$$

does not have $k + 1$ coprimes. The case $s = 1$, in which we have $\mathbb{N}_1(n) = [n]$, is of particular interest.

Conjecture 1

$$f(n, k, 1) = |\mathbb{E}(n, k, 1)| \quad \text{for all } n, k \in \mathbb{N}.$$

It seems that this conjecture of Erdös appeared for the first time in print in his paper [E62].

The papers [ESS69] and [ESS80] by Erdös, Sárközy, and Szemerédi and [ES92] by Erdös and Sárközy are centered around this problem. Whereas it is easy to show that the conjecture is true for $k = 1$ and $k = 2$, it was proved for $k = 3$ by Szabó and Tóth [ST85] only in 1985. Conjecture 1 can also be found in Section 3 of the survey [E73]. In the survey [E80] one finds the

General Conjecture

$$f(n,k,s) = |\mathbb{E}(n,k,s)| \quad \text{for all } n,k,s \in \mathbb{N}.$$

Erdös mentions in [E80] that he did not succeed in settling the case $k = 1$. We focus on this special case by calling it

Conjecture 2

$$f(n,1,s) = |\mathbb{E}(n,1,s)| \quad \text{for all } n,s \in \mathbb{N}.$$

Notice that

$$\mathbb{E}(n,1,s) = \{u \in \mathbb{N}_1(n) : p_s|u; p_1,\ldots,p_{s-1} \nmid u\}.$$

Our interest in these conjectures is motivated by an attempt to search for new combinatorial principles in this number theoretic environment. Consequently, we look for statements that do not depend on the actual distribution of primes. As we will see, these problems have some resemblance with the intersection problem. In fact, it gave ideas for solving the long-standing $4m$-conjecture of Erdös/Ko/Rado and marginally led to the Complete Intersection Theorem (Lecture 1). However, it turns out that the prime number distribution affects the behavior of Erdös' problem above in a very delicate way. We make use of refined estimates presented in the next section.

§2 Auxiliary Results: New Combinatorial and Known Number Theoretical Properties

The key tool in our proofs is a combinatorial result. For a subfamily $\mathcal{A} \subset \binom{[m]}{\ell}$, the lower shadow $\Delta\mathcal{A}$ is defined by

$$\Delta\mathcal{A} = \left\{ B \in \binom{[m]}{\ell-1} : B \subset A \text{ for some } A \in \mathcal{A} \right\}$$

and the upper shadow of $\mathcal{B} \subset \binom{[m]}{\ell-1}$ is

$$\nabla \mathcal{B} = \left\{ A \in \binom{[m]}{\ell} : B \subset A \text{ for some } B \in \mathcal{B} \right\}.$$

With any function $g : \mathcal{A} \to \mathbb{R}_+$ we associate the function $h : \Delta \mathcal{A} \to \mathbb{R}_+$, where $h(B) = \max_{A \in \nabla(B) \cap \mathcal{A}} g(A)$.

We formulate the following combinatorial

Theorem 54 (Ahlswede and Khachatrian 1995) *Let* $\mathcal{A} \subset \binom{[m]}{\ell}$ *have the property that no $k+1$ elements of $\mathcal{A}$ are disjoint. Then for any function $g : \mathcal{A} \to \mathbb{R}_+$ and its associated function $h : \Delta \mathcal{A} \to \mathbb{R}_+$*

$$\sum_{B \in \Delta \mathcal{A}} h(B) \geq \frac{1}{k} \sum_{A \in \mathcal{A}} g(A).$$

In particular,

$$|\Delta \mathcal{A}| \geq \frac{1}{k}|\mathcal{A}|.$$

We begin with a special case of Theorem 54.

Lemma 38 *Let* $\mathcal{A} \subset \binom{[m]}{\ell}$ *have the property that no $k+1$ of its members are pairwise disjoint. Then*

$$|\Delta \mathcal{A}| \geq \frac{1}{k}|\mathcal{A}|.$$

Proof. The left pushing operation preserves the "no $k+1$ disjoint"-property and only can decrease the shadow. We can assume therefore that $\mathcal{A}$ is left compressed. We distinguish two cases.

Case 1: $m \leq (k+1)\ell - 1$. Counting the pairs (A, B) with $B \subset A$ in two ways we get

$$|\Delta \mathcal{A}| \geq \frac{\ell}{m - \ell + 1}|\mathcal{A}| \geq \frac{\ell}{(k+1)\ell - 1 - \ell + 1}|\mathcal{A}| = \frac{1}{k}|\mathcal{A}|.$$

Case 2: $m \geq (k+1)\ell$. We consider the following partition of $[m]$:

$$I_1 = [k],$$
$$I_2 = [k+1, 2k+1], \ldots, I_j = [(j-1)(k+1), j(k+1) - 1], \ldots,$$
$$I_\ell = [(\ell-1)(k+1), \ell(k+1) - 1], I_{\ell+1} = [\ell(k+1), m].$$

First we show that for every $A \in \mathcal{A}$ there exists an index j, $1 \leq j \leq \ell$, for which

$$|A \cap (I_1 \cup \ldots \cup I_j)| = j. \tag{1}$$

To see this, assume that this does not hold for some $A \in \mathcal{A}$. Then, necessarily, $|A \cap I_{\ell+1}| \geq 1$, because otherwise $|A \cap (I_1 \cup \ldots \cup I_\ell)| = \ell$ since $|A| = \ell$. Therefore, we must have $|A \cap (I_1 \cup \ldots \cup I_\ell)| \leq \ell - 1$ and a fortiori $|A \cap (I_1 \cup \ldots \cup I_{\ell-1})| \leq \ell - 2,$

$|A \cap (I_1 \cup \ldots \cup I_{\ell-2})| \le \ell - 3, \ldots, |A \cap (I_1 \cup I_2)| \le 1, |A \cap I_1| = 0$. However, since $\mathcal{A}$ is also left compressed, we can choose $k + 1$ elements from $\mathcal{A}$ (including A) which are pairwise disjoint. This contradicts our assumption on $\mathcal{A}$.

Now, for every $A \in \mathcal{A}$ define j_A, $1 \le j_A \le \ell$, as the largest index j for which (1) holds. This can be used to partition $\mathcal{A}$ into disjoint subsets:

$$\mathcal{A} = \bigcup_{i=1}^{\ell} \mathcal{A}_i, \ \mathcal{A}_i = \{A \in \mathcal{A} : j_A = i\}. \tag{2}$$

Some of the subsets may be empty. Consider now the shadows $\Delta \mathcal{A}_i$, $1 \le i \le \ell$ and their subshadows $\Delta^* \mathcal{A}_i = \{B \in \Delta \mathcal{A}_i : \ |B \cap (I_1 \cup \ldots \cup I_i)| = i - 1\}$. It follows immediately from the definition of $\mathcal{A}_i$ that

$$\Delta^* \mathcal{A}_{i_1} \cap \Delta^* \mathcal{A}_{i_2} = \emptyset, \ i_1 \ne i_2. \tag{3}$$

Moreover, using the left compressedness of $\mathcal{A}$, it can be easily shown that

$$\Delta \mathcal{A} = \bigcup_{i=1}^{\ell} \Delta^* \mathcal{A}_i. \tag{4}$$

In the light of (2)–(4) it suffices to show that

$$|\Delta^* \mathcal{A}_i| \ge \frac{1}{k} |\mathcal{A}_i|, \ i = 1, \ldots, \ell. \tag{5}$$

We look therefore for fixed i at the intersections

$$\mathcal{U}_i = \{A \cap (I_1 \cup \ldots \cup I_i) : \ A \in \mathcal{A}_i\}$$

and partition $\mathcal{A}_i$ as follows:

$$\mathcal{A}_i = \bigcup_{U \in \mathcal{U}_i} \mathcal{A}_i^U, \ \mathcal{A}_i^U = \{A \in \mathcal{A}_i : \ A \cap (I_1 \cup \ldots \cup I_i) = U\}. \tag{6}$$

Also, introduce the intersections

$$\mathcal{V}_i = \{B \cap (I_1 \cup \ldots \cup I_i) : \ B \in \Delta^* \mathcal{A}_i\}$$

and partition $\Delta^* \mathcal{A}_i$ as follows:

$$\Delta^* \mathcal{A}_i = \bigcup_{V \in \mathcal{V}_i} (\Delta^* \mathcal{A}_i)^V, \tag{7}$$

$$(\Delta^* \mathcal{A}_i)^V = \{B \in \Delta^* \mathcal{A}_i : \ B \cap (I_1 \cup \ldots \cup I_i) = V\}.$$

Now counting for Δ^*-operation pairs again in two ways, we get the inequality

$$i \sum_{U \in \mathcal{U}_i} |\mathcal{A}_i^U| \le \sum_{V \in \mathcal{V}_i} (i(k+1) - 1 - (i-1)) |(\Delta^* \mathcal{A}_i)^V| \le ik \sum_{V \in \mathcal{V}_i} |(\Delta^* \mathcal{A}_i)^V|.$$

Together with (6) and (7) it implies (5). $\square$

The next result enables us to get Theorem 54 from Lemma 38. Let $G = (V, W, E)$ be a bipartite graph. Write $\sigma(s)$ for the set of vertices adjacent to vertex s, and $\sigma(S)$ for the set of vertices adjacent to vertices in S. We assume that

$$\sigma(V) = W.$$

Lemma 39 *Suppose that for some $\alpha \in \mathbb{R}_+$ we have that*

$$|S| \le \alpha |\sigma(S)| \text{ for every } S \subset V. \tag{8}$$

Then for every function $g \in V \to \mathbb{R}_+$ and the associated function $h : W \to \mathbb{R}_+$ defined by $h(b) = \max_{a \in \sigma(b)} g(a)$ for all $b \in W$ we have

$$\sum_{a \in V} g(a) \le \alpha \sum_{b \in W} h(b). \tag{9}$$

Proof. Let $\{\gamma_1 < \ldots < \gamma_r\}$ be the range of g. Then we have the partition $V = \bigcup_{i=1}^{r} V_r$, where

$$V_i = \{v \in V : g(v) = \gamma_i\}, \ 1 \le i \le r.$$

Clearly,

$$\sum_{a \in V} g(a) = \sum_{i=1}^{r} \gamma_i |V_i|. \tag{10}$$

By the definition of h, obviously

$$h(b) = \gamma_r, \ b \in \sigma(V_r). \tag{11}$$

We now proceed by induction on r.

$r = 1$: Here $h(b) = \gamma_1$ for all $b \in W$ and hence by (8),

$$\sum_{a \in V} g(a) = \gamma_1 |V| \le \gamma_1 \alpha |W| = \alpha \sum_{b \in W} h(b).$$

$r - 1 \to r$: We assume that (9) holds for every function $g' : V \to \mathbb{R}_+$ with $r - 1$ different values. With our g under consideration we associate the function $g^* : V \to \mathbb{R}_+$ defined by

$$g^*(a) = \begin{cases} \gamma_i, & a \in V_i, \ i \le r - 1, \\ \gamma_{r-1}, & a \in V_r. \end{cases}$$

Denote by $h^* : W \to \mathbb{R}_+$ the usual function associated with g^*. We verify that

$$\sum_{a \in V} g(a) = \sum_{a \in V} g^*(a) + (\gamma_r - \gamma_{r-1}) |V_r|, \tag{12}$$

$$\sum_{b \in W} h(b) = \sum_{b \in W} h^*(b) + (\gamma_r - \gamma_{r-1}) |\sigma(V_r)|. \tag{13}$$

From condition (8) and the induction hypothesis applied to g^* we know that

$$|V_r| \le \alpha |\sigma(V_r)| \text{ and } \sum_{a \in V} g^*(a) \le \alpha \sum_{b \in W} h^*(b).$$

these inequalities and (12), (13) give (9). $\square$

Proof of Theorem 54. Consider $G = (V, W, E) = (\mathcal{A}, \Delta\mathcal{A}, E)$, where $(A, B) \in E$ iff $A \supset B$, and $\mathcal{A}$ satisfies the hypothesis of Theorem 54 and hence also of Lemma 38. Since every subfamily $\mathcal{A}' \subset \mathcal{A}$ also satisfies this hypothesis, we know that

$$|\nabla(\mathcal{A}')| \geq \frac{1}{k}|\mathcal{A}'|. \tag{14}$$

Since $\nabla(\mathcal{A}') = \sigma(\mathcal{A}')$, (14) guarantees (8) for $\alpha = k$.

The conclusion (9) says now

$$\sum_{A \in \mathcal{A}} g(A) \leq k \sum_{A \in \Delta\mathcal{A}} h(A).$$

this completes the proof of Theorem 54.   ∎

Now suppose we have two counters at time 0: $x_0 = x$ and $y_0 = y$ where $x, y \in \mathbb{R}_+$. At any step $i \geq 1$, we arbitrarily remove a_i, b_i with $0 \leq a_i \leq x_{i-1}$, $0 \leq b_i \leq y_{i-1}$, and add $a_i^* \geq 0$, $b_i^* \geq 0$, where

$$a_i^* + b_i^* > \beta(a_i + b_i), \ \beta > 1.$$

the new counter values are

$$x_i = x_{i-1} - a_i + a_i^*, \ y_i = y_{i-1} - b_i + b_i^*.$$

Lemma 40 *If for some $\ell \in \mathbb{N}$ the event $y_\ell = 0$ (resp. $x_\ell = 0$) occurs, then we have $x_\ell > x + \beta y$ (resp. $y_\ell > y + \beta x$).*

Proving this lemma is the task of Exercise 1.

We will also need the following number theoretical properties.

Theorem 55 (Rosser and Schoenfeld 1962) *For the function*

$$\varphi(x, y) = \left| \left\{ a \leq x : \left(a, \prod_{p < y} p \right) = 1 \right\} \right|$$

there exist positive absolute constants c_1, c_2 such that

$$c_1 x \prod_{p < y} \left(1 - \frac{1}{p} \right) \leq \varphi(x, y) \leq c_2 x \prod_{p < y} \left(1 - \frac{1}{p} \right) \tag{15}$$

for all x, y satisfying $x \geq 2y \geq 4$. Furthermore, the RHS inequality in (15) remains valid also for $x < 2y$.

We are not going to give the proof of this theorem here. The reader can find it in [HR66] on pages 200–204.

Proposition 15 *For positive constants c_1, c_2, κ there exists a $t(c_1, c_2, \kappa)$ such that for $t > t(c_1, c_2, \kappa)$,*

$$\frac{c_1}{c_2}p_t \prod_{p<p_t}\left(1-\frac{1}{p}\right) > \kappa.$$

This is asked to be proved in Exercise 2.

§3 Maximal Sets Without $k+1$ Coprimes

It came as a surprise when Ahlswede and Khachatrian disproved Conjecture 1 of Erdös. Their counter example can be found in §6 below. However, they were able to prove the conjecture when, for given k, n is sufficiently large. For simplicity we make the convention $f(n,k) = f(n,k,1)$ and $|\mathbb{E}(n,k)| = |\mathbb{E}(n,k,1)|$

Theorem 56 (Ahlswede and Khachatrian 1995) *For every $k \in \mathbb{N}$ there is an $n(k)$ such that $f(n,k) = |\mathbb{E}(n,k)|$ for all $n > n(k)$ and the optimal set is unique.*

Inspection of the methods and proofs show that they apply also to the General Conjecture. Only some extra notation is needed and the proof is left to the reader. The result is as follows.

Theorem 57 (Ahlswede and Khachatrian 1995) *For every $k,s \in \mathbb{N}$ there exists an $n(k,s)$ such that for all $n \geq n(k,s)$,*

$$|\mathbb{E}(n,k,s)| = f(n,k,s).$$

We give first a few concepts from Number Theory and state the consequence of Theorem 56.

Define the upper and lower asymptotic densities

$$\overline{d}A = \limsup_{n\to\infty}\frac{|A\cap[n]|}{n}, \ \underline{d}A = \liminf_{n\to\infty}\frac{|A\cap[n]|}{n}.$$

In the case $\overline{d}A = \underline{d}A$ we write $dA = \overline{d}A = \underline{d}A$ for the asymptotic density.

Let $M(A)$ be the set of multiples of A. The set of nonmultiples of A we denote by

$$N(A) = \mathbb{N}\setminus M(A).$$

Thus $\mathbb{E}(n,k) = \mathbb{E}(n,k,1) = M(\{p_1,\ldots,p_k\}) \cap [n]$ and also for any finite $A = \{a_1,\ldots,a_t\} \subset \mathbb{N}$ and $a = \prod_{i=1}^t a_i$, $N(A)\cap[a]$ is the set of integers in $[a]$ not divisible by any member of A. It is easy to see that

$$|N(A)\cap[a]| = a\prod_{i=1}^{t}\left(1-\frac{1}{a_i}\right)$$

if the elements of A are pairwise relatively prime.

For general A, by inclusion–exclusion,

$$|N(A)\cap[a]| = a\left(1-\sum_{i=1}^{t}\frac{1}{a_i}+\sum_{i<j}\frac{1}{[a_i,a_j]}-\cdots\right)$$

and therefore

$$dN(A) = 1 - \sum_{i=1}^{t} \frac{1}{a_i} + \sum_{i<j} \frac{1}{[a_i, a_j]} - \cdots. \tag{16}$$

We introduce the family $\mathcal{S}(n,k,s)$ of all subsets of $\mathbb{N}_s(n)$ no $k+1$ elements of which are pairwise relatively prime. In case $s = 1$ we also write $\mathcal{S}(n,k)$ and $\mathcal{S}(\infty,k)$ in the unrestricted case $n = \infty$.

Let us mention that Theorem 56 implies

$$\sup_{A \in \mathcal{S}(\infty,k)} \overline{d}A = \sup_{A \in \mathcal{S}(\infty,k)} \underline{d}A = \sup_{A \in \mathcal{S}(\infty,k)} dA.$$

Reduction to Left-Compressed Sets

The operation "pushing to the left" is frequently used in Extremal Set Theory, but to our surprise seems not to be as popular in Combinatorial Number Theory, perhaps because its usefulness is less obvious. Anyhow, our first (but not only) idea is to exploit it.

We need the following definition and result. The set $A \subset \mathbb{N}_s$ is said to be left compressed if for any $a \in A$ of the form

$$a = p_r^i a_1, \ (a_1, p_r) = 1,$$

and any p_ℓ of the form

$$p_s \leq p_\ell < p_r, \ (p_\ell, a_1) = 1,$$

it follows that $a^* = p_\ell^i a_1 \in A$ as well.

For any $n \in \mathbb{N} \cup \{\infty\}$ we denote the family of all left compressed sets from $\mathcal{S}(n,k,s)$ by $\mathcal{C}(n,k,s)$. The next lemma shows that the family of optimal sets from $\mathcal{S}(n,k,s)$ has a nonempty intersection with $\mathcal{C}(n,k,s)$.

Lemma 41 *For* $n \in \mathbb{N}$,

$$\max_{A \in \mathcal{S}(n,k,s)} |A| = \max_{A \in \mathcal{C}(n,k,s)} |A| = f(n,k,s).$$

Proof. For any $A \in \mathcal{S}(n,k,s)$ and $p_s \leq p_\ell < p_r$ we consider the partition of A

$$A = A^1 \cup A^0,$$

where

$$A^1 = \{a \in A : a = p_r^i a_1 \ (i \geq 1), \ (a_1, p_r p_\ell) = 1; \ p_\ell^i a_1 \notin A\},$$
$$A^0 = A \setminus A^1.$$

Define $A_*^1 = \{u \in \mathbb{N}_s : u = p_\ell^i a_1, \ (a_1, p_\ell, p_r) = 1, \ p_r^i a_1 \in A^1\}$ and notice that by definitions $A_*^1 \subset \mathbb{N}_s(n)$. Consider now $A^* = (A \cup A_*^1) \setminus A^1$ and observe that $|A^*| = |A|$ and also that $A^* \in \mathcal{S}(n,k,s)$.

Finitely many iterations of this procedure (which is called left pushing operation) to primes $p_s \leq p_\ell < p_r$ give the result. $\qquad\square$

Moreover, by countably many left pushing operations one can transform every $A \in \mathcal{S}(\infty, k)$ into a left compressed set A' such that

$$|A_n| \leq |A'_n|$$

and therefore also

$$\underline{d}A \leq \underline{d}A', \ \overline{d}A \leq \overline{d}A'.$$

For the left compressed sets $\mathcal{C}(\infty, k)$ in $\mathcal{S}(\infty, k)$ we have shown the following.

Lemma 42

$$\sup_{B \in \mathcal{S}(\infty, k)} \underline{d}B = \sup_{B \in \mathcal{C}(\infty, k)} \underline{d}B$$

and

$$\sup_{B \in \mathcal{S}(\infty, k)} \overline{d}B = \sup_{B \in \mathcal{C}(\infty, k)} \overline{d}B.$$

Any optimal $B \in \mathcal{S}(n, k, s)$, that is, $|B| = f(n, k, s)$, is an "upset":

$$B = M(B) \cap \mathbb{N}_s(n)$$

and it is also a "downset" in the following sense:

$$b \in B, \ b = q_1^{\alpha_1} \ldots q_t^{\alpha_t}, \ \alpha_i \geq 1 \Rightarrow b' = q_1 \ldots q_t \in B. \tag{17}$$

For any $B \in \mathbb{N}$ the unique primitive subset $P(B)$ has the properties

$$b_1, b_2 \in P(B) \Rightarrow b_1 \nmid b_2, \ B \subset M(P(B)).$$

We know from (17) that for an optimal $B \in \mathcal{S}(n, k, s)$, $P(B)$ consists only of square-free integers.

We could use also the following concept of left compressedness: $A \subset \mathbb{N}_s$ is left compressed iff for any $a \in A$ of the form

$$a = p_i^{\alpha_i} a_1, \ a_i \geq 1, \ (a_1, p_i) = 1,$$

we have that for any p_j, $p_s \leq p_j < p_i$, in the case $\alpha_i \geq 2$,

$$a^* = p_j p_i^{\alpha_i - 1} a_1 \in A,$$

and in the case $\alpha_i = 1$,

$$a^* = p_j a_1 \in A \text{ if } (a_1, p_j) = 1.$$

Although the two definitions are different in general, it can be easily seen that when the considered set $A \subset \mathbb{N}_s$ is an "upset" and a "downset," both definitions of left compressedness coincide.

The number theoretic properties Theorem 55, Proposition 15, and Lemma 40 in §2 and Lemma 44 in this section are the technical ingredients of the proof of Theorem 56. Lemma 43 uses Theorem 54 to prove that the number of elements in $A \subset \mathcal{S}(n,k)$, which have a divisor from some (sufficiently large) initial interval of primes, in some sense, dominates the number $|A|$. The meaning of this domination will be clear later (Lemma 43).

Next we will make, in some sense, a similar thing as we make in the proof of the Complete Intersection Theorem, but for a set of integers: we take some integers from the primitive set of $A \in \mathcal{S}(n,k)$ (the definition of the primitive set can be found after Lemma 42), say $B \subset A$ which have the largest prime divisor $p > p_k$ and delete this divisor from the prime number decomposition of these integers. From the obtained subset we choose a sufficiently large set C, which has a common prime divisor (this divisor is less than the largest prime) and add the set of multiples of C to A instead of multiples of B. Proposition 16 helps to establish the fact that this procedure under some conditions increases $|A|$. To prove this we also use Lemmas 40 and 43. The obtained contradiction of maximality of $A \in \mathcal{S}(n,k)$ proves that the primitive set has prime divisors $p \leq p_k$ only.

§4 Proof of the Main Result

Now we turn to the proof of Theorem 56. We set $s = 1$ and introduce

$$\mathcal{O}(n,k) = \{B \in \mathcal{S}(n,k) : |B| = f(n,k)\}.$$

We know that for $A \in \mathcal{O}(n,k)$ we have properties (I) :

$$P(A) \subset \mathbb{N}^*,$$
$$A = M(P(A)) \cap \mathbb{N}(n),$$
$$\mathcal{O}(n,k) \cap \mathcal{C}(n,k) \neq \emptyset.$$

We now present first another auxiliary result. For every $S \in \mathcal{C}(n,k)$ with properties (I), S need not be optimal, that is, it can be in $\mathcal{C}(n,k) \setminus \mathcal{O}(n,k)$. Define

$$S_i = \{d \in S : p_i | d, \ (p_1, \ldots, p_{i-1}, d) = 1\}.$$

Clearly

$$S_i \cap S_j = \emptyset \ (i \neq j), \ S = \bigcup_{i \geq 1} S_i. \tag{18}$$

Lemma 43 *For every $k, n \in \mathbb{N}$ and every $S \in \mathcal{C}(n,k)$ with properties (I) we have*

(i) $|S_r| \geq \frac{1}{k} \sum_{i \geq r+1} |S_i|$ for every $r \in \mathbb{N}$,
(ii) for every $\alpha \in \mathbb{R}_+$ and for $k(\alpha) \geq k\alpha$ (independent of n)

$$\sum_{i=1}^{k(\alpha)} |S_{k+i}| \geq \alpha \sum_{j \geq k+k(\alpha)+1} |S_j|.$$

Proof. (ii) follows from (i), so we have to prove (i). We consider the set $\bigcup_{i \geq r+1} S_i$ and let, for every $\ell \in \mathbb{N}$,

$$T_\ell = \left\{ d \in \bigcup_{i \geq r+1} S_i : d \text{ has exactly } \ell \text{ different primes in factorization} \right\}. \quad (19)$$

Obviously,

$$\bigcup_{i \geq r+1} S_i = \bigcup_{\ell \geq 1} T_\ell \quad (20)$$

and for $d \in T_\ell$,

$$d = q_1^{\beta_1} \ldots q_\ell^{\beta_\ell}, \ p_r < q_1 < \ldots < q_\ell, \beta_i \geq 1. \quad (21)$$

Since $S \in C(n,k)$, we have

$$d_i = p_r^{\beta_i} q_1^{\beta_1} \ldots q_{i-1}^{\beta_{i-1}} q_{i+1}^{\beta_{i+1}} \ldots q_\ell^{\beta_\ell} \in S_r, \ i = 1, \ldots, \ell. \quad (22)$$

Define

$$\sigma(d) = \{d_1, \ldots, d_\ell\}, \ \sigma(T_\ell) = \bigcup_{d \in T_\ell} \sigma(d).$$

As $\sigma(T_\ell) \subset S_r$ and $\sigma(T_\ell) \cap \sigma(T_{\ell'}) = \emptyset$, $\ell \neq \ell'$, to prove (i) it is sufficient to show that

$$|\sigma(T_\ell)| \geq \frac{1}{k}|T_\ell|, \ \ell \in \mathbb{N}. \quad (23)$$

Let $T_\ell^* = T_\ell \cap \mathbb{N}^*$ be the square-free integers in T_ℓ. Then $\sigma(T_\ell^*) = \bigcup_{d \in T_\ell^*} \sigma(d)$ is the set of all square-free integers in $\sigma(T_\ell)$.

For an $a \in T_\ell^*$, $a = x_1 \ldots x_\ell, x_1 < \ldots x_\ell, x_i \in \mathbb{P}$, we consider

$$T(a) = \{d \in S : d = x_1^{\beta_1} \ldots x_\ell^{\beta_\ell}, \beta_i \geq 1\}$$

and for a $b \in \sigma(T_\ell^*)$, $b = p_r y_1 \ldots y_{\ell-1}, \ p_r < y_1 < \ldots < y_{\ell-1}, y_i \in \mathbb{P}$, we consider

$$U(b) = \{d \in S_r : d = p_r^{\gamma_r} y_1^{\gamma_1} \ldots y_{\ell-1}^{\gamma_{\ell-1}}, \gamma_i \geq 1,$$
$$y_1^{\gamma_1} \ldots y_{\ell-1}^{\gamma_{\ell-1}} x^{\gamma_\ell} \in T_\ell \text{ for some } x \in \mathbb{P}\}.$$

It is evident that

$$T_\ell = \bigcup_{a \in T_\ell^*} T(a), \ \sigma(T_\ell) = \bigcup_{b \in \sigma(T_\ell^*)} U(b)$$

are partitions. Next observe that for any $b \in \sigma(T_\ell^*)$,

$$|U(b)| = \max_{bx/p_r \in T_\ell^*} \left| T\left(\frac{bx}{p_r}\right) \right| \quad (24)$$

and this enables us to apply Theorem 54 to the sets $\mathcal{A} \sim T_\ell^*$ and $\Delta\mathcal{A} \sim \sigma(T_\ell^*)$, where $\sim$ is the canonical correspondence between square-free numbers and subsets. We indicate the correspondence by using small and capital letters such as $a \sim A$.

We define $g : \mathcal{A} \to \mathbb{R}_+$ by

$$g(A) = |T(a)|.$$

The associated function $h : \Delta\mathcal{A} \to \mathbb{R}_+$ is defined by $h(B) = |U(b)|$. We see from (24) that this definition is correct. Theorem 54 therefore yields (23) and thus (i).   □

Now we are ready to prove Theorem 56. Assume that $S \in \mathcal{C}(n,k) \cap \mathcal{O}(n,k)$ has properties (I). Notice that $P(S) = P(S \cap \mathbb{N}^*)$. Equivalent to Theorem 56 is the assertion that for large n always

$$\bigcup_{i \geq k+1} S_i = \emptyset. \tag{25}$$

Henceforth we assume to the contrary that

$$\bigcup_{i \geq k+1} S_i \neq \emptyset \tag{26}$$

for infinitely many n. Let $k_0 \in \mathbb{N}$, $k_0 > k$ be an integer to be specified later. By the disjointness property (18) we can write

$$S^0 = S \setminus \left(\bigcup_{i \geq k_0+1} S_i \right) = \left(\bigcup_{i=1}^{k} S_i \right) \cup \left(\bigcup_{i=k+1}^{k_0} S_i \right). \tag{27}$$

From (i) in Lemma 43 we know that

$$\left| \bigcup_{i=k+1}^{k_0} S_i \right| \geq \frac{k_0 - k}{k} \left| \bigcup_{i \geq k_0+1} S_i \right|$$

and hence also that

$$|S| \leq \left| \bigcup_{i=1}^{k} S_i \right| + \gamma \left| \bigcup_{i=k+1}^{k_0} S_i \right|, \tag{28}$$

where $\gamma = 1 + k/(k_0 - k)$.

Let $P(S^0)$ be the primitive subset of S^0, which generates S^0. We notice that by the properties of S,

$$P(S^0) \subset P(S), \tag{29}$$

because $d' \in P(S^0)$ and $d | d'$ for some $d \in S$ would by compressedness imply the existence of an $e' \in P(S^0)$ with $e' | d'$.

Let p_t be the largest prime occurring in any element of $P(S^0)$. In other words, $(p_t, d) = p_t$ for some $d \in P(S^0)$ and $(p_{t'}, d) = 1$ for all $t' > t$ and all $d \in P(S^0)$. By assumption (II) we have $p_t > p_k$.

Assume now $A \in \mathcal{O}(n,k) \cap \mathcal{C}(n,k)$ and $P(A) = \{a_1, \ldots, a_m\}$ is a primitive set, where a_i's are written in colexicographic order. It means that if $a_{m_1} =$

$p_1 p_2 \cdots p_{r_1}, a_{m_2} = q_1 \ldots q_{r_2}$ are the prime number decompositions of a_{m_1} and a_{m_2}, then $a_{m_1} < a_{m_2}$ iff either $p_{r_1} < q_{r_2}$ or $p_{r_1} = q_{r_2}$ and $p_{r_1-1} < q_{r_2-1}$ or $p_{r_1} = q_{r_2}$, $p_{r_1-1} < q_{r_2-1}$ and $p_{r_1-2} < q_{r_2-2}$ or... $p_i = q_i$, $i = 2, \ldots, r_1 - 1 = r_2 - 1$ and $p_1 < q_1$.

The set of multiples of $P(A)$ in $\mathbb{N}(n)$ can be written as a union of disjoint sets $B^i(n)$:

$$M(P(A)) \cap \mathbb{N}(n) = \cup B^i(n), \tag{30}$$

$$B^i(n) = \{x \in M(P(A)) \cap \mathbb{N}(n) : a_i | x, \ a_j \nmid x \text{ for } j = 1, \ldots, i-1\}. \tag{31}$$

As before, if $a_i = p_{j_1} \ldots p_{j_\ell}$ with $j_1 < \ldots < j_\ell$, then

$$B^i(n) = \left\{ x \in \mathbb{N}(n) : x = p_{j_1}^{\alpha_1} \ldots p_{j_\ell}^{\alpha_\ell} T, \ \alpha_i \geq 1, \ \left(T \prod_{p_i \leq p_{j_\ell}} p_i \right) = 1 \right\}.$$

Lemma 44 *Let* $a_i = q_1 \ldots q_r$ *and* $q_1 < \ldots, q_r$ *with* $q_j \in \mathbb{P}$. *Then*

(i)

$$B^i(n) = \{u \in \mathbb{N}(n) : a_i | u \text{ and } a_j \nmid u \text{ for } j = 1, \ldots, i-1\}.$$

(ii)

$$\lim_{n \to \infty} \frac{|B^i(n)|}{n} = \frac{1}{(q_1 - 1) \ldots (q_r - 1)} \prod_{p \leq q_r} \left(1 - \frac{1}{p}\right).$$

(iii) *For every* $\varepsilon > 0$, *and every* $a_i = q_1 \ldots q_r$, *there exists an* $n(\varepsilon)$ *such that for* $n > n(\varepsilon)$ *we have*

$$(1 - \varepsilon) \frac{n}{(q_1 - 1) \ldots (q_r - 1)} \prod_{p \leq q_r} \left(1 - \frac{1}{p}\right) < |B^i(n)|$$

$$< (1 + \varepsilon) \frac{n}{(q_1 - 1) \ldots (q_r - 1)} \prod_{p \leq q_r} \left(1 - \frac{1}{p}\right).$$

Proof.

(i) follows from the facts that A is an "upset," a "downset" and compressed.
(ii) We know that for $m \in \mathbb{N}$

$$d\mathbb{N}_m = \prod_{p \leq p_m} \left(1 - \frac{1}{p}\right)$$

and hence

$$\lim_{n \to \infty} \frac{|B^i(n)|}{n} = \sum_{\alpha_i \geq 1} \frac{1}{q_1^{\alpha_1} \ldots q_r^{\alpha_r}} \prod_{p \leq q_r} \left(1 - \frac{1}{p}\right)$$

$$= \frac{1}{(q_1 - 1) \ldots (q_r - 1)} \prod_{p \leq q_r} \left(1 - \frac{1}{p}\right).$$

(iii) follows from (ii), since the constant number of sequences converges uniformly.

$\square$

We now consider

$$P^t(S^0) = \{a \in P(S^0) : (a, p_t) = p_t\}.$$

We know that the contribution of every element $a \in P^t(S^0)$, $a = q_1 \ldots q_r p_t$ and $q_1 < \ldots < q_r < p_t$, to $M(P(S^0)) \cap \mathbb{N}(n)$ is the set of integers

$$B(a) = \left\{u = q_1^{\alpha_1} \ldots q_r^{\alpha_r} p_t^\beta Q \in \mathbb{N}(n) : a_i \geq 1, \ \beta \geq 1, \ \left(Q, \prod_{p \leq p_t} p\right) = 1\right\}. \quad (32)$$

We use the abbreviation

$$L_t = \bigcup_{a \in P^t(S^0)} B(a).$$

We also consider the partition

$$P^t(S^0) = \bigcup_{1 \leq i \leq k_0} P_i^t(S^0), \ P_i^t(S^0) = P^t(S^0) \cap S_i.$$

By the pigeon-hole principle, for some $\ell, 1 \leq \ell \leq k_0$,

$$\left| \bigcup_{a \in P_\ell^t(S^0)} B(a) \right| \geq \frac{|L_t|}{k_0}, \ t > k_0 \quad (33)$$

and for some $\ell, 1 \leq \ell \leq t - 1$,

$$\left| \bigcup_{a \in P_\ell^t(S^0)} B(a) \right| \geq \frac{|L_t|}{t - 1}, \ k < t \leq k_0. \quad (34)$$

We consider for this ℓ corresponding to t, the set (of square-free numbers)

$$\tilde{P}(S^0) = (P(S^0) \setminus P^t(S^0)) \cup R_\ell^t(S^0),$$

with

$$R_\ell^t(S^0) = \{u \in \mathbb{N} : up_t \in P_\ell^t(S^0)\}.$$

It can happen that $\tilde{P}(S^0)$ is not primitive, however, always $\tilde{P}(S^0) \subset S(n, k)$. Now we state the main result for $\tilde{P}(S^0)$.

Proposition 16 *For suitable* $n > n(k)$,

$$\left| \bigcup_{a \in R_\ell^t(S^0)} D(a) \right| > \gamma |L_t|, \quad (35)$$

where for an $a \in R_\ell^t(S^0)$, $a = q_1 \ldots q_r$, $q_1 < \ldots < q_r, p_t$, *we consider the set*

$$D(a) = \left\{ v = q_1^{\alpha_1} \ldots q_r^{\alpha_r} T_1 \in \mathbb{N}(n) : \alpha_i \geq 1, \left(T_1, \prod_{p \leq p_{t-1}} p \right) = 1 \right\}.$$

Proof. Since p_t was the largest prime, which occurred in $P(S^0)$, we observe that

$$M(P(S^0) \setminus P^t(S^0)) \cap D(a) = \emptyset \tag{36}$$

for all $a \in R_\ell^t(S^0)$. Moreover,

$$D(a) \cap D(a') = \emptyset$$

for $a, a' \in R_\ell^t(S^0)$, $a \neq a'$.

Hence, taking into account (33) and (34), to show (35) it is sufficient to prove that for $n > n(k)$ and

$$B(ap_t) = \left\{ u = q_1^{\alpha_1} \ldots q_r^{\alpha_r} p_t^\beta T \in \mathbb{N}(n) : \alpha_i \geq 1, \beta \geq 1, (T, \Pi_{p \leq p_t} p) = 1 \right\},$$

we have

$$|D(a)| > \begin{cases} \gamma k_0 |B(ap_t)|, & t > k_0, \\ \gamma(t-1)|B(ap_t)|, & t \leq k_0. \end{cases} \tag{37}$$

To prove (37), we consider three cases. We always have $a = q_1 \ldots q_r$, $q_1 < \ldots < q_r < p_t$.

Case 1: $n/(ap_t) \geq 2$ and $t > t(c_1, c_2, k_0)$. Using the RHS of (15), which is valid without restrictions, we get

$$|B(ap_t)| \leq c_2 \sum_{\alpha_i \geq 1, \beta \geq 1} \frac{n}{q_1^{\alpha_1} \ldots q_r^{\alpha_r} p_t^\beta} \prod_{p \leq p_t} \left(1 - \frac{1}{p} \right) \tag{38}$$

$$< c_2 \frac{n}{(q_1 - 1) \ldots (q_r - 1)} \prod_{p \leq p_t} \left(1 - \frac{1}{p} \right) \frac{1}{p_t - 1}.$$

For $D(a)$ we have

$$D(a) \supset D'(a) = \left\{ u = q_1 \ldots q_r T_1 \in \mathbb{N}(n) : \left(T_1, \prod_{p \leq p_{t-1}} p \right) = 1 \right\},$$

and since $n/(q_1 \ldots q_r) \geq 2p_t$, we can apply the LHS of (15) to get

$$|D(a)| > |D'(a)| \geq c_1 \frac{n}{q_1 \ldots q_r} \prod_{p \leq p_{t-1}} \left(1 - \frac{1}{p} \right) \tag{39}$$

$$= c_1 \frac{n}{q_1 \ldots q_r} \frac{p_t}{p_t - 1} \prod_{p \leq p_t} \left(1 - \frac{1}{p} \right).$$

Comparing (38) and (39) we get

$$\frac{|D(a)|}{|B(ap_t)|} > \frac{c_1}{c_2} p_t \frac{(q_1-1)\dots(q_r-1)}{q_1\dots q_r}$$

$$\geq \frac{c_1}{c_2} p_t \prod_{p<p_t}\left(1-\frac{1}{p}\right) > \kappa = \gamma k_0,$$

where in the last step we used Proposition 15. Thus we proved (37) in this case.

Case 2: $n/(ap_t) \geq 2$, $t < t(c_c, c_2, k_0)$. First let us specify k_0 and hence γ. We choose k_0 so large that

$$p_{k+i} > \gamma(k+i-1) = \left(1+\frac{k}{k_0-k}\right)(k+i-1), \ i \in \mathbb{N}. \tag{40}$$

Next we choose $\varepsilon > 0$ to guarantee

$$p_{k+i}\frac{1-\varepsilon}{1+\varepsilon} > \gamma(k+i-1). \tag{41}$$

Let $n(\varepsilon)$ be a positive integer so that for $n > n(\varepsilon)$ we can apply lemma 44 (iii). Then we have

$$|B(ap_t)| < (1+\varepsilon)\frac{n}{(q_1-1)\dots(q_r-1)(p_t-1)}\prod_{p\leq p_t}\left(1-\frac{1}{p}\right),$$

$$|D(a)| > (1-\varepsilon)\frac{n}{(q_1-1)\dots(q_r-1)}\prod_{p\leq p_{t-1}}\left(1-\frac{1}{p}\right)$$

$$= (1-\varepsilon)\frac{n}{(q_1-1)\dots(q_r-1)}\frac{p_t}{p_t-1}\prod_{p\leq p_t}\left(1-\frac{1}{p}\right),$$

and hence, by (41),

$$\frac{|D(a)|}{|B(ap_t)|} > \frac{1-\varepsilon}{1+\varepsilon}p_t > \gamma(t-1).$$

This establishes (37) in this case.

Case 3: $1 \leq n/(ap_t) < 2$. In this case $B(ap_t)$ consists of only one element, namely $q_1,\dots q_r p_t$. Let now $t_1 \in \mathbb{N}$ satisfy

$$p_{t_1} > (p_{k_0})^{\lceil \gamma k_0\rceil} \tag{42}$$

and let

$$n > \prod_{p\leq p_{t_1}} p. \tag{43}$$

Notice that in our case necessarily $p_t \geq p_{t_1}$, because $ap_t \leq \prod_{p\leq p_t} p$ and $p_{t_1} > p_t$ would imply

$$2ap_t < 2\prod_{p\leq p_t} p < \prod_{p\leq p_{t_1}} p < n$$

and this contradicts our case $2ap_t > n$.

Now by (42), $p_t \geq p_{t_1} > (p_{k_0})^{\lceil \gamma k_0 \rceil}$ and since $q_1 \leq p_{k_0}$, we get $q_1^{\lceil \gamma k_0 \rceil} < p_t$. Therefore

$$D(a) \supset \{q_1 \ldots q_r, \ q_1^2 q_2 \ldots q_r, \ldots, q_1^{\lceil \gamma k_0 \rceil} q_2 \ldots q_r, \ q_1 q_2 \ldots q_r p_t\},$$

$|D(a)| > \lceil \gamma k_0 \rceil$, and again (37) holds. k_0, γ, and ε are already fixed and depend only on k. Then for

$$n(k) = \max \left\{ \prod_{p \leq (p_{k_0})^{\lceil \gamma k_0 \rceil}} p, n(\varepsilon) \right\} \tag{44}$$

and $n > n(k)$, (37) holds in all three cases and the proof of Proposition 16 is complete. $\qquad \square$

We have already noticed that $\tilde{P}(S^0)$ may not be primitive. Moreover, $M(\tilde{P}(S^0))$ may even not be left compressed.

Let now $S^1 \subset \mathbb{N}(n)$ be any set that is obtained from $M(\tilde{P}(S^0))$ by left pushing and is left compressed. We know that

$$S^1 \subset \mathcal{C}(n,k), \ |S^1| \geq |M(\tilde{P}(S^0)) \cap \mathbb{N}(n)| \tag{45}$$

and we know from the Proposition 16 that

$$\left| \bigcup_{a \in R_\ell^t(S^0)} D(a) \right| > \gamma |L_t|. \tag{46}$$

We notice that $\left(a, \prod_{p \leq p_{k_0}} p \right) > 1$ for every $a \in S^1$ and the last prime p_{t^1}, which occurs as a factor of any primitive element of $P(S^1)$ is less than p_t.

If $S^1 \not\subset \mathbb{E}(n,k)$, then we repeat the whole procedure and get an S^2, for which

$$\left| \bigcup_{a \in R_{\ell_1}^{t^1}(S^0)} D(a) \right| > \gamma |L_{t^1}|.$$

Here L_{t^1} is defined analogously to L_t with respect to the largest prime p_{t^1} occurring in a member of $P(S^1)$.

By iteration we get an $S^i \subset \mathcal{C}(n,k)$ with

$$\left| \bigcup_{a \in R_{\ell_{i-1}}^{t^{i-1}}(S^0)} D(a) \right| > \gamma |L_{t^{i-1}}|. \tag{47}$$

and again, in analogy with the first step, we define S_j^i and the partition

$$S^i = \left(\bigcup_{j=1}^{k} S_j^i \right) \cup \left(\bigcup_{j=k+1}^{k_0} S_j^i \right)$$

and also the sets $R_{\ell_i}^{t^i}(S^i)$.

It is clear that the procedure is finite, i.e., there exists an $m \in \mathbb{N}$, for which

$$\bigcup_{j=k+1}^{k_0} S_j^m = \emptyset, \quad S^m \subset \mathbb{E}(n,k). \tag{48}$$

Now we do the counting via Lemma 40. The integers x, y are here

$$x = x_0 = \left| \bigcup_{j=1}^{k} S_j \right|, \quad y = y_0 = \left| \bigcup_{j=k+1}^{k_0} S_j \right|$$

and $\beta = \gamma > 1$. Furthermore,

$$x_i = \left| \bigcup_{j=1}^{k} S_j^i \right|, \quad y_i = \left| \bigcup_{j=k+1}^{k_0} S_j^i \right|,$$

$$a_i = \left| L_{t^{i-1}} \cap \left(\bigcup_{j=1}^{k} S_j^{i-1} \right) \right|, \quad b_i = \left| L_{t^{i-1}} \cap \left(\bigcup_{j=k+1}^{k_0} S_j^{i-1} \right) \right|,$$

$$a_i^* = \left| \left(\bigcup_{a \in R_{\ell_{i-1}}^{i-1}} D(a) \right) \cap \left(\bigcup_{j=1}^{k} S_j^{i-1} \right) \right|,$$

$$b_i^* = \left| \left(\bigcup_{a \in R_{\ell_{i-1}}^{i-1}} D(a) \right) \cap \left(\bigcup_{j=k+1}^{k_0} S_j^{i-1} \right) \right|,$$

and so

$$a_i + b_i = |L_{t^{i-1}}|, \quad a_i^* + b_i^* = \left| \bigcup_{a \in R_{\ell_{i-1}}^{i-1}} D(a) \right|$$

count the new elements in the i-th step.

We know from Proposition 16 that $a_i^* + b_i^* > \gamma(a_i + b_i)$ and from (48) that $y_m = 0$. Hence, by Lemma 40,

$$|\mathbb{E}(n,k)| \geq x_m = |S^m| > x + \gamma y \tag{49}$$

$$= \left| \bigcup_{j=0}^{k} S_j \right| + \left| \bigcup_{j=k+1}^{k_0} S_j \right| + (\gamma - 1) \left| \bigcup_{j=k+1}^{k_0} S_j \right| \geq |S|,$$

since

$$\gamma = 1 + \frac{k}{k_0 - k}, \quad S = \left| \bigcup_{j=1}^{k} S_j \right| + \left| \bigcup_{j=k+1}^{k_0} S_j \right| + \left| \bigcup_{j \geq k_0 + 1} S_j \right|,$$

and

$$\left| \bigcup_{j=k+1}^{k_0} S_j \right| \geq \frac{k_0 - k}{k} \left| \bigcup_{j \geq k_0} S_j \right|.$$

However, (49) says that $\mathbb{E}(n,k) > |S|$, which contradicts the optimality of S. Therefore (II) must be false and Theorem 56 is proved. ∎

§5 Maximal Sets Without Coprimes

From Theorem 57 it follows that

$$f(n,1,s) = |\mathbb{E}(n,1,s)| \tag{50}$$

is true for large n. However, we are going to prove Theorem 58, which says that 50 holds for all $n,s \in \mathbb{N}$. Moreover it contains a uniqueness statement. The proof is quite different from the proof of Theorem 56!
    Denote

$$\Pi(y) = |\{p \in \mathbb{P} :, p \leq y\}|, \ y \geq 0.$$

Let $\tilde{\mathbb{P}} = \{r_1 < r_2 < \ldots\} \subset \mathbb{P}$ be an infinite subset of primes and $\mathbb{X}$ be a set of integers whose prime decompositions contain only primes from $\tilde{\mathbb{P}}$ and $\mathbb{X}(z) = \{x \in \mathbb{X}, \ x \leq z\}$.
    Denote

$$\Pi(y) = |\{r \in \tilde{\mathbb{P}} : \ r \leq y\}|, \ y \geq 0$$

and

$$\Phi(u,y) = \{x \in \mathbb{X}(u) : \ (x,r) = 1 \ for \ all \ r < y\}.$$

Note that $1 \in \Phi(u,y)$ for all $u \geq y$, $u \geq 1$.
    A set $A \subset \mathbb{X}(z)$, $z \geq 1$, is said to be intersecting iff for all $a,b \in A$, we have $a = \prod_{i=1}^{\infty} p_i^{\alpha_i}$ and $b = \prod_{i=1}^{\infty} p_i^{\beta_i}$ with $\alpha_j \beta_j > 0$ for some j.
    We study $\mathcal{I}(z)$, the family of all intersecting $A \subset \mathbb{X}(z)$, and

$$f(z) = \max_{A \in \mathcal{I}(z)} |A|, \ z \in \mathbb{X}.$$

The subfamily $\mathcal{O}(z)$ consists of the optimal sets. We call $A \subset \mathbb{X}(z)$ a star if

$$A = M(\{p\}) \cap \mathbb{X}(z)$$

for some $p \in \tilde{\mathbb{P}}$.
    Clearly, any $A \in \mathcal{O}(z)$ is an "upset" and "downset" and for all $z \in \mathbb{N}$

$$f(z) = \max_{A \in \mathcal{C}(z)} |A|,$$

where $\mathcal{C}(z)$ is the family of left compressed (along $\mathbb{X}$) sets from $\mathcal{I}(z)$. The left pushing and left compressed sets here are considered along the set $\mathbb{X}$. It means that this notions apply to the set of primes from $\tilde{\mathbb{P}}$ only.

The following simple result is true.

Proposition 17 *For any $B \in \mathcal{I}(z)$ and $B' \subset \mathbb{X}(z)$ which is left compressed and obtained from B by left pushing, we have: B is a star iff B' is a star.*

To prove (50) we need the following

Lemma 45 *Suppose that for all $u \in \mathbb{R}_+$, r_ℓ, $\ell \geq 2$ the following relation is valid:*

$$2|\Phi(u, r_\ell)| \leq |\Phi(ur_\ell, r_\ell)|. \tag{51}$$

Then for all $z \in \mathbb{R}_+$, every optimal $A \in \mathcal{O}(z)$ is a star. In particular,

$$f(z) = |M(\{r_1\}) \cap N(z)|$$

for all $z \in \mathbb{N}$.

Proof. Let the elements of the primitive set $P(A) = \{a_1, \ldots, a_m\}$ be written in the colexicographic order, $m > 1$ and maximal prime $p^+(a_m)$ in the decomposition of a_m into primes is r_t,

$$p^+(a_m) = r_t, \ t \geq 2.$$

Write $P(A)$ in the form

$$P(A) = S_1 \cup \ldots \cup S_t, \ t \geq 2, \ S_t \neq \emptyset,$$

where

$$S_i = \{a \in P(A) : \ p^+(a) = r_i\}.$$

Since $A \in \mathcal{O}(z) \cap \mathcal{C}(z)$, we have the partition

$$A = M(P(A)) \cap \mathbb{N}(z) = \bigcup_{1 \leq j \leq t} B(S_j),$$

where $B(S_j) = \bigcup_{a_i \in S_j} B^i(z)$.

Now consider $S_t = \{a_\ell, a_{\ell+1}, \ldots, a_m\}$ for some $\ell \leq m$, and let $S_t = S_t^1 \cup S_t^2$, where

$$S_t^1 = \{a_i \in S_t : \ p_{t-1} | a_i\}, \ S_t^2 = S_t \setminus S_t^1.$$

We have

$$B(S_t) = B(S_t^1) \cup B(S_t^2), \tag{52}$$

where

$$B(S_t^j) = \bigcup_{a_i \in B_t^j} B^i(z), \ j = 1, 2.$$

Let $\tilde{B}_t = \{a_\ell/r_t, a_{\ell+1}/r_t, \ldots, a_m/r_t\}$ and similarly $\tilde{S}_t^j = \{a_i/r_t : \ a_i \in S_t^j\}, j = 1, 2$. It is clear that $a_i/r_t > 1$ for all $a_i \in S_t$.

Obviously, $\tilde{S}_t^1 \in \mathcal{I}(z)$, because all elements of $\tilde{S}_t^1$ have the common factor r_{t-1}. Let us show that $\tilde{S}_t^2 \in \mathcal{I}(z)$ as well. Suppose, to the contrary, that there exist $b_1, b_2 \in \tilde{S}_t^2$ with $(b_1, b_2) = 1$. We have $b_1 r_t, b_2 r_t \in S_t^2 \subset A$ and $(b_1 r_t, r_{t-1}) = 1$, $(b_2 r_t, r_{t-1}) = 1$. Since $A \in \mathcal{C}(z)$ and $r_{t-1} \nmid b_1 b_2$ (see the definition of S_t^2), we conclude that also $b_1 r_{t-1} \in A$. Hence $b_1 r_{t-1}, b_2 r_t \in A$ and at the same time $(b_1 r_{t-1}, b_2 r_t) = 1$, which is a contradiction. So we have $\tilde{S}_t^j \in \mathcal{I}(z)$, $j = 1, 2$, and therefore

$$A_j = M\left((P(A) \setminus S_t) \cup \tilde{S}_t^j \right) \cap \mathbb{X}(z) \in \mathcal{I}(z), \ j = 1, 2.$$

We now prove that either $|A_1| > |A|$ or $|A_2| > |A|$, and this will lead to a contradiction.

From (52) we know that $\max\{|B(S_t^1)|, |B(S_t^2)|\} \geq \frac{1}{2}|B(S_t)|$. W.l.o.g. let us assume that

$$|B(S_t^2)| \geq \frac{1}{2}|B(S_t)|, \tag{53}$$

and let us show that $|A_2| > |A|$ (if $|B(S_t^1)| \geq \frac{1}{2}|B(S_t)|$, the situation is symmetrically the same).

Let $b \in \tilde{S}_t^2$ and $b = r_{i_1} \ldots r_{i_s}$ with $r_{i_1} < \ldots < r_{i_s} < r_t$. We know that $a_i = br_t \in S_t^2$ for some $i \leq m$, and the contribution of $M(a_i)$ in $B(S_t)$ (and also in A) are the elements in the form

$$B^i(z) = \left\{ x \in \mathbb{X}(z) : x = r_{i_1}^{\alpha_1} \ldots r_{i_s}^{\alpha_s} r_t^{\alpha_t} T; \ \alpha_i \geq 1, \ \left(T, \prod_{i \leq t} r_i \right) = 1 \right\}.$$

We write $B^i(z)$ in the following form:

$$B^i(z) = \bigcup_{(\alpha_1, \ldots, \alpha_s), \alpha_i \geq 1} D(\alpha_1, \ldots, \alpha_s), \tag{54}$$

where

$$D(\alpha_1, \ldots, \alpha_s) \tag{55}$$
$$= \left\{ x \in \mathbb{X}(z) : x = r_{i_1}^{\alpha_1} \ldots r_{i_s}^{\alpha_s} r_t T_1, \ \left(T_1, \prod_{i \leq t-1} r_i \right) = 1 \right\}.$$

Now we look at the contribution of $M(b)$ in $A_2 = M((P(A) \setminus S_t) \cup \tilde{S}_t^2) \cap \mathbb{N}(z)$, namely at those elements in A_2 (denoted by $B(b)$) which are divisible by b, but not divisible by any element from $(P(A) \setminus S_t) \cup (\tilde{S}_t^2 \setminus b)$.

Since $A \in \mathcal{C}(z)$ and r_t is the largest prime in $P(A)$, we conclude that

$$B(b) \supseteq B^*(b)$$
$$= \left\{ x \in \mathbb{X}(z) : x = r_{i_1}^{\alpha_1} \ldots r_{i_s}^{\alpha_s} \tilde{T}, \ \alpha_i \geq 1, \ \left(\tilde{T}, \prod_{i \leq t-1} r_i \right) = 1 \right\},$$

and we can write

$$B^*(b) = \bigcup_{(\alpha_1,\ldots,\alpha_s),\ \alpha_i \geq 1} \tilde{D}(\alpha_1,\ldots,\alpha_s), \tag{56}$$

where

$$\tilde{D}(\alpha_1,\ldots,\alpha_s) \tag{57}$$

$$= \left\{ x \in \mathbb{X}(z) : x = r_{i_1}^{\alpha_1} \ldots r_{i_s}^{\alpha_s} \tilde{T}, \ \left(\tilde{T}, \prod_{i \leq t-1} r_i \right) = 1 \right\}, \tag{58}$$

Hence

$$|B(b)| \geq |B^*(b)| = \sum_{(\alpha_1,\ldots,\alpha_s),\ \alpha_i \geq 1} |\tilde{D}(\alpha_1,\ldots,\alpha_s)|. \tag{59}$$

First we prove that $|A_2| > |A|$. In the light of (53) and (54)–(59), for this it is sufficient to show that

$$|\tilde{D}(\alpha_1,\ldots,\alpha_s)| \geq 2|D(\alpha_1,\ldots,\alpha_s)|, \tag{60}$$

for all $(\alpha_1,\ldots,\alpha_s)$, $\alpha_i \geq 1$. But this is exactly the condition (51) for $u = z/(r_{i_1}^{\alpha_1} \ldots r_{i_s}^{\alpha_s} r_t)$ and $\ell = t$. Hence $|A_2| \geq |A|$. To prove that $|A_2| > |A|$, it is sufficient to show the existence of $(\alpha_1,\ldots,\alpha_s)$, $\alpha_i \geq 1$, for which strict inequality holds in (60). For this we take $\beta \in \mathbb{N}$ and $(\alpha_1,\ldots,\alpha_s) = (\beta,1,\ldots,1)$ such that

$$z/r_t < r_{i_1}^\beta r_{i_2} \ldots r_{i_s} \leq z.$$

This is always possible, because $r_{i_1} \ldots r_{i_s} r_t \leq z$ implies $r_{i_1} \ldots r_{i_s} \leq z/r_t$ and $r_{i_1} < \ldots r_{i_s} < r_t$.

We have $|\tilde{D}(\beta,1,\ldots,1)| = 1$ and $|D(\beta,1\ldots,1)| = 0$. Hence $|A_2| > |A|$, which is a contradiction, since $A_2 \in \mathcal{I}(z)$. This completes the proof of Lemma 45. $\qquad\Box$

Lemma 46 *Sufficient for condition (51) to hold is the condition*

$$2\Pi(v) < \Pi(r_2 v) \tag{61}$$

for all $v \in \mathbb{R}_+$.

Proof. Under condition (61) it is sufficient to prove for every $u \in \mathbb{R}_+$ and r_ℓ ($\ell \geq 2$) that $|\Phi(u,r_\ell)| \leq |\Phi_1(ur_\ell,r_\ell)|$, where $\Phi_1(ur_\ell,r_\ell) = \{x \in \Phi(ur_\ell,r_\ell) : u < x \leq ur_\ell\}$.

To avoid the trivial cases $u < 1$, for which $\Phi(u,r_\ell) = \emptyset$, and $1 \leq u < r_\ell$, for which $\Phi(u,r_\ell) = \{1\}$ and $r_\ell \in \Phi(ur_\ell,r_\ell)$, we assume $u \geq r_\ell$.

Let $F(u,r_\ell) = \{a \in \Phi(u,r_\ell), a \neq 1 : ap^+(a) \leq u\} \cup \{1\}$. On the one hand, it is clear that for any $b \in \Phi(u,r_\ell)$, $b \neq 1$, we have $b/p^+(b) \in F(u,r_\ell)$ and that

$$|\Phi(u,r_\ell)| = 1 + \sum_{a \in F(u,r_\ell)} |\tau(a)|, \tag{62}$$

where $\tau(a) = \{r \in \mathbb{P} : r_\ell \leq p^+(a) \leq r \leq u/a$ and the integer 1 in (62) stands to account for the element $1 \in \Phi(u,r_\ell)$.

On the other hand, we have

$$|\Phi_1(ur_\ell, r_\ell)| \geq \sum_{a \in F(u,r_\ell)} |\tau_1(a)|, \tag{63}$$

where

$$\tau_1(a) = \{r \in \mathbb{P} : u/a < r \leq ur_\ell/a\}.$$

We have

$$|\tau(a)| \leq \Pi\left(\frac{u}{a}\right) - \ell + 1 \leq \Pi\left(\frac{u}{a}\right) - 1 \ (\ell \geq 2)$$

and by condition (61)

$$|\tau_1(a)| = \Pi\left(\frac{up_\ell}{a}\right) - \Pi\left(\frac{u}{a}\right) \geq \Pi\left(\frac{u}{a}\right). \tag{64}$$

Hence $|\tau_1(a)| > |\tau(a)|$ for all $a \in F(u, r_\ell)$ and, since $F(u, r_\ell) \neq \emptyset$ $(u \geq r_\ell)$, from (62) - (64) we get $|\Phi_1(ur_\ell, r_\ell)| \geq |\Phi(u, r_\ell)|$. $\qquad\square$

Theorem 58 (Ahlswede and Khachatrian 1994)

(i) *Equation (50) is true for all $n, s \in \mathbb{N}$.*
(ii) *For all $s, n \in \mathbb{N}$, every optimal configuration is a star.*
(iii) *The optimal configuration is unique iff*

$$|M(p_s) \cap \mathbb{N}_s(n)| > |M(p_{s+1}) \cap \mathbb{N}_s(n)|,$$

which is equivalent to the inequality

$$\left|\Phi\left(\frac{n}{p_s}, p_s\right)\right| > \left|\Phi\left(\frac{n}{p_{s+1}}, p_s\right)\right|.$$

Proof. We prove (ii). Since $M(p_s) \cap \mathbb{N}_s(n)$ is not smaller than any competing star, this implies (i) and (iii). In the light of Lemmas 45 and 46, it is sufficient to show that

$$2\Pi(v) \leq \Pi(p_{s+1}v) \tag{65}$$

for all $v \in \mathbb{R}_+$. Since $v < p_s$, $\Pi(v) = 0$, we can assume $v \geq p_s$. Now (65) is equivalent to

$$2(\pi(v) - s + 1) \leq \pi(p_{s+1}v) - s + 1, \tag{66}$$

where $\pi(z) = |\mathbb{P}(z)|$ is the counting function for primes. To show (66), it is sufficient to prove that for all $v \in \mathbb{R}_+$,

$$2\pi(v) \leq \pi(3v). \tag{67}$$

It suffices to show (67) only for $v \in \mathbb{P}$.

Next we use the estimates on the distribution of primes [RS62]:

$$\frac{v}{\log v - 1/2} < \pi(v) < \frac{v}{\log v - 3/2}, \ v \geq 67. \tag{68}$$

From (68) we get inequality (67) for all $v > 298$. The cases $v < 298$, $v \in \mathbb{P}$, are verified by inspection. $\qquad\blacksquare$

§6 Counterexamples for Small n

Next we present an example that shows that the equality

$$f(n,k,s) = |\mathbb{E}(n,k,s)|$$

is not true in general, i.e., for small n.

Theorem 59 (Ahlswede and Khachatrian 1994) *For any $t \in \mathbb{N}$ with the properties*

$$p_{t+7}p_{t+8} < p_t p_{t+9}, \ p_{t+9} < p_t^2 \tag{69}$$

and every n in the interval $I_t = [p_{t+7}p_{t+8}, p_t p_{t+9})$ we have for $k = t+3$

$$f(n,k,1) > |\mathbb{E}(n,k,1)|. \tag{70}$$

Then we show that (69) holds for $t = 209$. Let

$$\mathbb{E}(n,t+3,1) = \left\{ u \in \mathbb{N}_1(n) : \ \left(u, \prod_{i=1}^{t+3} p_i \right) > 1 \right\}.$$

As a competitor we suggest $A_n(t+3) = B \cup C$, where

$$B = \left\{ u \in \mathbb{N}_1(n) : \ \left(u, \prod_{i=1}^{t-1} p_i \right) > 1 \right\}$$

and

$$C = \{ p_{t+i}p_{t+j} : 0 \le i < j \le 8 \}.$$

Note that by (69), $C \subset \mathbb{N}_1(n)$ for $n \in I_t$, that $B \cap C = \emptyset$, and that $|C| = \binom{9}{2} = 36$. Therefore

$$|A_n(t+3)| = |B| + 36. \tag{71}$$

Furthermore, no $k+1 = t+4$ numbers of $A_n(t+3)$ are coprimes, because we can take in B at most $t-1$ and in C at most 4 pairwise relatively prime integers.

For comparison, we write $\mathbb{E}(n,t+3,1)$ in the form $\mathbb{E}(n,t+3,1) = B \cup D$, where

$$D = \{ p_t, p_{t+1}, p_{t+2}, p_{t+3} \} \cup \{ p_t^2, p_{t+1}^2, p_{t+2}^2, p_{t+3}^2 \}$$
$$\cup \ \{ p_{t+i}p_{t+j} : 0 \le i \le 3, \ 1 \le j \le 8, \ i < j \}.$$

Notice that by (69), for $n \in I_t$, p_t^3 (and a fortiori $p_{t+1}^3, \ldots$) exceeds n and so does $p_t p_{t+9}$ (and a fortiori $p_{t+1}p_{t+9}, \ldots$).

Since $|D| = 4+4+8+7+6+5 = 34$, we conclude with (71) that

$$|A_n(t+3)| - |\mathbb{E}(n,t+3,1)| = |B| + 36 - (|B| + 34) = 2 > 0.$$

The hypothesis (69) remains to be verified. It is interesting that among the prime numbers less than 5,000 there is only one t which satisfies (69), namely $t = 209$. The respective primes $p_t, \ldots, p_{t+9}$ are

p_{209}	p_{210}	p_{211}	p_{212}	p_{213}	p_{214}	p_{215}	p_{216}	p_{217}	p_{218}
1289	1291	1297	1301	1303	1307	1319	1321	1327	1361

We calculate $p_{209}p_{218} = 1289 \cdot 1361 = 1754329 > p_{216}p_{217} = 1321 \cdot 1327 = 1752967$ and that $p_{209}^2 = 1289^2 > 1361 = p_{218}$. Hence for $k = 212$ and for all n with $p_{209}p_{218} = 1754329 > n \geq 1752967 = p_{216}p_{217}$ one has $f(n,k,1) \geq |\mathbb{E}(n,k,1)| + 2$. ∎

Notes to Chapter VI

Consider the following problem posed by Erdös and Graham [E73], [E80]. Let $1 < a_1, \ldots < a_k = n$, $(a_i, a_j) \neq 1$. What is the maximal value of k?

We generalize this problem in the following way. Let $Q = \{q_1 < \ldots < q_r\} \subset \mathbb{P}$ be any finite set of primes and let $A = \{a_1 < \ldots < a_k\} \subset \mathbb{N}(n)$ be a set such that for all $1 \leq i, j \leq k$

$$(a_i, a_j) \neq 1$$

and

$$\left(a_i, \prod_{j=1}^{r} q_j \right) \neq 1,$$

Denote by $I(n, Q)$ the family of all such sets.

The following is true.

Theorem 60 (Ahlswede and Khachatrian 1996) *For* $n \geq \prod_{i=1}^{r} q_i$,

$$\max_{A \in I(n,Q)} |A| = \max_{1 \leq j \leq r} |M(2q_1, \ldots, 2q_j, q_1 \ldots q_j) \cap \mathbb{N}(n)|,$$

where for $B \subset \mathbb{N}$, $M(B)$ *denotes the set of multiples of* B.

This theorem uses the same considerations as the proof of (50) and we refer to [AK96b] for the proof. There also was shown that the condition in the theorem saying that n should be sufficiently large, cannot be omitted.

Theorems 56 and 57 first were proved in [AK95b]. Equation (50) was proved in [AK96a]. Theorem 59 we took from [AK94b].

Exercises

1. Prove Lemma 40.
2. Prove Proposition 15. *Hint:* use the inequality

$$\ln\left(1 - \frac{1}{p} \right) > -\frac{2}{p}$$

and the RHS estimate from (68) (Lecture 16).

More precise calculations (see [A76], Theorem 13.13) show that

$$\prod_{p\in\mathbb{P}(t)}\left(1-\frac{1}{p}\right) \overset{t\to\infty}{\sim} \frac{e^{-C}}{\ln t},$$

where C is the Euler constant.

3. For the set of square-free natural numbers $\mathbb{N}^*$ one is naturally led to the sets $\mathbb{N}_s^* = \mathbb{N}_s \cap \mathbb{N}^*$, $\mathbb{N}_s^*(n) = \mathbb{N}_s(n) \cap \mathbb{N}^*$, $\mathbb{E}^*(n,k,s) = \mathbb{E}(n,k,s) \cap \mathbb{N}^*$ etc. and to the function $f^*(n,k,s)$.
   Prove that for all $s, n \in \mathbb{N}$,

$$f^*(n,1,s) = |\mathbb{E}^*(n,1,s)|.$$

   Hint: Use the Marica/Schönheim inequality.

4. Even for square-free numbers "Erdős sets" are not always optimal, that is, $f^*(n,k,1) \neq |E^*(n,k,1)|$ can occur. Show that the set $\mathbb{N}^* \cap A_n(t+3)$ (defined in §6) is an example.

5. Show that $f^*(n,2,s) \neq |E^*(n,2,s)|$ for $p_s = 101$ and $n \in [109 \cdot 113, 101 \cdot 127]$.

Research Problems

1. **Conjecture** The relation

$$|\mathbb{E}(n,k,s)| = f(n,k,s)$$

   is true for $k = 2, 3$ (see [AK96a]).

2. By considering instead of $\mathbb{N}_s$ the set $\mathbb{N}_{p'}$ of those natural numbers which do not have any prime of the finite set of primes $\mathbb{P}'$ in their prime number decomposition. We put $\mathbb{N}_{\mathbb{P}'}(n) = \mathbb{N}_{\mathbb{P}'} \cap [n]$ and consider sets $A \subset \mathbb{N}_{\mathbb{P}'}(n)$ of noncoprimes. We are again interested in cardinalities and therefore introduce

$$f(n,1,\mathbb{P}') = \max\{|A| : A \subset \mathbb{N}_{\mathbb{P}'}(n) \text{ has no coprimes}\}.$$

   In analogy to the set $\mathbb{E}(n,1,s)$ in the case $\mathbb{P}' = \{p_1, \ldots, p_{s-1}\}$, we introduce

$$\mathbb{E}(n,1,\mathbb{P}') = \{u \in \mathbb{N}_{\mathbb{P}'}(n) : q_1 | u\},$$

   where $\{q_1, q_2, \ldots\} = \{p_1, p_2, \ldots\} \backslash \mathbb{P}'$ and $q_1 < q_2 < \ldots$ and $Q_{\mathbb{P}'} = \prod_{p\in\mathbb{P}'} p$.

   Conjecture For any finite set of primes $\mathbb{P}'$ we have

$$f(n,1,\mathbb{P}') = |\mathbb{E}(n,1,\mathbb{P}')|.$$

   The equality has been proved for $n \geq \frac{q_1 q_2}{q_2 - q_1} Q_{\mathbb{P}'}$ in [AK94b].

Appendix:
Supplementary Material and Research Problems

Whereas previously there are notes given after every chapter, now we add notes to Lectures. A new ingredient is whole Research Programs. The style of the presentation of results is less streamlined, but always clear.

Notes to Lecture 1

Further Intersection Problems

1. A Higher Level Extremal Problem

First consider the following intersection problem for the family of sets $\mathcal{A}_1, \ldots, \mathcal{A}_M \subset \binom{[n]}{k}$ with the property that for any i, j there is an $A_{ij} \in \mathcal{A}_i$ intersecting all $A \in \mathcal{A}_j$. It is easy to see that the maximal M with this property satisfies

$$M \geq \binom{n-1}{k-1}.$$

It is shown in [AAERS96] that here equality holds for $k = 2, 3, 4$ and does not hold for $k \geq 8$.

2. A Sharpening of EKR

A generalization of EKR was given in [HR73]. The following intersection problem was considered in [A06a]. Let $n \geq 2k$, then for any $\mathcal{A} \subset \binom{[n]}{k}$ with the "triangle property": $\forall A, B, C \in \mathcal{A}, A \cap B \neq \emptyset, B \cap C \neq \emptyset \Rightarrow A \cap C \neq \emptyset$ we have

$$|\mathcal{A}| \leq \begin{cases} n, & \text{if } k = 2 \text{ and } n \equiv 0 \mod 3, \\ \binom{n-1}{k-1}, & \text{otherwise} \end{cases}$$

and this bound is best possible.

3. A Pushing–Pulling Method

It came as a surprise to Ahlswede and Khachatrian that at first they did not succeed to derive Katona's Intersection Theorem for the unrestricted case, that is in the space $2^{[n]}$, by the method of "generated sets." This led to the discovery of another method in [AK99], which yields Theorem 3 and Katona's Intersection Theorem. Subsequently, they found a way to derive Katona's Theorem from Theorem 3 via another comparison lemma. This is the most complicated proof, where several simple proofs exist, but it teaches something about methods, which made progress possible on the t-intersection problem in the truncated Boolean lattice covering the restricted and the unrestricted intersection problem as special cases [ABEK02]. Although there are vertex and edge-isoperimetric theorems, it went unsaid that diametric theorems are vertex-diametric theorems. We complete the story by introducing edge-diametric theorems into combinatorial extremal theory. Using again the pushing–pulling method such a result for $\mathcal{V} = \{0,1\}^n$ and $\mathcal{E} = \{(a^n, b^n) : a^n, b^n \in \mathcal{V}\}$ are established. Results and methods of this section are discussed in the survey [N], 45–74.

Left shifting is used in proofs of the Kruskal/Katona/Lindström and Zetterström ([LZ67]) Shadow Theorem. An essential progress on this method is made in [AAK03c]. Starting with any optimal configuration, it is possible by suitable left and right shifting, alternatively, to come to the beginning in the squashed order (see [AAK03d]). A first application led to the result on shadows of intersecting families of [AAK04a]. Useful formulas for cardinalities of squashed families were given in [M95].

4. Maximizing Pairs of Intersecting Sets: Excess Problems

Pioneers are the intersection problems considered by Ahlswede and Katona in [AK78]. For every extremal problem concerning the maximal cardinality M of sets whose elements have a property in question (like pairwise distances, intersections, etc.), one can study sets of a cardinality $N > M$ and ask for the maximal number of satisfied relations $S(N)$.

For $\mathcal{A} \subset \binom{[n]}{k}$ let $I(\mathcal{A}) = |\{(A_1, A_2) \in \mathcal{A}^2 : A_1 \cap A_2 \neq \phi\}|$, $G(\mathcal{A}) = |\{(A_1, A_2) \in \mathcal{A}^2 : |A_1 \cap A_2| \geq k - 1\}|$. Determine $f(N) = \max_{|\mathcal{A}| = N} I(\mathcal{A})$ or $g(N) = \max_{|\mathcal{A}| = N} G(\mathcal{A})$. The problems are the same for $k = 2$ and here these authors described two configurations, quasi-ball and quasi-star, one of which is always optimal.

The decision about the winner becomes number theoretically tricky. It is settled for

$$N \in \left\{ 1, 2, \ldots, \frac{1}{2}\binom{n}{2} - \frac{n}{2} \right\} \cup \left\{ \binom{n}{2} + \frac{n}{2}, \binom{n}{2} + \frac{n}{2} + 1, \ldots, \binom{n}{2} \right\}.$$

For

$$N \in \left\{ \frac{1}{2}\binom{n}{2} - \frac{n}{2}, \ldots, \frac{1}{2}\binom{n}{2} + \frac{n}{2} \right\}$$

there are at most three switches for the winner as N increases from the smallest to the largest value within the specified range. Depending on the number n, there can be also only two switches or even only one.

For $k \geq 3$ none of the problems is solved. The second has also been called the Kleitman/West Problem. Ahlswede and Cai (see [N], 1-16) disproved a conjecture of Kleitman, which generalizes the construction of [AK78], for this problem already for $k = 3$.

For the unrestricted case $\mathcal{A} \subset 2^{[n]}$ Frankl has results in [F77].

5. On Dense Sets

A family $\mathcal{F} \subset \binom{[n]}{k}$ is called k-dense, if there exists an $F \in \mathcal{F}$, such that for $\mathcal{F}|_F = \{F \cap F_1 : F_1 \in \mathcal{F}\}$ $|\mathcal{F}|_F| = 2^k$.

Frankl and Pach proved in [FP84] that for $|\mathcal{F}| \geq \binom{n}{k-1}$ $\mathcal{F}$ is k-dense and they conjectured that every $\mathcal{F}$ with $|\mathcal{F}| > \binom{n-1}{k-1}$ is k-dense.

Ahlswede and Khachatrian [AK97c] provide a counterexample to this conjecture by the construction of a set $\mathcal{F} \subset \binom{[n]}{k}$, $|\mathcal{F}| = \binom{n-1}{k-1} + \binom{n-4}{k-3}$, which is not k-dense.

These authors propose a stronger restriction. Loosely speaking, instead of forbidding a subset of A, they forbid for some x all subsets of A of cardinality x and ask how large $|\mathcal{F}|$ can be under this restriction.

Let $\mathcal{F}$ be a set system. For $A \in \mathcal{F}$, we say that A has x-intersection if there exists $B \in \mathcal{F}$ such that $|A \cap B| = x$ (here we do not assume $B \neq A$, therefore each A trivially has $|A|$-intersection).

6. On Chvátal's Conjecture

Next we discuss an interesting conjecture due to Chvátal [C72].

For a family $\mathcal{Q} \subset 2^{[n]}$ and $i \in [n]$ denote

$$\mathcal{Q}(i) = \{A \in \mathcal{Q} : i \in A\}.$$

Clearly $\mathcal{Q}(i)$ is an intersecting family.

The Chvátal's conjecture is formulated in Research Problem 3 below. A partial result was obtained by Ahlswede and Khachatrian.

Proposition 18 *Let $\mathcal{S} \subset 2^{[n]}$ be a downset such that*

$$L_{1,j}(\mathcal{S}) = \mathcal{S}, j = 2, \ldots, n$$

(family $\mathcal{S}$ is left-compressed with respect to $1 \in [n]$). Then Chvátal's conjecture is true.

Proof. Let $\mathcal{A} \subset \mathcal{S}$ be an intersecting system. Consider the partition $\mathcal{A} = \mathcal{A}_1 \cup \mathcal{A}_2$, where $\mathcal{A}_1 = \mathcal{A}(1) = \{B \in \mathcal{A} : 1 \in B\}$, $\mathcal{A}_2 = \mathcal{A} \setminus \mathcal{A}_1$. We can assume, that $\mathcal{A}_2 \neq \emptyset$, otherwise $A = A_1 = \mathcal{A}(1)$. Consider the set

$$\mathcal{F} = \{A_2 - A_2\} = \{A \setminus A_1 : A, A_1 \in \mathcal{A}_2\}$$

and the set

$$\mathcal{F}^* = \{F \cup \{1\} : F \in \mathcal{F}\}.$$

Clearly $|\mathcal{F}| = |\mathcal{F}^*|$.

Claim.

$$\mathcal{F}^* \subset \mathcal{S}.$$

Prove this using the compressedness of $\mathcal{S}$.

Claim.

$$\mathcal{F}^* \cap \mathcal{A}_1 = \emptyset.$$

Assume that there is a $C \in \mathcal{F}^* \cap \mathcal{A}_1$. Then we have $C = \{1\} \cup (A \setminus A_1)$ for some $A, A_1 \in \mathcal{A}_2$, and consequently $C \cap B_1 = \emptyset$, a contradiction to the conditions that $C, B_1 \in \mathcal{A}$ and $\mathcal{A}$ is intersecting.

Consider now $\mathcal{A}' = (\mathcal{A} \setminus \mathcal{A}_2) \cup \mathcal{F}^*$. We have

(i) $\mathcal{A}' \subset \mathcal{S}$,
(ii) $\mathcal{A}' \subset \mathcal{S}(1)$ and A' is intersecting,
(iii) $\mathcal{A}' = \mathcal{A}_1 \cup \mathcal{F}^*$ and hence

$$|\mathcal{A}'| = |\mathcal{A}_1| + |\mathcal{F}^*| = |\mathcal{A}_1| + |A_2 - A_2|$$

Next we use the Marica/Schönheim inequality (see Lecture 15): for any $\mathcal{A} \subset 2^{[n]}$

$$|\mathcal{A} - \mathcal{A}| \geq |\mathcal{A}|$$

holds.

Using this inequality we have

$$A' = |\mathcal{A}_1| + |A_2 - A_2| \geq |\mathcal{A}_1| + |\mathcal{A}_2| = |\mathcal{A}|$$

and $\mathcal{A}' \subset \mathcal{S}(1)$. $\square$

7. Cross-Disjoint Pairs of Clouds in the Interval Lattice

In this and the next section we present work of [AC96] and some consequences. Consider the set $[n] = \{1, 2, \ldots, n\}$, the set of all its subsets $\mathcal{L}_n$, and the lattice of intervals $\mathcal{I}_n = \{I = [A, B] : A, B \in \mathcal{L}_n\}$, where $[A, B] = \{C \in \mathcal{L}_n : A \subset C \subset B\}$, if $A \subset B$, and $[A, B] = I_\phi$ (the empty interval), if $A \not\subset B$. The lattice operations $\wedge$ and $\vee$ are defined by

$$[A, B] \wedge [A', B'] = [A, B] \cap [A', B'], \tag{1}$$

$$[A, B] \vee [A', B'] = [A \cap A', B \cup B']. \tag{2}$$

Here the empty interval I_ϕ is represented by $\big[[n],\phi\big]$. The pair $(\mathcal{A},\mathcal{B})$ with $\mathcal{A},\mathcal{B}\subset \mathcal{I}_n \smallsetminus \{I_\phi\}$ is cross-disjoint, if

$$I \wedge J = I_\phi \ \text{ for } \ I \in \mathcal{A}, J \in \mathcal{B}.$$

Let us denote the set of those pairs by $\mathcal{D}_n$.

Theorem 61 *For $n = 1,2,\dots$*

$$\max\big\{|\mathcal{A}||\mathcal{B}| : (\mathcal{A},\mathcal{B}) \in \mathcal{D}_n\big\} = 3^{2n-2}.$$

Equality is assumed for

$$\mathcal{A}^* = \big\{I \in \mathcal{I}_n : I = [A,B], 1 \notin B\big\}, \mathcal{B}^* = \big\{I \in \mathcal{I}_n : I = [A,B], 1 \in A\big\}.$$

All optimal pairs are obtained by replacing 1 in the definition of $\mathcal{A}^*$ and $\mathcal{B}^*$ by any element m of $[n]$, and by exchanging the roles of these two sets.

We shall relate *cross-disjoint* pairs of clouds from $\mathcal{I}_n$ to *cross-intersecting* pairs of clouds from $\mathcal{L}_n$ with a suitable weight.
$(\mathcal{U},\mathcal{V})$ with $\mathcal{U},\mathcal{V} \subset \mathcal{L}_n$ is cross-intersecting, if

$$U \cap V \neq \phi \ \text{ for } \ U \in \mathcal{U} \text{ and } V \in \mathcal{V}.$$

We denote the set of these pairs by $\mathcal{P}_n$. Furthermore, we introduce the weight $w : \mathcal{L}_n \to \mathbb{N}$ by

$$w(A) = 2^{n-|A|} \ \text{ for } \ A \in \mathcal{L}_n.$$

Theorem 62 *For $(\mathcal{U},\mathcal{V}) \in \mathcal{P}_n$*

$$W(\mathcal{U})W(\mathcal{V}) \triangleq \sum_{U \in \mathcal{U}} w(U) \cdot \sum_{V \in \mathcal{V}} w(V) \leq 3^{2(n-1)}$$

and the bound is best possible.

Next we give a common generalization of Theorem 62 and a Theorem of Erdös/Schönheim [ES69]. In deriving their intersection theorem for multisets, they established first an intersection theorem with weights for $\mathcal{L}_n$. Those weights $w(A), A \in \mathcal{L}_n$, are *increasing* in $|A|$, whereas our weights $w(A) = 2^{n-|A|}$ used in Theorem 62 are *decreasing* in $|A|$. The latter does not allow to just choose the "heavier" one of A and $A^c = [n] \smallsetminus A$ in order to construct an optimal configuration. This difference makes things more difficult in our case. Nevertheless, we can give an unified approach.

Let $\mathcal{W} = \{w_i : 1 \leq i \leq n\}$ be positive reals, which give rise to the weight w on $\mathcal{L}_n$:

$$w(A) = \prod_{t \in A} w_t \ \text{ for } \ A \subset [n]$$

and

$$W(\mathcal{A}) = \prod_{A \in \mathcal{A}} w(A) \ \text{ for } \ \mathcal{A} \subset \mathcal{L}_n.$$

Define

$$\alpha(n,w) = \max\{W(\mathcal{A}) : \mathcal{A} \subset \mathcal{L}_n \text{ is intersecting}\}$$

(i.e. $A \cap B \neq \phi$ for $A, B \in \mathcal{A}$).

Theorem 63 (Erdös and Schönheim 1969)

$$\alpha(n,w) \leq \frac{1}{2} \sum_{A \subset [n]} \max\left(w(A), w(A^c)\right)$$

and the bound is best possible when $w_i \geq 1$ for $i \in [n]$.

Next by relabeling we can always assume that

$$w_1 \geq w_2 \geq \cdots \geq w_n$$

and define

$$\mathcal{A}^*(n,w) = \begin{cases} \{A \subset [n] : A \cap [m] \in \mathcal{A}(m,w')\}, & \text{if } m \geq 1 \\ \{A \subset [n] : 1 \in A\}, & \text{if } m = 0. \end{cases}$$

Theorem 64

$$\alpha(n,w) = W\left(\mathcal{A}^*(n,w)\right).$$

8. Incomparability and Intersection Properties of Boolean Interval Lattices and Chain Posets

We can define on $\mathcal{I}_n$ a partial order "$\leq$" by

$$[A,B] \leq [A',B'] \Leftrightarrow [A,B] \subset [A',B'] \text{ or (equivalently) } A' \subset A \subset B \subset B'.$$

We define a rank function $\rho : \mathcal{I}_n \to \mathbb{N} \cup \{0\}$ by

$$\rho\left([A,B]\right) = \begin{cases} 0, & \text{if } [A,B] = I_\phi \\ |B \setminus A| + 1, & \text{if } [A,B] \neq I_\phi. \end{cases}$$

Let us introduce $\mathcal{L}_n^k = \binom{[n]}{k}$ and let us denote by $\mathcal{I}_n^k$ the set of intervals from $\mathcal{I}_n$ of rank $k (0 \leq k \leq n+1)$.

Observe first that for all $I \in \mathcal{I}_n^k$

$$|\{I' \in \mathcal{I}_n^{k+1} : I' \supset I\}| = n - k + 1$$

and that

$$|\{I' \in \mathcal{I}_n^{k-1} : I' \subset I\}| = 2(k-1).$$

This regularity property of a lattice is sufficient for the LYM-inequality to hold. We move directly to the AZ-identity. For any $\mathcal{A} \subset \mathcal{I}_n$ and any $I = [A,B] \in \mathcal{I}_n$ with

$$\mathcal{A}_I = \{K \in \mathcal{A} : K \subset I\} \neq \phi \tag{3}$$

write

$$\mathcal{A}_I = \{[A_i, B_i] : 1 \leq i \leq \alpha\}$$

and define

$$W_{\mathcal{A}}(I) = \left(|B| - |\bigcup_{i=1}^{\alpha} A_i|\right) + \left(|\bigcap_{i=1}^{\alpha} B_i| - |A|\right).$$

If (3) does not hold, set $W_{\mathcal{A}}(I) = 0$.

Theorem 65 (AZ-identity) *For any $\mathcal{A} \subset \mathcal{I}_n$*

$$\sum_{I \in \mathcal{I}_n} \frac{W_{\mathcal{A}}(I)}{2^{n-\rho(I)+2}(\rho(I) - 1)\binom{n}{\rho(I)-1}} \equiv 1.$$

The Whitney numbers w_k of $\mathcal{I}_n$ are defined by

$$w_k = |\mathcal{I}_n^k| \text{ for } 0 \leq k \leq n+1.$$

They can be evaluated.

Lemma 47 *We have*

(i) $w_k = \binom{n}{k-1} 2^{n-k+1}$ *for $0 < k \leq n+1$ and $w_0 = 1$*
   and consequently
(ii) $|\mathcal{I}_n| = \sum_{k=0}^{n-1} w_k = 3^n + 1$.

LYM-Inequality (See Exercise 2)

For any antichain $\mathcal{A} \subset \mathcal{I}_n$

$$\sum_{k=0}^{n+1} \frac{|\mathcal{A} \cap \mathcal{I}_n^k|}{w_k} \leq 1.$$

We now consider intersecting systems of intervals of rank k. This is analogous to the case of k element sets considered originally in [EKR61]. It is remarkable that in the new situation we have uniqueness in the sense that only the $\mathcal{I}_n^k(C)$'s appear as optimal systems.

Theorem 66 *For every intersecting system $S \subset \mathcal{I}_n^k$*

$$|S| \leq \binom{n}{k-1}$$

and the $\mathcal{I}_n^k(C)$ $(C \subset [n], 1 \leq |C| \leq n - k + 1)$ are exactly the intersecting systems achieving equality.

A well-known inequality of Bollobás [B65] states that for any intersecting antichain $\mathcal{F} \subset \mathcal{L}_n$

$$\sum_{k=1}^{\lfloor n/2 \rfloor} \frac{|\mathcal{F} \cap \mathcal{L}_n^k|}{\binom{n-1}{k-1}} \leq 1.$$

What is the Bollobás-type inequality for $\mathcal{I}_n$? The answer follows by simple reasoning. For an intersecting antichain $\mathcal{S} = \{[A_i, A_i \cup D_i] : A_i \cap D_i = \emptyset, 1 \leq i \leq m\}$ in $\mathcal{I}_n$ necessarily $\{D_i : 1 \leq i \leq m\}$ is an antichain in $\mathcal{L}_n$. We get the following inequality.

Theorem 67 *For an intersecting antichain $\mathcal{S}$ in $\mathcal{I}_n$*

$$\sum_{k=1}^{n+1} \frac{|\mathcal{S} \cap \mathcal{I}_n^k|}{\binom{n}{k-1}} \leq 1.$$

Conversely, we can translate this inequality backwards. Thus the LYM inequality for the Boolean lattice is exactly the Bollobás-type inequality for the Boolean interval lattice.

A strictly increasing sequence of subsets of $[n]$ and of length k is called a k-chain. $\mathcal{C}_n^k$ denotes the set of all those chains and we define $\mathcal{C}_n = \bigcup_{k=1}^{n+1} \mathcal{C}_n^k$. $M(n,k)$ denote the maximum size of intersecting families of k-chains in $[n]$.

Theorem 68 *The intersecting family of all k-chains in $\mathcal{L}_n$ starting with the empty set ϕ has the maximal cardinality $M(n,k)$.*

Remark We have started to think about families of d-intersecting k-chains. Here the chains $A_1 \subset A_2 \subset \cdots \subset A_m$ and $A'_1 \subset A'_2 \subset \cdots \subset A'_{m'}$ are d-intersecting if there are indices $i_1 < \cdots < i_d$ and $j_1 < \cdots < j_d$ with $A_{i_\ell} = A'_{i_\ell}$ for $\ell = 1, 2, \ldots, d$.

Applying a shifting operator as in [ESS94], one can show that there is an optimal family $\mathcal{F}$ with a *strong d-intersection property* saying that for any $\mathbb{C}, \mathbb{C}' \in \mathcal{F}$ there is a subset $S \subset \{0, 1, \ldots, n\}$, $|S| \geq d$, such that all $X_j = \{1, 2, \ldots, j\}$ with $j \in S$ are contained in both, $\mathbb{C}$ and $\mathbb{C}'$. Here $X_0 = \phi$.

This means that there is a set $\mathcal{S} = \{E(F) : F \in \mathcal{F}\}$ of subsets of $\{0, 1, \ldots, n\}$ with

$$|S(F) \cap S(F')| \geq d$$

associated with F, such that for $j \in E(F)$ X_j is contained in F.

It seems natural to *conjecture* that for some $\varepsilon > 0$ and n large for $d \leq n(1 - \varepsilon)$, there is an optimal family of d-intersecting chains all containing $X_0, X_1, \ldots, X_{d-1}$. The restriction on d is essential, because otherwise the guess is false.

To see this, let us assume that $n - d$ is bounded by a constant b. Then the number of chains in the family just specified is bounded by a function of b only. However, the family of chains containing $\{X_i\}_{i \in S}$, where S runs through all subsets of $\{0, 1, \ldots, n\}$ with cardinality at least $\frac{n+d}{2}$, is d-intersecting and increases with n. In the spirit of this construction is our last contribution in Research Problem 5.

9. An Application of the Complete Intersection Theorem (Here Theorem 1) in Complexity Theory

Here Theorem 1 is used only in the special case $t = 2$ to yield a continuous version. More precisely, in [DS05] an inequality is derived for the p-biased product distribution μ_p, $0 < p < 1$, on $\{0,1\}^n$ or, equivalently on $2^{[n]}$, where for $\mathcal{S} \subset [n]$

$$\mu_p(\mathcal{S}) = p^{|\mathcal{S}|}(1-p)^{n-|\mathcal{S}|}.$$

Lemma 48 (Dinur and Safra 2005) *For* $\mathcal{A}_i = \{F \in 2^{[n]} : |F \cap [2+2i]| \geq 2+i\}$ *with* $0 \leq i \leq \lfloor \frac{n}{2} \rfloor - 1$, $0 < p < \frac{1}{2}$, *and 2-intersecting* $\mathcal{F} \subset 2^{[n]}$

$$\mu_p(\mathcal{F}) \leq \max_i \mu_p(\mathcal{A}_i).$$

Proof. Denote $\mu = \max \mu_p(\mathcal{A}_i)$. Assume that for a 2-intersecting $\mathcal{F}_0 \subset 2^{[n_0]}$ the inequality is false: $a = \mu_p(\mathcal{F}_0) - \mu > 0$. Thus consider $\mathcal{F} = \mathcal{F}_0 \cup 2^{[n]-[n_0]}$ for $n > n_0$ large, to be determined later.

Clearly $\mathcal{F}$ is 2-intersecting and in natural notation $\mu_p^{[n]}(\mathcal{F}) = \mu_p^{[n_0]}(\mathcal{F}_0)$. Consider for $\theta < \frac{1}{2} - p$, to be determined later,

$$S = \{k \in \mathbb{N} : |k - pn| \theta n\},$$

and for every $k \in S$ define $\mathcal{F}_k = \mathcal{F} \cap \binom{[n]}{k}$. It is clear from elementary Probability Theory that most of $\mathcal{F}$'s probability comes from $\bigcup_{k \in S} \mathcal{F}_k$. Therefore, there must be at least one $\mathcal{F}_n$ violating the bound in Theorem 1. Indeed,

$$\mu + a = \mu_p(\mathcal{F}) = \sum_{k \in S} p^k (1-p)^{n-k} |\mathcal{F}_k| + o(1).$$

Hence, there exists a $k \in S$ for which $\frac{|\mathcal{F}_k|}{\binom{n}{k}} \geq \mu + \frac{1}{2}a$. We are left to show that $\mu\binom{n}{k}$ is close enough to $\max_i |\mathcal{A}_i \cap \binom{[n]}{k}|$. This follows from the usual tail bounds. Subsets in $\binom{[n]}{k}$ for large enough i (depending only on $\frac{k}{2}$ but not on k or n) have roughly $\frac{k}{2}(2i+2)$ elements in the set $[2i+2]$.

Moreover, the subsets in $\mathcal{A}_i$ have at least $i+2$ elements in $[2i+2]$, thus are very few (compared with $\binom{n}{k}$), because $\frac{i+2}{2i+2} > \frac{1}{2} > p + \theta \geq \frac{k}{n}$. In other words, there exists some constant $C_{p+\theta,\mu}$, for which $|\mathcal{A}_i \cap \binom{[n]}{k}| < \mu\binom{n}{k}$ for all $i \geq C_{p,\mu}$ as long as $\frac{k}{n} \leq p + \theta$.

Additionally, for every $i < C_{p,\mu}$ taking n to be large enough we have for all $k \in S$

$$\frac{|\mathcal{A}_i \cap \binom{[n]}{k}|}{\binom{n}{k}} = \mathcal{M}_{\frac{k}{n}}(\mathcal{A}_i) + o(1) = \mu_p(\mathcal{A}_i) + o(1) < \mu + o(1)$$

where the first inequality follows from a straightforward computation.

Exercises

1. Derive Theorem 61 from Theorem 62 (ad 7).
2. (LYM-inequality) Using the AZ-identity (Theorem 65) and Lemma 47, prove that for any antichain $\mathcal{A} \subset \mathcal{I}_n$

$$\sum_{k=0}^{n+1} \frac{|\mathcal{A} \cap \mathcal{I}_n^k|}{w_k} \leq 1. \qquad (\text{ad } 8)$$

3. (Sperner property) Prove that for every antichain $\mathcal{A} \subset \mathcal{I}_n$

(i)

$$|\mathcal{A}| \leq \max_{0 < k \leq n+1} w_k = \binom{n}{\lceil \frac{n+1}{3} \rceil - 1} 2^{n - \lceil \frac{n+1}{3} \rceil + 1}$$

(ii)

$$\max_{0 < k \leq n+1} w_k = \begin{cases} w_{\ell+1}, & \text{if } n+1 = 3\ell+1 \\ w_{\ell+1}, & \text{if } n+1 = 3\ell+2 \\ w_{\ell+1} = w_\ell, & \text{if } n+1 = 3\ell. \end{cases}$$

(iii) The antichains $\mathcal{A}$ of maximal length are

$$\mathcal{A} = \begin{cases} \mathcal{I}_n^{\ell+1}, & \text{if } n+1 = 3\ell+m; m = 0,1,2, \\ \mathcal{I}_n^\ell, & \text{if } n+1 = 3\ell. \end{cases}$$

Thus, if $3 \mid n+1$, then there are two optimal antichains (ad 8).

Research Problems

1. Find the bound for the cardinality of $\mathcal{A} \subset \binom{[n]}{k}$ with the property $\forall A, B, C \in \mathcal{A}$, $|A \cap B| \geq t$, $|B \cap C| \geq t \Rightarrow |A \cap C| \geq t$ (ad 2).
2. **Conjecture** Let $\mathcal{F} \subseteq \binom{[n]}{k}$, $(1 \leq k < n/2)$. Suppose that for each $A \in \mathcal{F}$ there exists $x = x(A) \in \{0, 1, \ldots, k\}$ such that A has no x-intersection. Then $|\mathcal{F}| \leq \binom{n-1}{k-1}$.
   It is clear that the conjecture is not true for $n < 2k$: in $\mathcal{F} = \binom{[n]}{k}$ each member has no 0-intersection.
   The following example shows that the conjecture is wrong also for $n = 2k$, for each $k > 1$.

   Example Construction: let $A, B \in \binom{[n]}{k}$ be a pair of disjoint sets. For odd k: let

   $$\mathcal{F} = \left\{ F \in \binom{A \cup B}{k} : F = B \text{ or } |F \cap A| > \frac{k}{2} \right\}.$$

   For even k: let a be fixed point of A, and let

$$\mathcal{F} = \left\{ F \in \binom{A \cup B}{k} : F = B \text{ or } |F \cap A| > \frac{k}{2}, \text{ or } |F \cap A| = \frac{k}{2}, \, a \in F \right\}.$$

The system $\mathcal{F}$ satisfies the hypothesis of the conjecture:
B has no x-intersection for all $x : k/2 < x < k$,
A has no x-intersection for all $x : 0 < x < k/2$.
Each $F \in \mathcal{F} \setminus \{A, B\}$ has no 0-intersection.

The size of $\mathcal{F}$: it is easy to see that $|\mathcal{F} \setminus \{B\}| = \frac{1}{2} \left| \binom{A \cup B}{k} \right|$. Therefore,

$$|\mathcal{F}| = \frac{1}{2}\binom{2k}{k} + 1 = \binom{2k-1}{k-1} + 1. \tag{ad 5}$$

3. **Conjecture** (Chvátal (1974)) Let $\mathcal{S} \subset 2^{[n]}$ be a downset and let $\mathcal{A} \subset \mathcal{S}$ be an intersecting system. Then

$$|\mathcal{A}| \le \max_{i \in [n]} |\mathcal{S}(i)|. \tag{ad 6}$$

4. **Conjecture** ([ESS94]) $M(n,k)$ is assumed by the simple intersecting family $_\phi \mathcal{C}_n^k$ of all k-chains meeting ϕ (or $[n]$). As

$$_\phi \mathcal{C}_n^k = \bigcup_{I=[\phi,B], B \in \binom{[n]}{r-1}, r \ge k} \mathcal{C}_n^k(I)$$

and therefore

$$|_\phi \mathcal{C}_n^k| = \sum_{r \ge k}^{n+1} q(r) \binom{n}{r-1},$$

the *conjecture* can be restated as

$$M(n,k) = \sum_{r \ge k}^{n+1} q(r) \binom{n}{r-1}. \tag{ad 8}$$

5. **Conjecture** For all n, d and some $w \ge d$ there is an optimal d-intersecting family of chains, which contain at least $\lceil \frac{w+d}{2} \rceil$ members of $X_0, X_1, \ldots, X_{w-1}$ (ad 8).

Notes to Lectures 2–4

1. Vertex Isoperimetric Theorems (VIP)

The senior author, while aiming at strong converse proofs in Multi-user Information Theory, came to conjecture a result, which he named in analogy to results in classical geometries **isoperimetric theorem** in binary Hamming space. Then

Katona [K75] proved the conjecture, and it was noticed later that it had already been proved by Harper [H64], who was also motivated by Information Theory. For general Hamming spaces carrying even a product probability distribution P^n an inequality sufficient for the information theoretical purposes was also established with the help of a 0-1-law by Margulis [M74].

We define the k-Hamming-neighborhood $\Gamma^k A$ of a set $A \subset \mathcal{X}^n$ as

$$\Gamma^k A \triangleq \{y^n \in \mathcal{X}^n : d(y^n, x^n) \leq k \text{ for some } x^n \in A\}.$$

Lemma 49 (Blowing up Lemma, Ahlswede, Gács, and Körner [AGK76]) *Let $\mathcal{P}$ be a finite set of probability distributions on $\mathcal{X}$. For any $P^n = \prod\limits_{i=1}^{n} P_i$ with $P_i \in \mathcal{P}$, there is a constant $c(\mathcal{P}) : A \subset \mathcal{X}^n$ with $P^n(\Gamma^k A) \geq \Phi(\Phi^{-1}(P^n(A))) + n^{-1/2}(k-1)c$ if $\Phi(t) = \int_{-\infty}^{t} (2\pi)^{-1/2} e^{-u^2/2} du$.*

On the one hand, it became the forerunner of great activities in probability theory concerning concentration of measures ([M96], [M98], [T91], [T95]). On the other hand, still open is the Research Problem 1.

For the Taxi space the vertex isoperimetric problem was easier and quickly solved independently by Furlmeier/Kalus [FK77] in their joint diploma thesis and by Wang/Wang [WW77]. Some additional information about isoperimetric problems in discrete space one can find in [B94].

2. Vertex Diametric Theorems (VDP)

Somewhat earlier systematic investigations were made in [AK77], where stimulated by the "Spiegelungstheorem" of E. Schmidt ([S48a], [S48b], [S49]) the following two family result was found.

We define

$$\Gamma_l(A) \triangleq \{s : s \in \mathcal{S}, \mu(s, a) \leq l \text{ for all } a \in A\}$$

and

$$\Gamma_{-l}(A) \triangleq \{s : s \in \mathcal{S}, \mu(s, b) \geq l \text{ for all } b \in A^c\}$$

in a metric space $(\mathcal{S}, \mu)$.

Lemma 50 (Ahlswede and Katona [AK77]) *In the binary Hamming space,*

$$\min_{|A|=N} |\Gamma^l(A)| = 2^n - \max_{|A|=N} |\Gamma_{-(u-r)}(\bar{A})|,$$

where

$$\bar{A} = \{b^n \in \{0.1\}^n : b^n = \bar{a}^n = (1 - a_1, \dots, 1 - a_n), a^n \in A\}.$$

Also the equivalence of a result of Kleitman [K66b], now termed diametric theorem for binary Hamming spaces, with Katona's Unrestricted Intersection Theorem 4 was shown.

3. Updating Memories with Cost Constraints: Optimal Anticodes

That Combinatorics and Information Sciences often come together is of no surprise, because they were born as twins (Leibniz in Ars Combinatoria gives credit to Raimundus Lullus from Catalania, who wanted to create a formal language).

In the example

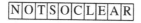

$d_H = 7$ letters have to be changed for an updating, where d_H is the Hamming distance, measuring the cost.

How many messages can be updated into each other, if cost $\leq c$? This is equivalent to the diametric problem in Hamming spaces.

For a Hamming space $(\mathcal{H}_q^n, d_H)$, the set of n-length words over the alphabet $\mathcal{X}_q = \{0, 1, \ldots, q-1\}$ endowed with the distance d_H, we determine the maximal cardinality of subsets with a prescribed diameter d or, in another language, anticodes with distance d. We refer to the result as Diametric Theorem.

In a sense anticodes are dual to codes, which have a prescribed *lower* bound on the pairwise distance. It is a hopeless task to determine their maximal sizes exactly.

We find it remarkable that the Diametric Theorem (for arbitrary q) can be derived from the Complete Intersection Theorem, which can be viewed as a Diametric Theorem (for $q = 2$) in the constant weight case, where all n-length words considered have exactly k ones.

This model for updating memories gave further motivation to study VDP for other metrics like the Taxi metric (Lecture 3), Lee metric, and general sum-type functions $\varphi^n(x^n, y^n) = \sum_{t=1}^{n} \varphi(x_t, y_t)$, where $\varphi : \mathcal{X} \times \mathcal{X} \to \mathbb{R}$.

For the Hamming space the problem was much harder (Lecture 2). Optimal are certain Cartesian products of a ball and a suitable subcube (or cylinder set). Depending on the parameters this configuration can degenerate into a ball and up to isometrics (with one exception of two solutions), and there is only one solution.

Notice that the Complete Intersection Theorem for parameters (n, k, t) can be viewed as the Diametric Theorem on the Johnson space of binary n-tuples of given Hamming weight k for diameter $D = 2k - 2t$.

Its proof uses essentially the new concept of generated sets, and the result is used to prove the Diametric Theorem. This connection is the reason for presenting in this book diametric theorems and not isoperimetric theorems, which are proved by other methods and have already been presented in books as for instance [H04].

It is our aim to have the striking breakthrough in problem solving show up in the first chapter of these Advances. Later other proofs for both theorems were given by Ahlswede and Khachatrian [AK99]. The most advanced method in this direction is the shifting technique of [AAK03d], where the Shadow Theorem ([K63],[K68]) is proved by a sequence of steps not reducing the size of the shadows.

There are several essential contributions to VDP (and VIP) in Lee spaces mostly by Bollobás and Leader ([BL90], [BL91], [BL93], and [BL04]) and a start in Koppenrade [K91]. However, the cases of odd q are not completely solved.

A complete solution of an EIP has been given in [AB95] for integer arrays after work of [BL91] for good parameters.

4. Edge Diametric Theorems (EDP)

Linked to the Complete Intersection Theorem is Theorem 9 with a new concept concerning edges vs. diameter. The solution is for $\mathcal{H}_2^n$. We propose the

Research Program A

Analyze EDP for $\mathcal{H}_q^n$ and other metric sequence spaces.

We now sketch a wider scope: rate-wise optimality, average performance criteria, additional algebraic structure, and more diametric and also isoperimetric theorems in sequence spaces to give the reader a broader view. The essence is the message that advanced information-theoretical methods give "rate-wise" optimal solutions. We also present a generalization of Harper's vertex-isoperimetric problem and finally a novel vertex-diametric theorem in direct products of Abelian groups.

For some distance function d (like Hamming, Lee or Taxi metrics) the surface $\Gamma_d(A)$ is the set of points in the complement of A and with distance 1 to A.

Harper's solution of the isoperimetric problem in the Hamming space $\mathcal{H}_2^n$ can be read now in [H04]. However, a "rate-wise" optimal solution was found in [A99] for the t-th surface $\Gamma_{\rho_n}^t(A) = \{x^n \in \mathcal{X}^n : x^n \notin A, \ \rho_n(x^n, a^n) \leq t \text{ for some } a^n \in A\}$ with $t = \tau n$, where $\rho_n(x^n, a^n) = \sum_{i=1}^n \rho(x_i, a_i)$ and $\rho : \mathcal{X} \times \mathcal{X} \to \mathbb{R}$ is any symmetric function, i.e. found was the rate

$$R(\lambda, \tau) = \lim_{n \to \infty} \frac{1}{n} \min_{|A| \geq \exp(\lambda n)} \log |\Gamma_{\rho_n}^{\tau n}(A)|,$$

using the Inherently Typical Subset Lemma by Ahlswede/Yang/Zhang [AYZ97].

Another kind of diametric theorem is for *average diameter* constraint (see below). We have now gained by examples an understanding of the following classification:

$$\text{restricted case} \leftrightarrow \text{unrestricted case}$$
$$\text{worst case} \leftrightarrow \text{average case}$$
$$\text{vertex-isoperimetric} \leftrightarrow \text{edge-isoperimetric}$$
$$\text{vertex-diametric} \leftrightarrow \text{edge-diametric}$$
$$\text{exact solution} \leftrightarrow \text{rate-wise optimal solution}$$

5. Rate-Wise Optimal Solutions for the Average Case (Vertex)-Diametric Problem

Exact solutions for the worst case vertex-diametric problem have been discussed in connection with intersection theorems in Lecture 2 for the Hamming distance. A worst case diametric problem for edges in the binary case was also considered in Lecture 2.

The Diametric Theorem for the *average* was obtained in [AA94] with rate-wise optimal solution: $\mathcal{A} \subset \mathcal{X}^n$ has an average diameter not exceeding D, if

$$D_{ave} = \frac{1}{|\mathcal{A}|^2} \sum_{x^n, y^n \in \mathcal{A}} \rho_n(x^n, y^n) \leq D.$$

In [AA94] was proved the following "rate-wise" statement.

Theorem 69 (Diametric Theorem in Average) *For the Hamming space $\mathcal{H}_q^n$ and rate $0 \leq R \leq \log q$ the smallest average diameter*

$$\overline{d}_n(R) = \min_{\mathcal{A}_n \subset \mathcal{H}_q^n, \, |\mathcal{A}_n| \geq exp(Rn)} D_{ave}(\mathcal{A}_n), \, n \in \mathbb{N},$$

satisfies

$$\overline{d}(R) = \lim_{n \to \infty} \frac{1}{n} \overline{d}_n(R)$$

$$= \min \left[\sum_{x,y \in \mathcal{H}_q^1} \left(\lambda d_H(x,y) P(x) P(y) + (1-\lambda) d_H(x,y) P'(x) P'(y) \right) \right],$$

where "min" is taken over $\lambda \in [0,1]$, and probability distributions on $\mathcal{H}_q^1$ with $\lambda H(P) + (1-\lambda) H(P') \geq R$. Here $H(P) = -\sum_{x \in \mathcal{H}_q^1} P(x) \log P(x)$ stands for the entropy of the distribution P.

Writing $R = -\beta \log \beta - (1-\beta) \log(1-\beta)$ for $q = 2$ we get $\overline{d}(R) = 2\beta(1-\beta)$. For $\alpha = 3$ calculations show that $P \neq P'$ occurs in the optimization.

For general sum-type functions ρ_n instead of d_H the same result was proved in [AC97a]. There are also extensions to several sets with some pairwise mutual average distances (or costs), and some internal average distances all simultaneously valid are treated. The proofs use a tool from Information Theory to bound the cardinality of ranges of auxiliary random variables:

Lemma 51 (Support Lemma, [AK75]) *Let $\mathcal{P}(\mathcal{Z})$ be the set of all probability distributions on the finite set $\mathcal{Z}$, let f_j $(j \in [k]) : \mathcal{P}(\mathcal{Z}) \to \mathbb{R}$ be continuous functions, and let μ be a probability distribution on $\mathcal{P}(\mathcal{Z})$ with Borel σ-algebra, then there exist elements $P_i \in \mathcal{P}(\mathcal{Z})$ and $\alpha_1, \ldots, \alpha_k \geq 0$, $\sum_{i=1}^k \alpha_i = 1$ such that*

$$\int_{\mathcal{P}(\mathcal{Z})} f_j(P) \mu(dP) = \sum_{i1}^k \alpha_i f_i(P_i), \, j = 1, 2, \ldots, k.$$

Research Program B

Find sum-type functions for which the diametric problem in average can be solved exactly.

6. Edge-Isoperimetric Inequalities Rate-Wise Optimal

In this book we have focused on solving difficult problems (Chapters II, VI), introducing new concepts (such as higher level extremal problems in Chapter IV), and developing new methods (to establish AZ-identities and AD-inequalities in Chapter V). Perhaps most important here as a novelty is the local-global principle, which plays a key role in the recent book by Harper [H04]. It also stimulated work for vertex-isoperimetric problems [BS02]. Furthermore, applications and motivations for finding methods are discussed.

The lexicographical order $\mathcal{L}$ on the sequence space $[0, q-1]^n$ is defined by $x_{\mathcal{L}}^n < y_{\mathcal{L}}^n$ iff there exists a t such that $x_t < y_t$ and $x_s = y_s$ for $s < t$.

Let $G = (V, E)$ be a finite graph. For any $A \subset V$, define the set of all boundary edges

$$B(G, A) = \{\{x, y\} \in E : |\{x, y\} \cap A| = 1\}.$$

The edge-isoperimetric problem is to find the minimum of $B(G, A)$ over all sets $A \subset V$ with given $|A|$. Harper's result ([H64]) of an edge-isoperimetric problem, correctly proved by Lindsey [L64] (see also Eggleston [E58], Harper [H64], [H67b], Bernstein [B67], Clements/Lindström [CL69], Clements [C71], Kleitman/Krieger/Rothschild [KKR71], Hart [H76], and Frankl [F87]), in the binary Hamming space says that the first segments in $\mathcal{L}$ are optimal.

Now we study the edge-isoperimetric problem in the product G^n of a given graph $G = (V, E)$. For finite sets $\mathcal{X}_i$, $i = 1, 2$ and two functions $\varphi_i : 2^{\mathcal{X}_i} \to \mathbb{R}$, $i = 1, 2$ the product $\varphi_1 \times \varphi_2 : 2^{\mathcal{X}_1 \times \mathcal{X}_2} \to \mathbb{R}$ is defined by

$$\varphi_1 \times \varphi_2(A) = \sum_{x \in \mathcal{X}_2} \varphi_1(A_1(x)) + \sum_{x \in \mathcal{X}_1} \varphi_2(A_2(x)), \ A \subset \mathcal{X}_1 \times \mathcal{X}_2,$$

where $A \subset A_1 \times A_2$ and

$$A_1(x) = \{x_1 \in \mathcal{X}_1 : (x_1, x) \in A\},$$
$$A_2(x) = \{x_2 \in \mathcal{X}_2 : (x, x_2) \in A\}.$$

The n-power φ^n is defined as

$$\varphi^n = ((\varphi \times \varphi) \times \varphi) \times \ldots \times \varphi),$$

It is easy to see that if $\varphi(G, \cdot) = -B(G, \cdot)$, then $\varphi^n(\cdot) = -B(G^n, \cdot)$ and the edge-isoperimetric problem reduces to the problem of maximizing $\varphi^n(A)$ over $A \subset V^n$ with fixed $|A|$.

W.l.o.g. we assume that $V = [0, q-1]$. We need the following properties.

1. (Nestedness): For $k \in \mathcal{X}$, $A \subset [0, q-1]$, $|A| = k+1$

$$\varphi(A) \leq \varphi([0,k]).$$

(Hamming, Lee and Taxi metrics satisfy this property).

2. (Submodularity): For $A, B \subset [0, q-1]$

$$\varphi(A) + \varphi(B) \leq \varphi(A \cup B) + \varphi(A \cap B).$$

3. We can always assume that $\varphi(\emptyset) = 0$, otherwise we replace $\varphi(A)$ by $\varphi(A) - \varphi(\emptyset)$.

The problem is to maximize (for given φ) value $\varphi^n(A)$ over all $A \subset \mathcal{X}^n$ with fixed $|A|$.

In [AC97c] it is proved that if $q > 2$, then for any set function $\varphi : 2^{[0,q-1]} \to \mathbb{R}$, satisfying the above three conditions it holds that for any integer $n \geq 2$, sets in the initial order $\mathcal{L}$ on $\mathcal{X}^n$ are optimal for φ^n iff initial sets in order $\mathcal{L}$ on $[0, q-1]^2$ are optimal for φ^2. One can easily conclude from here that this theorem (called local–global principle) gives solutions of edge-isoperimetric problems in Hamming spaces and in some other cases ([H04]).

Define

$$\Delta_\varphi(k) = \varphi([0,k]) - \varphi([0, k-1]).$$

A pair (R, δ) is achievable, if for all $\varepsilon_1, \varepsilon_2 > 0$ there exists an $n(\varepsilon_1, \varepsilon_2)$ such that for every $n > n(\varepsilon_1, \varepsilon_2)$ there is an $A_n \subset \mathcal{X}^n$ with $\left| \frac{1}{n} \log |A_n| - R \right| < \varepsilon_1$ and $\frac{1}{n|A_n|} \varphi^n(A) < \delta - \varepsilon_2$. Let $\mathcal{R}_\varphi$ is the set of achievable pairs (R, δ).

Theorem 70 (Ahlswede and Cai [AC97b]) *The following is true:*

$$\mathcal{R}_\varphi = \{(H(X|U), \mathbb{E}\Delta_\varphi(X)) : X, U \text{ satisfy conditions below}\}.$$

Conditions:

(i) *Random variable X takes values in a finite set $\mathcal{X}$ and random variable U takes values in $\mathcal{U}$.*

(ii) $|\mathcal{U}| \leq |\mathcal{X}| + 1$.

(iii) $Pr(X = 0 | U = u) \geq Pr(X = 1 | U = u) \geq \ldots (X = q-1 | U = u)$.

7. Boundaries with Intensity

Let $B(x^n, t) = \{z^n \in \mathcal{H}_2^n : d_H(x^n, z^n) \leq t\}$ be the ball in Hamming space $\mathcal{H}_2^n$ of radius t with the center in x^n. For the set $V \subset \mathcal{H}_2^n$ denote

$$B_k(V) = \left\{ x^n \in \mathcal{H}_2^n \setminus V, \left| B(x^n, 1) \cap V \right| \geq k \right\}.$$

8. A Vertex-Diametric Theorem in $\mathcal{G}^n$

Next we consider the vertex-diametric problem with group structure in the direct product $\mathcal{G}^n = \sum_{i=1}^n \mathcal{G}$ of finite Abelian groups $\mathcal{G} = \{a_0 = 0, \ldots, a_{q-1}\}$. We impose the condition that the set in $\mathcal{G}^n$ of diameter d, which contains the maximal number of elements, should be a subgroup of $\mathcal{G}^n$. We write the elements $g^n \in \mathcal{G}^n$ as concatenations $u^n = u_1 u_2 \ldots u_n$, $u_i \in \mathcal{G}$, while the operation in group $\mathcal{G}$ will be addition. We need the following definitions.

The zero word of length ℓ is denoted 0^ℓ.

For $\mathcal{U} \subset \mathcal{G}^n$ we define $S \subset \mathcal{G}$, $S \neq \emptyset$,

$$\mathcal{U}_S = \{u_1 \ldots u_{n-1} : u_1 \ldots u_{n-1}s \in \mathcal{U} \; \forall \, s \in S,$$
$$\text{and } u_1 \ldots u_{n-1}s \notin \mathcal{U} \; \forall \, s \in \mathcal{G} \setminus S\}.$$

clearly
$$\mathcal{U}_S \cap \mathcal{U}_{S'} = \emptyset, \; S \neq S'.$$

For $\mathcal{U} \subset \mathcal{G}$ we define $\mathcal{U}_{(n)} = \{u_n \in \mathcal{G} : \text{there exists a } u_1 \ldots u_{n-1} \text{ with } u_1 \ldots u_n \in \mathcal{U}\}$.
For $A \subset \mathcal{G}^m$, $B \subset \mathcal{G}^\ell$ we write

$$AB = \{a^m b^\ell : a^m \in A, \, b^\ell \in B\}.$$

We assume that on $\mathcal{G}^n$ is defined (Hamming) metrics: $d(u^n, v^n) = |\{i : v_i \neq u_i\}|$ and for $\mathcal{U} \subset \mathcal{G}^n$ denote $D(\mathcal{U}) = \max_{v^n, u^n \in \mathcal{U}} d(v^n, u^n)$. We consider

$$A\mathcal{G}(n, d) = \max\{|\mathcal{U}| : \mathcal{U} \text{ is a subgroup of } \mathcal{G}\}.$$

We are going to show the proof of the following

Theorem 71 (Ahlswede [A06b]) *For any finite Abelian group $\mathcal{G}$ and $n \geq d$*

$$A\mathcal{G}(n, d) = |\mathcal{G}|^d.$$

Proof. We need some propositions.

Proposition 19 *For a subgroup $\mathcal{U} \subset \mathcal{G}^n$ a nonempty $\mathcal{U}_{S_0} 0$ is a subgroup of $\mathcal{U}$.*

Proof. If for $u, v \in \mathcal{U}_{S_0}$ we have $u0 + v0 \notin \mathcal{U}_{S_0} 0$, then $u0 + v0 \in \mathcal{U}_S 0$ and $u + v \in \mathcal{U}_S$, where $S \neq S_0$ and $S \supset S_0$, because $u0 \in \mathcal{U}$, $(u + v)su0 + vs \in \mathcal{U}$. Now for $x \in S \setminus S_0$, $(u + v)s - u0 = vx \in \mathcal{U}$, but this contradicts $v \in \mathcal{U}_{S_0}$, and hence $u0 + v0 \in \mathcal{U}_{S_0} 0$. It remains to be seen that $u0$ has an inverse in $\mathcal{U}_{S_0} 0$.

There is a $v0 \in \mathcal{U}$ with $u0 + v0 = 0^n$. If $v0 \notin \mathcal{U}_{S_0} 0$, then $v0 \in \mathcal{U}_S 0$, where $S \neq S_0$ and $S \supset S_0$, because for all $s \in S_0$, $us \in \mathcal{U}$ and since $u0 + vsus + v0 \in \mathcal{U}$ also $vs \in \mathcal{U}$.

Now for $x \in S \setminus S_0$ we have $u0 + vx \in \mathcal{U}$ and therefore $ux + v0 \in \mathcal{U}$ and $ux \in \mathcal{U}$ in contradiction to $u \in \mathcal{U}_{S_0}$. Thus $\mathcal{U}_{S_0} 0$ is a subgroup. $\square$

Proposition 20 (i) *There is exactly one S_0 with $\mathcal{U}_{S_0} \neq \emptyset$*
(ii) *This S_0 is a subgroup of $\mathcal{G}$.*
(iii) *$\mathcal{U}_{S_0} S_0$ is a subgroup of $\mathcal{U}$.*

Proof.

(i) As $0^n \in \mathcal{U}_{S_0} S_0$ for all sets of type S_0 (by Proposition 19), disjointness of these sets gives the statement of the Proposition.

(ii) As $0^n \in \mathcal{U}_{S_0} S_0$, also $0^{n-1}s \in \mathcal{U}_{S_0} S_0$ for all $s \in S_0$, and for all $s, s' \in S_0$

$$0^{n-1}s + 0^{n-1}s' = 0^{n-1}s'' \in \mathcal{U}.$$

If $s'' \notin S_0$, then this contradicts that $0^{n-1} \in \mathcal{U}_{S_0}$. Therefore $s + s's'' \in S_0$. Concerning the inverse of s in S_0 use that $0^{n-1}s$ has an inverse $0^{n-1}(-s) \in \mathcal{U}$. Again by definition of $\mathcal{U}_{S_0}$ we have $-s \in S_0$.

(iii) $\mathcal{U}_{S_0} S_0$ is a subgroup of $\mathcal{U}$ because it is a direct sum of groups and contained in $\mathcal{U}$. $\qquad\square$

Next consider the decomposition of $\mathcal{U}$ into set of cosets of $\mathcal{U}_{S_0} S_0$ and with representatives of these cosets of the form $0^{n-1}\alpha_i$, $(i = 1, \dots, I)$ such that

$$S_0 + \alpha_i \text{ are disjoint for } i = 1, \dots, I.$$

This gives cosets in $\mathcal{U}$:

$$\mathcal{U}_{S_0}(S_0 + \alpha_i), \ i = 1, \dots, I.$$

However, necessarily $I = 1$. We can choose $\alpha_1 = 0$. Generally, using $\mathcal{U}_{S_0} S_0$ we can make the decomposition into cosets

$$\mathcal{U} = \bigcup_{\gamma}(\mathcal{U}_{S_0} + \gamma)(S_0 + \psi(\gamma)) \tag{4}$$

or equivalently

$$\mathcal{U} = \bigcup_{\beta}(\mathcal{U}_{S_0} + \psi^{-1}(\beta))(S_0 + \beta).$$

Proposition 21 *If for a subgroup $\mathcal{U} \subset \mathcal{G}^n$, $|S_0| > 1$, then the transformation*

$$L : \bigcup_S \mathcal{U}_S S \to \left(\bigcup_S \mathcal{U}_S\right) \cdot \mathcal{G}$$

results in a group of diameter $\leq d$ and not a decreased cardinality.

Proof. From (4) it follows that every u^{n-1} occurring in some $\mathcal{U}_{S_0} + \gamma$ has multiplicity $|S_0 + \psi(\gamma)| = |S_0|$ and gets by the transformation multiplicity $|\mathcal{G}| \geq |S_0|$. So the cardinality does not decrease. Furthermore $D(\mathcal{U}_{S_0} + \gamma) \leq d - 1$ and also

$$D(\mathcal{U}_{S_0}\gamma, \mathcal{U}_{S_0}\gamma') \leq d - 1$$

and the transformation L is appropriate. $\qquad\square$

It remains to analyze the case $S_0 = \{0\}$. Here, due to the definition of $\mathcal{U}_{S_0}$, the decomposition

$$\mathcal{U} \bigcup_{\beta \in \mathcal{U}(n)} (\mathcal{U}_{S_0} + \psi^{-1}(\beta))\beta \tag{5}$$

holds. The terms $(\mathcal{U}_{S_0} + \psi^{-1}(\beta))$ are disjoint or equal. If $\mathcal{U}_{S_0} + \psi^{-1}(\beta) = \mathcal{U}_{S_0} + \psi^{-1}(\alpha)$, then $\psi^{-1}(\alpha) - \psi^{-1}(\beta) \in \mathcal{U}_{S_0}$. Hence $D(\mathcal{U}_{S_0} + \psi^{-1}(\alpha)) = D(\mathcal{U}_{S_0}) \leq d - 1$ and we can make the same procedure L as in the Proposition 21 replacing all last symbols β in the decomposition (5) by $\mathcal{G}$.

So it remains to consider the case, when

$$\mathcal{U}_{S_0} + \psi^{-1}(\alpha) \neq \mathcal{U}_{S_0} + \psi^{-1}(\beta)$$

for all $\alpha \neq \beta$. In this case we replace all last β's in the decomposition (5) by 0.

We can continue this process until we come to the set $\mathcal{U} = \mathcal{G}^d$, which has the desired properties. This completes the proof of the theorem. ∎

Remark Inspection shows that for all n $A\mathcal{G}(n,d) = |\mathcal{G}|^{\min(n,d)}$.

Exercises

1. Derive Kleitman's result from Lemma 50 (ad 2).

Research Problems

1. Generalize Harper's isoperimetric theorem for binary to general Hamming spaces $\mathcal{H}_q^n$. The aim here is of course an exact result (ad 1)!
2. Convert the incomplete diametric theorems for Lee spaces to a Complete Diametric Theorem (ad 3)!
3. Determine
$$b_k(n,N) = \min_{V \in \mathcal{H}_2^n: |V|=N} |B_k(V)|.$$

In the case $k = 1$ this problem reduces to the case which was solved by Harper (we have already mentioned this earlier) (ad 7).
4. Establish vertex-diametric theorems for Krotov-distances ([K00], [K01], [K07]) (ad 8).

Notes to Lecture 5

In [E05] sets $F \subset W_q^n$ are studied by considering relations different from but closely related to the relation $\nearrow\searrow$. For each position $t \in [n]$, let $L_t \subset [n] \setminus \{t\}$ be a list of positions. Furthermore, let the words in F be ordered, say $F = \{u^n(1), u^n(2), \ldots, u^n(|F|)\}$. We consider the family $\mathcal{F}_q^n(L_1, L_2, \ldots, L_n)$ of ordered sets F, which satisfy for every $i < j$ the condition that there exists a $t \in [n]$ and an $s \in L_t$ such that $u_t(i) = u_s(j)$. Then

$$f_q^n(L_1, L_2, \ldots, L_n) = \max\{|F| : F \in \mathcal{F}_q^n(L_1, L_2, \ldots, L_n)\} \leq \prod_{t=1}^n (|L_t| + 1)$$

From here follows the inequality

$$f_q^n([n] \setminus \{1\}, \ldots, [n] \setminus \{n\}) \leq n^n.$$

In [T89] a better bound is proved:

$$f_q^n([n] \setminus \{1\}, \ldots, [n] \setminus \{n\}) \leq \frac{\binom{q}{n}}{\lceil q/n \rceil^r \lfloor q/n \rfloor^{n-r}}.$$

Note, that if n divides q, then

$$\frac{\binom{q}{n}}{\lceil q/n \rceil^r \lfloor q/n \rfloor^{n-r}} = n^n \frac{\binom{q}{n}}{q^n}$$

and the last bound is tight for $q = n + 1$ giving the bound $q!/2$.

Let us just also draw attention to another direction. There is a wide area of extremal problems concerning words of various length. For instance with relations of sequence–subsequence, sequence–supersequence, for which many problems considered in this book are also meaningful. We draw the reader's attention to the papers [AC97d], [DD97a], and [DD97b].

Also of interest are extremal problems for special families of sets like downsets and upsets, also with weight assignments, and of course also diametric and isoperimetric problems for abstract graphs and hypergraphs.

Theorem 72 (Ahlswede and Katona [AK77]) *Denote by $K_i = K_i(\mathcal{U})$, $i = (0, 1, \ldots, n)$, the number of i-element members of a downset. Then*

$$\min_{\mathcal{U}:|\mathcal{U}|=N} \sum_{i=1}^n K_i w_i, 1 \leq N \leq 2^n,$$

is assumed

(i) *in case $w_0 \leq w_1 \leq \cdots \leq w_M \geq w_{M+1} \geq \cdots \geq w_n$ by a union of a quasi-cylinder and a quasi-sphere,*

(ii) *in case $w_0 \geq w_1 \geq \cdots \geq w_M \leq w_{M+1} \leq \cdots \leq M_n$ by an intersection of a quasi-cylinder and a quasi-sphere.*

Research Problems

1. Find the value of $f_q^n([n] \setminus \{1\}, \ldots, [n] \setminus \{n\})$ for all q.

Notes to Lecture 6

For the value

$$M_q(n) = \max_{0 \le \delta \le n} M_q(n, \delta)$$

when $q = 2$ the bound

$$M_2(n) \le \begin{cases} 2^n, & 2|n, \\ 2^{n-1}, & 2 \nmid n. \end{cases} \tag{6}$$

was first proved in [AEP84].

Equality in (6) is achieved for example on the sets $A = \{01, 10\}^m$, $B = \{11, 00\}^m$, $n = 2m$ or $A = \{01, 10\}^m \times \{0\}$, $B = \{11, 00\}^m \times \{0\}$ $n = 2m + 1$. Simple (and closely related) proofs of (6) based on elementary Linear Algebra and slight generalizations were given in [DP85], [HL67].

One can replace the property (A, B) to be a constant distance pair by the one-sided constant distance property:

$$d_H(a^n, b^n) = d_H(a^n, b'^m), \ a^n \in A, \ b^n, b'^m \in B.$$

Obviously from the property of the pair to be constant distance follows the one-sided property, from which in turn follows $4-$WP.

Next consider the family $A \subset \{0, 1\}^n$ which for a prescribed set $L \subset \{0, 1, \dots\}$ has the property that $\lambda(a^n, b^n) \in L$ for distinct members $a^n, b^n \in A$. In [FW81] it is proved that

$$|A| \le \sum_{i=0}^{|L|} \binom{n}{i}.$$

This inequality implies as a corollary the following fact. Given integers p and $c < n$, let $A \subset \{0, 1\}^n$ be such that for all $a^n \ne b^n \in A$ one has

$$\lambda(a^n, b^n) \equiv c \mod p, \tag{7}$$

then

$$|A| \le \sum_{i=0}^{\lfloor (n-c-1)/p \rfloor} \binom{n}{i}.$$

Moreover, Frankl and Rödl [FR87] extended this result to two families, thus obtaining a generalization of Ahlswede, El Gamal, and Pang [AEP84].

Theorem 73 (Frankl and Rödl 1987) *If for $A, B \subset 2^{[n]}$ (7) holds for all $a^n \in A$ and $b^n \in B$ then either*

(i) $c = 0$ and $|A||B| \le 2^n$

or

(ii) $0 < c < p$ and $|A||B| \le 2^{n-1}$.

Another generalization of (6) for T families is given by Aydinian in [A88].

Theorem 74 *Let $A_1, \dots, A_T \subset \mathcal{H}_2^n$ be distinct families with the property*

$$d_H(a_i^n, a_j^n) = \delta(i, j), \ a_i^n \in A_i, \ a_j^n \in A_j, i \ne j \in [1, T],$$

then

$$|A_1| \cdot \ldots \cdot |A_T| \leq 2^n. \tag{8}$$

Proof. Consider instead of vectors $a^n = (a_1, \ldots, a_n) \in \mathcal{H}_2^n$ their bipolar images $\bar{a}^n = (\bar{a}_1, \ldots, \bar{a}_n)$, $\bar{a} = (-1)^a$. If $A_1, A_2 \subset 2^{[n]}$ are such that $d_H(a_1^n, a_2^n) = \delta$, $a_1^n \in A_1$, $a_2^n \in A_2$, then we say that vectors $\bar{c}^k, \bar{d}^k \in \{-1,1\}^k$ are orthogonal if $(\bar{c}^n, \bar{d}^n) = \sum_{i=1}^n \bar{c}_i, \bar{d}_i = 0$. Let now $\alpha^m(i) = (\alpha_1(i), \ldots, \alpha_m(i))$, $i = 1, \ldots, T$ be T orthogonal vectors of length $m \geq T$ with components from $\{-1, +1\}$.

Note that such vectors always exist (use e.g., Hadamard matrices of order $m = 2^r$, $r \in \mathbb{N}$).

Consider new sets $B_1, \ldots, B_T \subset \{-1,1\}^n$: $B_i = \{(\alpha_1(i) \cdot \bar{a}_i^n(1), \ldots, \alpha_m(i) \cdot \bar{a}_i^n(m)), a_i^n(j) \in A_i, j = 1, \ldots, |A_i|\}$. It is easy to see that $|B_i| = |A_i|^m$ and $span(B_i) \perp span(B_j)$. Thus

$$\sum_{i=1}^T \dim span(B_i) \leq mn.$$

As $|\{-1,1\}^n \cap span(B_i)| \leq 2^{\dim span(B_i)}$ we have $|B_1| \cdots |B_T| \leq 2^{mn}$. Since $|B_i| = |A_i|^m$, from the last inequality we obtain (8). □

Notes to Lecture 7, 8

Recall that for product hypergraphs $\mathcal{H}^n$ the minimal covering number $c(\mathcal{H}^n)$, the minimal partition number $\pi(\mathcal{H}^n)$, and the maximal packing number $p(\mathcal{H}^n)$ satisfy

$$c(\mathcal{H}^n) \leq \pi(\mathcal{H}^n) \leq p(\mathcal{H}^n),$$

if $c(\mathcal{H}^n)$ and $p(\mathcal{H}^n)$ are well defined, whereas the zero error capacity $\lim_{n \to \infty} \frac{1}{n} \log p(\mathcal{H}^n)$ is known only for very few cases, a nice formula exists for $\lim_{n \to \infty} \frac{1}{n} \log c(\mathcal{H}^n)$ (see Theorem 17).

Even in the case of nonidentical factors $\mathcal{H}_i = (\mathcal{V}_i, \mathcal{E}_i)$, $i \in \mathbb{N}$, with $\max_i |\mathcal{E}_i| < \infty$, the asymptotics of $c(\mathcal{H}^n)$ is known (pages 762–771 in [G]):

$$\lim_{n \to \infty} \frac{1}{n} \left(\log c(\mathcal{H}^n) - \sum_{t=1}^n \log \left(\max_{q \in Prob(\mathcal{E}_t)} \min_{v \in \mathcal{E}_t} \sum_{E \in \mathcal{E}_t} 1_E(v) q_E \right)^{-1} \right) = 0,$$

where $Prob(\mathcal{E}_t)$ is the set of all probability distributions on $\mathcal{E}$, 1_E is the indicator function of the set E.

The main difficulty in determining the zero error capacity comes from odd cycles of length ≥ 5. Cycle length 5 was settled by Lovász [L79b]. The case of length 7 is already unsolved. Haemers [H79] and Alon [A98] disproved conjectures by Shannon [S56] and Lovász: neither is the zero error capacity additive for two parallel channels, that is, $\mathcal{V}_t = \mathcal{V} \times \mathcal{V}'$ and $\mathcal{E}_t = \mathcal{E} \times \mathcal{E}'$ for $t = 1, 2, \ldots$, nor has the "sum channel" an "equivalent number of letters" equal to the sum of the "equivalent number of letters" for the individual channels. These phenomena are a good entrance into concepts and methods of this fascinating study of independence numbers of graphs.

The difficulties in analyzing $\pi(\mathcal{H}^n)$ are similar to those for $p(\mathcal{H}^n)$. For the case of graphs with edge set $\mathcal{E}$ including all loops, we prove that $\pi(\mathcal{H}^n) = \pi(\mathcal{H})^n$ ([AC93b]). This result is derived from the corresponding result for complete graphs with the help of Gallai's Lemma in Matching Theory. Another interesting quantity is $\mu(\mathcal{H}^n)$, the maximal size of a partition of V^n into sets that are elements of $\mathcal{E}^n$ (again only hypergraphs $(V, \mathcal{E})$ with a partition are considered). We also call μ the maximal partition number. It behaves more like the packing number. Clearly, $\pi(\mathcal{H}^n) \leq \mu(\mathcal{H}^n) \leq p(\mathcal{H}^n)$. It seems to us that an understanding of these partition problems would be a significant contribution to an understanding of the basic, and seemingly simple, notion of Cartesian products.

We present now a hypergraph covering lemma useful for deriving capacity results in the theory of Identification (for classical [A06c] in [G], 926–937 and quantum channels [AW02]), in the theory of Common Randomness [AC98] (where actually also balanced partitions, started in [A79] and [A80], play a basic rule nowadays) and helpful for the analysis of number theoretical complexity measures useful for Cryptography ([AKMS03], [AMS06a] in [G], 293–307, [AMS06b] in [G], 308–325).

Lemma 52 *Let* $\Gamma = (V, \mathcal{E})$ *be a hypergraph, with a measure* Q_E *on each edge* E, *such that* $Q_E(v) \leq \eta$ *for all* E, $v \in E$. *For a probability distribution* P *on* $\mathcal{E}$ *define*

$$Q = \sum_{E \in \mathcal{E}} P(E) Q_E,$$

and fix $\varepsilon, \eta > 0$. *Then there exist vertices* $V_0 \subset V$ *and edges* $E_1, \ldots, E_L \in \mathcal{E}$ *such that with*

$$\bar{Q} = \frac{1}{L} \sum_{i=1}^{L} Q_{E_i}$$

the following holds:

$$Q(V_0) \leq \eta,$$

$$\forall v \in V \setminus V_0 \ (1 - \varepsilon) Q(v) \leq \bar{Q}(v) \leq (1 + \varepsilon) Q(v),$$

$$L \leq 1 + \eta |V| \frac{2 \ln 2 \log(2|V|)}{\varepsilon^2 \eta}.$$

Basic covering problems are investigated by Dumer, Pinsker, and Prelov in [G], 738–761: the problem of thinnest coverings of spheres and ellipsoids with balls and ellipsoids in Hamming and Euclidean spaces. New bounds in terms of the ε-entropy of Hamming balls and spheres are established. The derived upper and lower bounds are optimal up to an additive logarithmic term on the dimension.

Notes to Lecture 11

1. On Pairs of Families with Mutually Incomparable Sets

We consider the partially ordered set with elements from $[0, k-1]^n$, which is defined as the direct product of the chain $[0, k-1] = \{0, 1, \ldots, k-1\}$, and study the set $CAC(n, k)$ of pairs $(\mathcal{A}, \mathcal{B})$ of incomparable sets $\mathcal{A}, \mathcal{B} \subset [0, k-1]^n$, that is, $A \not\leq B$, $A \not\geq B$ for all $A \in \mathcal{A}$, $B \in \mathcal{B}$.

In [AZ90b] for the case $k = 2$ it is proved that $|\mathcal{A}||\mathcal{B}| \leq 2^{2n-4}$. Equality here occurs for instance if

$$\mathcal{A} = \{X \in 2^{[n]} : 1 \in X, 2 \notin x\},$$
$$\mathcal{B} = \{X \in 2^{[n]} : 1 \notin X, q \in X\}.$$

We are concerned with the growth of the functions

$$f(k, n, m) = \max\{|\mathcal{B}| : (\mathcal{A}, \mathcal{B}) \in CAC(n, k), |\mathcal{A}| = m\}.$$

In [AK96c] for $(\mathcal{A}, \mathcal{B})$ the inequality

$$|\mathcal{A}|^{1/2} + |\mathcal{B}|^{1/2} \leq k^{n/2} \tag{9}$$

is proved.

In the same paper it is shown that for $0 \leq m \leq 2^{n-1}$ we have

$$f(2, n, m) = 2^{n-1} + 2f(2, n-2, m) - m.$$

2. A Higher Level Extremal Problem for Hamming Distance

Consider the following problem: find the maximal number $m(n)$ of subsets $\mathcal{A}_i$ such that $\{\mathcal{A}_i, i = 1, \ldots m(n)\}$ is a partition of the Hamming space $\mathcal{H}_2^n$ and the distance between each pair of elements from different subsets is equal to 1. In [ABBMM93] the bounds

$$\frac{\sqrt{2}}{2} \sqrt{n2^n} \leq m(n) \leq \sqrt{n2^n} + 1 \tag{10}$$

were derived.

3. Other Kinds of Dimensions

For a finite set $[n]$ we studied the family $2^{[n]}$ of all its subsets by assigning to $A \in 2^{[n]}$ a 0-1-vector in the canonical way and to a family $\mathcal{A} \subset 2^{[n]}$ a set of 0-1-vectors.

As subset of the vector space $\{0,1\}^n$ it has a dimension, which is the vector space dimension of the subspace it generates.

According to [FK06] Erdös communicated to them in 1993 that for a set system $\mathcal{A} \subset 2^{[n]}$ he considers set systems $\mathcal{B} \subset 2^{[n]}$ with the property that every $A \in \mathcal{A}$ can be written as a (nonempty) union of at most $b = 2$ members of $\mathcal{B}$. Calling such a $\mathcal{B}$ a 2-base he asked for determining

$$d_2(\mathcal{A}) := \min\{|\mathcal{B}| : \mathcal{B} \text{ is a 2-base of } \mathcal{A}\},$$

which we call 2-dimension of $\mathcal{A}$. He was primarily interested in the cases

$$\mathcal{A}(n) = 2^{[n]}, \mathcal{A}(n, \le k) = \binom{[n]}{\le k} = \bigcup_{l=0}^{k} \binom{[n]}{l}, \quad \text{and} \quad \mathcal{A}(n,k) = \binom{[n]}{k}.$$

He conjectured that

$$d_2(\mathcal{A}(n)) = 2^{\lfloor \frac{n}{2} \rfloor} + 2^{\lceil \frac{n}{2} \rceil} - 1 \tag{11}$$

and that an extremal family consists of all subsets of V_1 and V_2, where $V_1 \cup V_2 = [n]$ is a partition of $[n]$ into two almost equal parts. A lower bound $d_2(\mathcal{A}(n)) \ge (1 + o(1))2^{\frac{(n+1)}{2}}$ is a consequence of the inequality $|\mathcal{A}| \le \binom{|\mathcal{B}|}{2} + |\mathcal{B}|$, which holds for any 2-base $\mathcal{B}$ of $\mathcal{A}$.

Although the determination of $d_2(\mathcal{A}(n, \le k))$ is trivial for $k \le 2$ for $k = 3$, it can be obtained from Turan's theorem. The next case $k = 4$ was settled in [FK06]. From $d_2(\mathcal{A}(n, \le k))$ also $d_2(\mathcal{A}(n,k))$ can be derived.

Recently [FLS07] modified the question of Erdös by allowing only (nonempty) disjoint unions and generalized it from $b = 2$ to any value of b by allowing at most b members of $\mathcal{B}$. This leads to a b-dis-basis and the b-dis-dimension d_b^*. For $\mathcal{A}(n)$ they conjectured optimality of $\mathcal{B}$ having all subsets as members, which are contained in the atoms of balanced partitions of $[n]$ with b atoms. That is, writing n in the form

$$n = ab + c \quad \text{with} \quad 0 \le c < a,$$

they conjecture
$$d_b(\mathcal{A}(n)) = (b+c)2^a + 1 - b, \tag{12}$$

which they proved among others for the cases $n \le 3b$.

There are further reasonable generalizations. We emphasize the extensions of the sequence space $\{0,1\}^n$ to $\{0,1,\ldots,q-1\}^n$ with the operation $a^n \vee b^n = (a_1 \vee b_1, \ldots, a_n \vee b_n)$.

We also draw attention to a complexity problem, 3.10 on pages 57–58 of [CG87].

Finally, the area of extremal problems under dimension constraints [ACZ96] becomes a wide field, if we include the dimension concepts mentioned earlier.

Research Problems

1. For every k describe all $CAC(n,k)$ with equality in (9) (ad 1).
2. How does $f(k,n,m)$ behave asymptotically in k,n and m (ad 1)?

3. Reduce the gap between the bounds in (10) (ad 2).
4. Prove the conjecture (11) of Erdös (ad 3).
5. Prove the conjecture (12) of [FLS07] (ad 3).

Notes to Lecture 12

1. On Security of Statistical Databases

It is interesting that results and methods developed for systematic study of extremal problems under dimension constraints in this lecture (see also the survey paper [AAK03e]) could be useful for applications, in particular for the study of database security problems. The results presented in [AA07] are either direct consequences of results in [AAK03e] or can be easily derived using tools from that paper.

A Statistical Database (SDB) is a database that returns statistical information derived from the records to user queries for statistical data analysis. An important problem is to provide security to SDB against the disclosure of confidential information. Examples of confidential information stored in a SDB might be salaries or data concerning the medical history of individuals. A statistical database is said to be *secure* if no protected data can be inferred from the available queries, otherwise it is called *compromised*. One of the security-control methods suggested in the literature (see [D82] and [D-F02] for more information) consists of query restriction: the security problem is to limit the use of the SDB, introducing a control mechanism, such that no protected data can be obtained from the available queries. Such a control mechanism for query restriction was proposed in [CO82], where only SUM queries, that is only certain sums of individual records, are available for the users.

As an example consider a company N with n employees. Suppose that for each member of N is recorded the sex, age, rank, length of her/his employment with N, salary, etc. The salaries $\{z_1, \ldots, z_n\}$ of the individual employees are confidential. Only SUM queries are allowed, i.e., the sum of the salaries of the specified people is returned. For example one might pose the query: What is the sum of salaries for males above 50?

How large can be the number of SUM queries, preventing compromise (i.e., no individual salary z_i can be inferred using the outcomes from the list of allowed SUM queries)?

More generally, let $z_1, \ldots, z_n$ be nonzero real numbers, which are n confidential records stored in a database. A possible SUM query for users is $S_A := \sum_{i \in A} z_i$ for some $A \subset [n] := \{1, \ldots, n\}$ with $|A| > 1$. A natural problem is to maximize the number of SUM queries, possibly with some other side constraints, without compromise. This problem was originally stated in [CO82] and is studied (in different settings) in [BHM00], [DKM04], [E65], [G99], [HBM99], and [MS94].

In particular, consider the problem (without constraints): Maximize the number M of subsets (answerable query sets) $A_1, \ldots, A_M \subset [n]$ such that the knowledge of the corresponding sums $S_{A_1}, \ldots, S_{A_M}$ does not enable one to determine any of records z_i.

This problem can be reduced to the following one. Given nonzero real numbers $a_1, \ldots, a_n$, determine the maximum possible number of subsets with a zero sum, that is determine the maximum number of $(0,1)$-solutions of an equation

$$a_1 x_1 + \ldots + a_n x_n = 0 \quad (a_i \in \mathbb{R} \setminus \{0\}). \tag{13}$$

Theorem 75 (Miller, Roberts, and Simpson 1991) *(i) The maximum number of answerable SUM queries without compromise, from a database of n real entries z_i is $\binom{n}{\lfloor n/2 \rfloor}$.*

(ii) ([G99]) The maximum is achieved iff the set of entries is partitioned into two parts, of sizes $\lfloor \frac{n}{2} \rfloor$ and $\lceil \frac{n}{2} \rceil$, and all query sets have equal number of elements from each part. Equivalently, the maximum number of $(0,1)$-solutions of $(1,1)$, assumed for $a_i = -a_{i+1}$ $(i = 1, \ldots, n-1)$, is unique up to permutations of the coordinates.

In [BHM00], [G99], [HBM99], [MRS91] other models of compromise were introduced and studied. Among them so called *relative compromise* where either some record z_i or some difference $z_i - z_j$ $(i \neq j)$ can be inferred from available queries. This model leads to the Erdös-Moser problem [E65] (determine the largest possible number of subsets of a set of real numbers $\{a_1, \ldots, a_n\}$ having a common sum of elements) and its generalizations ([G99]). In an excellent survey paper by Griggs, further fundamental models of security: *group-security, internal-security, etc.*, were proposed. It was shown that they lead to challenging combinatorial, number theoretic, and geometric problems.

All these problems can be formulated in terms of $(0,1)$ – solutions of some linear equations (over real numbers) with certain restrictions.

In the model called *group-security model* suggested in [G99] not only individual data but also subset sums of subsets $I \subset [n]$ with small size, say $0 < |I| \leq g$, must be protected. This problem is equivalent to the following one.

Determine the maximum number $G(n,g)$ of $(0,1)$-solutions of equation (13) provided there are no nonzero solutions of Hamming weight less than $g + 1$. Assume that the number of elements in the SUM queries are restricted by the size constraint: only sums of m (or at most m) elements are considered. Then the problem is equivalent to finding the maximal number of $(0,1)$-solutions of weight m (or weight not exceeding m) of equation (13), provided there are no (nonzero) solutions of weight less than $g + 1$. We denote these quantities by $G(n, m, g)$ (resp. $G(n, \leq m, g)$). The problem for $g = 1$ was considered and solved for $n \gg m$ in [DKM04].

Theorem 76 (Demetrovich, Katona, and Miklos 2004) *For integers $1 < m \leq n$ and $t := \lfloor n/m \rfloor$, holds*

(i) $G(n, m, 1) = t \binom{n-t}{m-1}$, if $n \gg m$.
(ii) $G(n, \leq m, 1) = t \binom{n-t}{m-1}(1 + o(1))$, as $n \to \infty$.

In [AA07] $G(n, m, g)$ is determined for all parameters. Surprisingly the answer is the same as for 1-security, that is $G(n, m, g) = G(n, m, 1)$. Also $G(n, g)$ is determined (as well as $G_k(n,g)$), within a constant factor less than $1/2$. It turns out that for all $1 < g < \frac{n}{2}$, the number of answerable queries decreases less than two times when compared with 1-security, that is $\frac{1}{2}G(n,1) < G(n,g) < G(n,1)$.

Given $n,m,w \in \mathbb{N}$ let $F(n,m,w)$ denote the maximum number of (0,1)-vectors $X \subset \mathbb{R}^n$ of weight m such that the $span(X)$ does not contain (0,1)-vectors of weight w. Similarly the function $F(n,w)$ is defined where again vectors of weight w are forbidden but we have no restriction on the weights of (0,1)-vectors corresponding to the query sets (the unrestricted case). In [AAK03b] $F(n,m,w)$ is determined for all parameters $1 \leq w < m < n$. Results for $F(n,w)$ are presented in [AAK05]. It is clear that $F(n,m,1) = G(n,m,1)$ and $F(n,m,g) \geq G(n,m,g)$. Surprisingly, one has also equality here.

Theorem 77 (i) *For integers* $1 \leq g < m \leq n$ *let* $t \in \{\lfloor n/m \rfloor, \lfloor (n+1)/m \rfloor\}$. *Then we have*

$$F(n,m,g) = G(n,m,g) = t\binom{n-t}{m-1}. \tag{14}$$

(ii) *An optimal set of SUM queries corresponds to the set of (0,1)-solutions of weight* m *of equation* $(m-1)x_1 + \ldots + (m-1)x_t - x_{t+1} - \ldots - x_n = 0$ *and is unique (up to the permutations of the elements).*

Clearly $F(n,g) \geq G(n,g)$, note however that $F(n,g) \neq G(n,g)$, unless $g = 1$. For example, observe that $F(n,n-1) \geq |\{0,1\}^{n-2} \times \{(0,0)\}| = 2^{n-2}$, while clearly $G(n,n-1) = 1$. Finding the exact value of $G(n,g)$ seems to be more difficult. The first open case is $g = 2$.

Let us consider the following more general problem, which clearly makes sense theoretically and hopefully also practically. For integers $1 \leq g,k \leq n$ let $G_k(n,g)$ denote the maximum number of (0,1)-solutions of equation (18) such that $rank(B) = n - k$, provided there are no solutions of the weight g or less. Note that $G(n,g) = G_{n-1}(n,g)$.

A simple upper bound for $G_k(n,g)$ is 2^k.

Proposition 22 *For integers* $1 \leq k,g < n$ *holds* $G_k(n,g) = 2^k$ *if and only if* $n \geq k(g+1)$.

The next result is a generalization of Theorem 75.

Theorem 78 (i) *For integers* $\frac{n}{2} \leq k < n$ *we have*

$$G_k(n,1) = \binom{2k-n+2}{\lfloor \frac{2k-n+2}{2} \rfloor} 2^{n-k-1} \tag{15}$$

(ii) *An optimal set of SUM queries corresponds to the following set of vectors* $X \subset \{0,1\}^n$ $X = X_1 \times X_2$ *with* $X_1 := \{(x_1,\ldots,x_{2k-n+2}) : x_1 + \ldots + x_s - x_{s+1} - \ldots - x_{2k-n+2} = 0\}$ *and* $X_2 := \{00,11\}^{n-k-1}$, *where* $s := \lfloor (2n-k+2)/2 \rfloor$.

Note that in case $k = n-1$ we have $G(n,1) = \binom{n}{\lfloor n/2 \rfloor}$ (Theorem 75).

Theorem 79 *For* $2g < n < (g+1)k$ *we have*

$$\frac{1}{2}G_k(n,1) < G_k(n,g) \leq G_k(n,1). \tag{16}$$

2. On Bohman's Conjecture Related to a Sum Packing Problem of Erdös

Let H be a hyperplane in $\mathbb{R}^n$ so that $H \cap \{0, \pm 1\}^n = \{0, \dots, 0\}$. How large can an intersection with $\{0, \pm 1, \pm 2\}^n$ be? Let us denote

$$f(n) = \max_H |H \cap \{0, \pm 1, \pm 2\}^n|.$$

This problem was raised in [B96] in connection with a famous subset sum problem in [E56]. A set of positive integers $S \subseteq \{1, \dots, N\}$ has distinct subset sums, if all sums of subsets are distinct. Erdös has asked for the value of

$$g(n) := \min\{N : \exists S \subseteq N, |S| = n, S \text{ has distinct subset sums }\}.$$

In [B96] the relationship between functions $f(n)$ and $g(n)$ is explained, and it is noticed that studying the function $f(n)$ might be useful for better understanding the problem of Erdös. Suppose a hyperplane H defined by the equation $\sum_{i=0}^{n-1} a_i x_i = 0$; $a_0, \dots, a_{n-1} \in \mathbb{N}$ satisfies $H \cap \{0, \pm 1\}^n = \{0^n\}$. This means that $\{a_0, \dots, a_{n-1}\}$ has distinct subsets sums.

A simple example of such a set with $a_{n-1} \leq 2^{n-1}$ is the set of powers $\{2^i\}$, $i = 0, \dots, n-1$. The first nontrivial upper bound for $g(n) < 2^{n-2}$ is due to Conway and Guy. They defined a sequence A_n of sets of integers ([CG68]) and conjectured that A_n has distinct subset sums for every n. They showed that the conjecture is true for $1 \leq n \leq 40$. The set A_{21} in the Conway-Guy sequence has a largest element smaller than 2^{19}. This gives the upper bound 2^{n-2} for all $n > 21$. This bound was later improved to $0.2246(2^n)$ in [L88]. In [B96] the conjecture of Conway and Guy is proved for all n.

For $f(n)$ Bohman conjectured that $f(n) = \frac{1}{2}(1 + \sqrt{2})^n + \frac{1}{2}(1 - \sqrt{2})^n$, showing that this number can be achieved, taking $a_i = 2^i$ $(i = 0, \dots, n-1)$ as coefficients.

In [AAK04b] lower and upper bounds for $f(n)$ are established. It is shown that

$$(2,538)^n \ll f(n) \ll (2,723)^n. \tag{17}$$

The lower bound in (17) disproves the conjecture of Bohman. The construction giving this bound is as follows. Let H be the hyperplane defined by

$$(2S_{n-2} + 1)x_0 + 2x_1 + \cdots + 2^{n-1}x_{n-1} = 0.$$

where for $n \geq 3$

$$S_{n-2} := \begin{cases} 2^{n-3} + 2^{n-5} + \cdots + 2^3 + 2, & \text{if } 2 \mid n \\ 2^{n-3} + 2^{n-5} + \cdots + 2^2 + 1, & \text{if } 2 \nmid n. \end{cases}$$

Let $h(n) = |H \cap \{0, \pm 1, \pm 2\}^n|$. It is shown that for some constant c holds $h(n) \geq c\beta^n$, where β ($\beta = 2,5386\ldots$) is the greatest real root of the equation $z^8 - 8z^6 + 10z^4 + 1 = 0$.

3. Problems in GF$(2)^n$

In §3 of Lecture 12 the function

$$\Gamma(n,k,w) = \max_{U_k^n \subset \mathbb{R}} |U_k^n \cap \binom{[n]}{w}|,$$

where U_k^n is a k-dimensional subspace of $\mathbb{R}^n$, was determined.

It seems important from a coding theoretical viewpoint to replace $\mathbb{R}^n$ by vector spaces over finite fields. Already the first case is unsolved. We define

$$\gamma(n,k,w) = \max_{U_k^n \subset GF(2)^n} |U_k^n \cap \binom{[n]}{w}|,$$

where U_k^n is a k-dimensional subspace of $GF(2)^n$.

One can easily verify that

$$\gamma(n,k,w) \geq \Gamma(n,k,w).$$

Note that $\gamma(n,k,w)$ depends on the parity of w. On the one hand, for example one can easily see that for $k < w$ and odd w we have

$$\gamma(n,k,w) \leq 2^{k-1}, \text{ if } n \geq w+k-1.$$

On the other hand, for suitable even w's we can have

$$\gamma(n,k,w) = 2^k - 1.$$

It can be easily shown that this bound can be achieved iff $w = t2^{k-1}, n \geq t(2^k - 1)$, $t \in \mathbb{N}$. In this case we just take t copies of the simplex code (of length $2^k - 1$) well known in Coding Theory ([MS77]).

Note also that here we have no symmetry like for $\Gamma(n,k,w)$. That is, in general $\gamma(n,k,w) \neq \gamma(n,k,n-w)$. However, for the odd w's and even n's $\gamma(n,k,w) = \gamma(n,k,n-w)$ holds.

Research Problems

1. Given $2 \leq g < n$, determine $G(n, g)$ (ad 1).
2. Determine $G(n, \leq m, g)$ (ad 1).
3. Under similar restrictions as in Research Problems 1,2 determine the maximal number of (0,1)-solutions of a linear equation

$$B(x_1, \ldots, x_n)^T = 0, \tag{18}$$

   where B is a real $r \times n$ matrix of rank r (ad 1).
4. **Conjecture** *For an integer $n \geq 3$ and $t := \lfloor n/3 \rfloor$ holds*

$$G(n, 2) = \sum_{i=0}^{t} \binom{t}{i} \binom{n-t}{2i}. \tag{19}$$

   This number is achieved for the family $\mathcal{A}$ corresponding to the set of (0,1)-solutions of equation (13) where $a_1 = \ldots = a_t = 2, a_{t+1} = \ldots = a_n = -1$ (ad 1).
5. Prove the long-standing conjecture of Erdös that $g(n) \geq c2^n$ for some constant c (ad 2).
6. In the paper [AAK04b] the following conjecture was stated. **Conjecture** For some constant c we have $f(n) \sim c\beta^n$. (ad 2)
7. A more general problem as Research Problem 6 is the following. Let $Q \subset \mathcal{Z}$ be finite and $F = \{0, \pm 1, \ldots, \pm k\}$, then

$$f(n, Q, F) := \max\{|H \cap Q^n| : H \text{ is a hyperplane and } H \cap F^n = \{0^n\}\}.$$

   Conjecture For $Q = \{0, \pm 1, \ldots, \pm(k+1)\}$, $F = \{0, \pm 1, \ldots, \pm k\}$ and $k \geq 2$ (or a weaker condition: for $k > k_0$) one has

$$f(n, Q, F) = \frac{1}{2}(1 + \sqrt{2})^n + \frac{1}{2}(1 - \sqrt{2})^n. \tag{ad 2}$$

8. Determine $\gamma(n, k, w)$ (ad 3).

Notes to Lecture 13

Sperner's Lemma answers a number theoretical question: How many square-free numbers with prime factors $p_1, p_2, \ldots, p_n$ are there so that no two divide each other? It was one of the beginnings of Extremal Set Theory as a separate subject. Our point is that Extremal Number Theory and Extremal Set Theory should continue to be studied in a parallel manner. EKR for instance has a number theoretical interpretation: How many square-free numbers with k factors and one common factor for any two are there?

1. Uniqueness by the AZ-Identity and its Dual

It has been explained in this lecture that Theorem 27 gives immediately what LYM does not, namely the uniqueness part in Sperner's Lemma and in Erdös's result (Theorem 53). We suggest two further uniqueness proofs from [AZ90b] and [AC93a] as Exercises 1,2.

In [DT94], a dual to the AZ-identity was presented and for related identities see [T97].

2. Posets with a Rank Function

In a finite POS $(\mathcal{X}, <), r : \mathcal{X} \to \mathbb{N} \cup \{0\}$ is a rank function if

1. $r(x) = 0$ for the normal elements $x \in \mathcal{X}$ (x is normal, if there is no $y \in \mathcal{X}$ with $y < x$)
2. $r(x) = r(y) + 1$, where x is a direct successor of y (i.e., $y < x$ and there is no $z \in \mathcal{X}$ such that $y < z < x$)

In the POS $(2^{[n]}, <)$ the cardinality $r(S) = |S|$ for $S \in 2^{[n]}$ is a rank function.

In a q-regular tree $r(x) = $ length of the path from the root to x is a rank function.

A code $\mathcal{C} \subset \mathcal{X}^* = \bigcup_{n=0}^{\infty} \mathcal{X}^n$ is a prefix (suffix) code, if no word $c_i \in \mathcal{C}$ is the prefix (suffix) of another word $c_j \in \mathcal{C}$. $\mathcal{X}^*$ can be identified with a $q = |\mathcal{X}|$ regular tree in an obvious way, where words correspond to nodes.

Kraft Inequality *For every prefix (suffix) code $\mathcal{C} \subset \mathcal{X}^*$ for the length function L*

$$\sum_{c \in \mathcal{C}} q^{-L(c)} \leq 1$$

holds and, conversely, for every set of numbers $L_1, L_2, \ldots, L_k \in \mathbb{N}$ with

$$\sum_{i=1}^{k} q^{-L_i} \leq 1$$

there exists a prefix (suffix) code $\{c_1, \ldots, c_k\}$ such that $L(c_i) = L_i$ for $i = 1, 2, \ldots, k$.

This is an analogon to the LYM-inequality.

The two posets have an important difference. The antichain $\{\{1\}, \{2,3\}\} \subset 2^{[3]}$ cannot be extended, but is not blocking, that is, it does not meet every maximal chain, in the other POS every nonextendable antichain is blocking.

3. Forbidden Configurations in the Families of Subsets

There are many generalizations of Sperner's Lemma ([EK86] and Chapter III of [E97b]). Let us consider one more kind of generalization. Let $\mathcal{F} \subset 2^{[n]}$. We mention some results when the condition on $\mathcal{F}$ excludes certain configurations that can be expressed by inclusion only. That is, no intersections, unions, etc. are involved. For a poset P in the Boolean lattice B_n, let $\text{La}(n,P)$ denote the maximal number of elements in B_n such that the poset induced by these elements does not contain P as a subposet. The first such generalization was obtained by Erdős (see Theorem 53).

Let V_r denote the *r-fork*, that is, the following family of distinct sets: $F \subset G_1, F \subset G_2, \ldots F \subset G_r$. The quantity $\text{La}(n,V_r)$, that is, the largest family on n elements containing no V_r, was first (asymptotically) determined for $r = 2$.

Theorem 80 (Katona and Tarján [KT83])

$$\binom{n}{\lfloor \frac{n}{2} \rfloor}\left(1 + \frac{1}{n} + \Omega\left(\frac{1}{n^2}\right)\right) \le La(n,V_2) \le \binom{n}{\lfloor \frac{n}{2} \rfloor}\left(1 + \frac{2}{n}\right).$$

A more general result for $r \ge 2$ is obtained in

Theorem 81 (Thanh [T98])

$$\binom{n}{\lfloor \frac{n}{2} \rfloor}\left(1 + \frac{r}{n} + \Omega\left(\frac{1}{n^2}\right)\right) \le La(n,V_{r+1}) \le \binom{n}{\lfloor \frac{n}{2} \rfloor}\left(1 + 2\frac{r^2}{n} + O\left(\frac{1}{n}\right)\right).$$

The upper bound in this result was recently improved.

Theorem 82 (De Bonis and Katona [BK07])

$$\binom{n}{\lfloor \frac{n}{2} \rfloor}\left(1 + \frac{r}{n} + \Omega\left(\frac{1}{n^2}\right)\right) \le La(n,V_{r+1}) \le \binom{n}{\lfloor \frac{n}{2} \rfloor}\left(1 + 2\frac{r}{n} + O\left(\frac{1}{n^2}\right)\right).$$

Following the definition of the r-fork, let us define the *r-brush* Λ_2 (in a poset), which contains $r+1$ elements: $a, b_1, \ldots, b_r$ where $a > b_1, \ldots a > b_r$ and is the "dual" of the r-fork. It is easy to see that Theorem 81 gives the same solution for Λ_2. However, the result is different when both of them are excluded. The notation $\text{La}(n,R)$ is extended to the case when two subposets R_1 and R_2 are excluded: $\text{La}(n,R_1,R_2)$.

Theorem 83 (Katona and Tarján [KT83])

$$La(n,V_2,\Lambda_2) = 2\binom{n-1}{\lfloor \frac{n-1}{2} \rfloor}.$$

The construction giving the equality is the following:

$$\left\{ F \subset [n] : 1 \notin F, |F| = \left\lfloor \frac{n-1}{2} \right\rfloor \right\} \cup \left\{ F \subset [n] : 1 \in F, |F| = \left\lfloor \frac{n+1}{2} \right\rfloor \right\}.$$

The poset N contains four distinct elements a, b, c, d satisfying $a < c, b < c, b < d$. In the Boolean lattice, a subposet N consists of four disticts subsets satisfying $A \subset C$, $B \subset C, B \subset D$. It is somewhat surprising that excluding N the result is basically the same as in the case of V_2.

Theorem 84 (Griggs and Katona [GK08])

$$\binom{n}{\lfloor \frac{n}{2} \rfloor} \left(1 + \frac{1}{n} + \Omega \left(\frac{1}{n^2} \right) \right) \le La(n,N) \le \binom{n}{\lfloor \frac{n}{2} \rfloor} \left(1 + \frac{2}{n} + O \left(\frac{1}{n^2} \right) \right)$$

holds.

Let us also mention a result when the excluded poset contains one more relation. The *butterfly* $\bowtie$ contains four elements: a, b, c, d with $a < c, a < d, b < c, b < d$.

Theorem 85 (De Bonis, Katona, and Swanepoel [BKS05]) *Let $n \ge 3$. Then*

$$La(n, \bowtie) = \binom{n}{\lfloor n/2 \rfloor} + \binom{n}{\lfloor n/2 \rfloor + 1}.$$

4. Sharpening of LYM

Theorem 86 (Bey [B05a]) *Let $\mathcal{A} \subseteq 2^{[n]}$ be an antichain. Then we have*

$$\sum_{i=0}^{n} \frac{f_i}{\binom{n}{i}} + \sum_{I = \{i_1, \ldots, i_s\} \subseteq [n-1], s \ge 2} \left(\prod_{j=1}^{s-1} \frac{n(i_{j+1} - i_j)}{i_j (n - i_{j+1})} \right) \left(\prod_{i \in I} \frac{f_i}{\binom{n}{i}} \right) \le 1.$$

In fact, the inequality sharpens LYM with inclusion of additional nonnegative terms in LYM, which are polynomials in f_i's.

Exercises

1. Prove the uniqueness of an optimal configuration of unrelated antichains of subsets due to [GST84].
2. In [KS92] the following LYM-type inequality was observed:
   For $\mathcal{A} = \{A_1, \ldots, A_N\}, \mathcal{B} = \{B_1, \ldots, B_N\} \subset 2^{\Omega}$ with $A_i \cap B_i = \emptyset$, $A_i \not\subseteq A_j \cup B_j$,
   $B_i \not\subseteq A_j \cup B_j$ for $i \ne j$ $\sum_{i=1}^{N} \binom{n - |A_i|}{|B_i|}^{-1} + \binom{n - |B_i|}{|A_i|}^{-1} - \binom{n}{|A_i| + |B_i|}^{-1} \le 1$ and they
   asked "Is this inequality ever tight?"

This rather modest question was a challenging test of the power of the identities above or, more precisely, of the procedure to produce new identities described in [AZ90b].

The outcome is the Ahlswede–Zhang type identity of Theorem 29. From a special case of this identity derive a full characterization of the cases with equality (even for a generalized version of the inequality above). In other words characterize the cases with deficiency zero.

3. Formulate the AZ identity for the regular tree.

4. C is called fix-free if it is simultaneously a prefix and a suffix code.
   Show that for every $\varepsilon > 0$ there are numbers $L_1, \ldots, L_k$ with

$$\sum_{i=1}^{k} q^{-L_i} \leq \frac{3}{4} + \varepsilon$$

for which there is no fix-free code with word lengths $L_1, \ldots, L_k$.

5. If for $L_1, \ldots, L_k$

$$\sum_{i=1}^{k} q^{-L_i} \leq \frac{1}{2}$$

then show that there exists a fix-free code with word lengths $L_1, \ldots, L_k$.

Research Problems

1. ($\frac{3}{4}$-conjecture of Ahlswede, Balkenhol, and Khachatrian [ABK96]) For every k and integers $L_1, \ldots, L_k \in \mathbb{N}$ with

$$\sum_{i=1}^{k} q^{-L_i} \leq \frac{3}{4}$$

there exists a fix-free code with word length $L_1, \ldots, L_k$.

After 12 years of strong efforts involving also long computer calculations the problem is still open. There are several cases that have been settled ([DS06], [HK99], [HKG08] [Y01], [Y04], [YY01], and the survey [S05]).

Notes to Lecture 14

The conjecture of [AEG95] that every countable strongly dense partially ordered set has the splitting property was disproved with the poset of square-free integers in [N], 29–44, where the reader also finds several open problems.

Notes to Lecture 15

Further Important Correlation Inequalities

1. Janson Inequality

An application of the FKG inequality arose in the following estimations. Let Ω be a finite set with probabilities $P(\omega) = p_\omega$, $\omega \in \Omega$. Let R be a random subset of Ω with $P(\omega \in R) = p_\omega$. Let A_i, $i \in I$, $A_i \subset \Omega$ and let B_i be the event that $A_i \subset R$.) Let 1_{B_i} be the indicator function of the event B_i. Then $X = \sum_{i \in I} 1_{B_i}$ is the number of A_i's which belong to R and $\bigcap_{i \in I} \overline{B}_i, X = 0$ are identical events. Denote $\Delta = \sum_{i \sim j, \, i > j} P(B_i \cap B_j)$, where $i \sim j$ means $i \neq j$, $A_i \cap A_j \neq \emptyset$. Denote also $M = \prod_{i \in I} P(\overline{B}_i)$.

Lemma 53 *If* $P(B_i) \leq \varepsilon < 1$, *then*

$$M \leq P\left(\bigcap_{i \in I} \overline{B}_i\right) \leq M \exp(\Delta/(2(1-\varepsilon))).$$

This inequality is called Janson inequality. The key observation in the proof of this inequality by R. Boppana and J. Spencer is the following inequality, which easily follows from FKG

$$P\left(B_i \bigg| \bigcap_{i \in I} \overline{B}_j\right) \leq P(B_i).$$

For the complete proof of the Janson inequality see [AS92].

2. Suen Inequality

Another estimation of the probability is due to [S90]. Assume that B_i, $i \in I$ are arbitrary events in the probability space. We write $i \sim j$, $i, j \in I$ iff the following holds: For $I_1, I_2, \subset I$, $I_1 \cap I_2 = \emptyset$ so that $i \not\sim j$ for all $i \in I_1$, $j \in I_2$. Then any Boolean combinations of the event B_i, $i \in I_1$ and B_j, $j \in I_2$ are independent.

Lemma 54 *The following inequality is valid*

$$\left| P\left(\bigcap_{i \in I} \overline{B}_i\right) - M \right| \leq M\left(\exp\left(\sum_{i \sim j} d(i,j)\right) - 1\right),$$

where
$$d(i,j) = (P(B_i \cap B_j) + p(B_i)P(B_j)) / \prod_{k \sim i \text{ or } k \sim j} (1 - p(B_k)).$$

This and other related inequalities one can find in [AS92].

3. Baston Inequality

We take from [B81] an inequality which differs from Theorem 36 in the choice of functions f_i.

Theorem 87 *Let L be a ring of subsets of the finite set (family of subsets closed under the operations $\cap, \Delta$) and let $f_1, f_2, f_3, f_4 : L \to \mathbb{R}_+$ with f_3 and at least one of f_1, f_2 monotone nondecreasing functions, such that*

$$f_1(a)f_2(b) \leq f_3(a\Delta b)f_4(a \cap b), \; a, b \in L.$$

Then

$$f_1(A)f_2(B) \leq f_3(A\Delta B)f_4(A \wedge B), \; A, B \subset L.$$

4. The BKR Inequality

Let P be a product probability on $\mathbb{Z}_2^n$ of binary n-tuples with mod 2 addition in coordinates. For $x = (x_1, \ldots, x_n) \in \mathbb{Z}_2^n$ and $K \subset [n]$ define the cylinder $[x]_K = \{y \in \mathbb{Z}_2^n : y_i = x_i, \, \forall i \in K\}$. Define also

$$A \square B = \{x \in \mathbb{Z}_2^n : \exists K \subset [n] \text{ such that } [x]_K \subset A, \text{ and } [x]_{[n]\setminus K} \subset B\}.$$

Van den Berg and Kesten in [BK85] conjectured that for all $A, B \subset \mathbb{Z}_2^n$,

$$P(A \square B) \leq P(A)P(B). \tag{20}$$

In [BF87] it was shown that this inequality is equivalent to the following statement, which was finally proved in [R91].

Theorem 88 *For any $A, B \subset \mathbb{Z}_2^n$ holds*

$$2^n |A \square B| \leq |A||B|.$$

Actually, Reimer proved an equivalent version of this theorem due to [FS91]. An attempt to simplify Reimer's proof is made in [BCR99].

The RBK-inequality is a breakthrough in correlation inequalities. Instead of binary operations $\varphi : S \times S \to S$, which induce a set operation $\varphi(A, B) = \{\varphi(a, b) : a \in A, b \in B\}$, the operation $\square$ produces $A \square B$ without being based on a map $S \times S \to S$. More generally, in forthcoming work [A234] "Higher level correlation inequalities" maps of the type $2^S \times 2^S \to 2^S$ are investigated.

Two known consequences of RBK are stated as exercises 1,2. Here and later we adopt the notation $A \uparrow$ (resp. $A \downarrow$), if A is upset (resp. if A is downset).

Previously, known correlation inequalities such as the FKG-inequality and the AD-inequality go into the opposite direction, bounding for instance $|A||B|$ or $\alpha(A)\beta(B)$ from above. In [AD79b] the authors abstractly looked for "reversed" inequalities for pairs of operations $\varphi, \psi : S \times S \to S$. Although the notion "$\mathcal{M}$-expansive" for those operations turned out to be very useful for finding and (in conjunction with a product theorem) also for establishing new direct inequalities, its natural dual for reversed inequalities, namely the notion "$\mathcal{M}$-contractive" *turned out to be essentially vacuous.* By adding the smallest possible constant factor that make (φ_1, ψ_1) and (φ_2, ψ_2) mass contractive the product theorem applies and gives the best constant factor for $((\varphi_1, \varphi_2), (\psi_1, \psi_2))$ (Compare §5 in Lecture 15). In the light of these results the seemingly *surprising fact,* that the *BK-inequality holds without* an auxiliary factor, must be understood. We shall explain that it is related to a "loss of mass."

Motivated by certain percolation problems van den Berg and Kesten found their interesting correlation inequality involving a non-Boolean binary operation of the type

$$S \times S \rightarrow S^* \qquad \text{with} \qquad S^* \supset S.$$

This adds a new direction to the theory of correlation inequalities presented in [AD79b].

More specifically, for $S = \{0, 1\}$, $S^* = \{0, 1, 2\}$ they consider as operation the addition "+" for integers. Further, for sequences $a = (a_1, \ldots, a_n), b = (b_1, \ldots, b_n) \in S^n = \{0, 1\}^n$ the addition is defined component-wise

$$a + b = (a_1 + b_1, \ldots, a_n + b_n)$$

and for sets of sequences $A, B \subset S^n$ addition is understood in the sense of Minkowski

$$A + B = \{a + b : a \in A, b \in B\}.$$

Here we consider only probability distributions P on S^n, which are of product type:

$$P \triangleq \prod_{t=1}^n p_t. \tag{21}$$

S will always be a **finite** subset of the set of nonnegative integers $\mathbb{N}_0$.

The BK-Inequality States:

$$P((A + B) \cap S^n) \leq P(A)P(B) \tag{22}$$

for all product distributions P on $S^n = \{0, 1\}^n$ and all $A \uparrow, B \uparrow \subset \{0, 1\}^n$.

We show how the operation $(A, B) \rightarrow (A + B) \cap S^n$ can be based on the operation $* : S \times S \rightarrow S^*$ and also derive new inequalities for product distributions on chains.

We call $I = \{0, 1, 2, \ldots, T\}$ a *chain* with the operations $\wedge, \vee$:

$$a \wedge b = \min(a, b), \quad a \vee b = \max(a, b), \quad a, b \in I. \tag{23}$$

The chain has the structure of a distribution lattice. However, this lattice is *not* complementary:

$$\text{to } a \in I \quad \exists a^c : \quad a \vee a^c = I, \quad a \wedge a^c = 0.$$

We can define analogies of other Boolean operations:

$$a - b = (a - b)^+ = (a - b) \vee 0, \quad a \triangle b = |a - b|. \tag{24}$$

Furthermore,

$$a * b = a + b, \quad \text{if} \quad a + b \leq T, \quad \text{erasure otherwise} \tag{25}$$

$$a \triangle b = \overline{(\bar{a} * \bar{b})}, \quad \text{where} \quad \bar{a} = T - a \quad \text{and} \quad \overline{\text{erasure}} = \text{erasure}. \tag{26}$$

All operations have product extensions to $I^n = \prod_{i=1}^{n} I_i$.

In [A234] also the following results are proved:

Theorem 89 *For product distributions* $\alpha, \beta, \gamma, \delta : I^n \to \mathbb{R}_+$

$$\alpha(A)\beta(B) \leq \gamma(A \triangle B)\delta(A \wedge B) \qquad \text{for all} \quad A, B \downarrow \subseteq I^n.$$

This supplements Baston's work. Notice also that we have no monotonicity assumptions on $\alpha, \beta, \gamma, \delta$. Instead we assume product distributions and monotonicity ($\downarrow$) on B.

Theorem 90 *For product distributions* $\alpha, \beta, \gamma, \delta : I^n \to \mathbb{R}_+$

$$\alpha(A)\beta(B) \leq \gamma(A * B)\delta(A \wedge B) \qquad \text{for all} \quad A, B \downarrow \subseteq I^n.$$

For $\alpha = \beta = \gamma = \delta = P$ this gives

$$P(A)P(B) \leq P(A * B)P(A \wedge B) \qquad \text{for} \quad A, B \downarrow \subset I^n,$$

which is by the factor $P(A \wedge B) \leq 1$ *sharper* than a converse to the BK-inequality

$$P(A)P(B) \leq P(A * B) \qquad \text{for} \quad A \downarrow B \downarrow,$$

and more *general*, because monotonicity is only assumed for one set.

Applications of AD

5. On Representation of Posets

Here we explain an application of the AD-inequality from the paper [E84]. A representation of a finite poset $\mathcal{P}$ is a function $f : \mathcal{P} \to \mathbb{R}$, such that $a >_{\mathcal{P}} b$ implies $f(a) - f(b) \geq 1$. The variance D_f of the representation f is defined by

$$D_f = \frac{1}{|\mathcal{P}|} \sum_{P \in \mathcal{P}} f^2(P) - \left(\frac{1}{|\mathcal{P}|} \sum_{P \in \mathcal{P}} f(P) \right)^2.$$

A representation D_f is called optimal iff $D_f = \min_\phi D_\phi$. r is the rank of poset $\mathcal{P}$ iff there exists a function $r : \mathcal{P} \to \mathbb{N}$ such that $r(a) = 0$ if a is minimal in $\mathcal{P}$ and $r(b) = r(a) + 1$ if $b >_{\mathcal{P}} a$ and there is no $c \in \mathcal{P}$ with the property $b >_{\mathcal{P}} c >_{\mathcal{P}} a$. For subset $A \subset \mathcal{P}$ denote $r(A) = \sum_{a \in A} r(a)$. We will use the following simple (see also [A74])

Proposition 23 *Let r be the rank function of poset $\mathcal{P}$. Then r is optimal representation of $\mathcal{P}$ iff for all upsets $U \subset \mathcal{P}$*

$$\frac{r(U)}{|U|} \geq \frac{r(\mathcal{P})}{|\mathcal{P}|}.$$

Theorem 91 *If $\mathcal{P}$ is a distributive lattice, then its rank function is an optimal representation of $\mathcal{P}$.*

Indeed let U be an upset. Then taking $f_1 = f_3 = r$, $f_2, f_4 = 1$, $A = \mathcal{P}$, we see that from inequality (7) in Lecture 15 follows

$$r(\mathcal{P})|U| \leq r(U)|\mathcal{P}|.$$

Using Proposition 23 we conclude that r is an optimal representation.

6. Coloring in Random Graphs

The following example of an application of the AD inequality we take from [F70]. Let $G = (E, V)$ be a graph on n vertices. Assume that we choose each vertex to our collection independent with equal probability p. Denote by $A(G, p)$ the probability that the chosen set is stable, i.e., no edge of G has both endpoints in it.

Another notion that we need is that of a chromatic polynomial. Assume that each vertex of G is independently given a color from the set $\{1, 2, \ldots, \lambda\}$ with equal probabilities, then $P(G, \lambda)/\lambda^n$ is the probability that this assignment of colors is a coloring of G (for no edge both ends belong to the same color).

Theorem 92 *For all positive integers n and λ*

$$\frac{P(G, \lambda)}{\lambda^n} \leq A(G, \lambda^{-1})^\lambda. \tag{27}$$

Sketch of the Proof. For each $i \in \{1, \ldots, \lambda\}$ let C_i be the random set of vertices of G receiving color i. Then

$$\frac{P(G, \lambda)}{\lambda^n} = P(C_1, \ldots, C_\lambda \text{ are stable}).$$

Suppose that there are k colors and that for each vertex $v \in V(G)$ and each color i, v receives color i with probability p. The key observation, from which (27) follows, is the following inequality (which will be used for $p = 1/\lambda$)

$$P(C_1, \ldots, C_k \text{ are stable}) \leq P(C_1, \ldots, C_{k-1} \text{ are stable}) P(C_k \text{ is stable}) \tag{28}$$

and the fact that

$$\prod_{i=1}^{\lambda} P(C_i \text{ is stable}) = A^{\lambda}(G, \lambda^{-1}).$$ (29)

It is convenient now to denote by w, x, y, z subsets of $V(G)$ and by c_i the realization of random set C_i. Relation (28) follows from the AD inequality if we choose

$$f_1(w) = p^{|w|}(1-p)^{n-|w|}$$
$$f_2(x) = P(c_1, \ldots, c_{k-1} \text{ all stable and } c_k = x)$$
$$= p^{|x|} \sum_{\substack{c_1, \ldots, c_{k-1} \subset V(G) \setminus x, \\ \text{pairwise disjoint, stable}}} p^{|c_1| + \ldots + |c_{k-1}|} q^{n-|x|-|c_1|-\ldots-|c_{k-1}|},$$

where $q = 1 - p$

$$f_3(z) = f_2(z),$$
$$f_4(y) = f_1(y)$$

and

$$W = \{w : w \text{ is not stable in } G\}, \ X = 2^{V(G)}.$$

Indeed

$$W \vee X = X, \ W \wedge X = X$$

and it can be seen that

$$f_1(w)f_2(x) \leq f_3(w \cup x)f_4(w \cap x)$$

and thus we have

$$f_1(W)f_2(X) \leq f_3(W)f_4(X).$$

The last inequality is just what we need, because

$$f_1(W) = 1 - P(C_k \text{ stable})$$
$$f_2(X) = P(C_1, \ldots, C_{k-1} \text{ all stable})$$
$$f_3(W) = P(C_1, \ldots, C_{k-1} \text{ all are stable}) - P(C_1, \ldots, C_k \text{ all are stable})$$
$$f_4(X) = 1.$$

$\square$

7. Negatively Correlated or Associated Random Variables

In the following not the full power of AD is needed – already FKG suffices ([DPR96] and [DR98]). Along with negative correlation, which says that for two random variables

$$\mathbb{E}(XY) - \mathbb{E}(X)\mathbb{E}(Y) \leq 0,$$

consider the negative association of random variables $X_1, \ldots, X_n$, which says that for an arbitrary index set $I \subset [n]$ and arbitrary nondecreasing functions $f: \mathbb{R}^{|I|} \to \mathbb{R}$, $g: \mathbb{R}^{n-|I|} \to \mathbb{R}$

$$\mathbb{E}(f(X_i, \ i \in I)g(X_i, \ i \in [n] \setminus I)) \le \mathbb{E}(f(X_i, \ i \in I))\mathbb{E}(g(X_i, \ i \in [n] \setminus I)).$$

Function $f: \mathbb{R}^k \to \mathbb{R}$ is nondecreasing if $f(x^k) \le f(y^k)$ for all $x^k, y^k \in \mathbb{R}^k$, $x_i \le y_i$, $i = 1, \ldots, k$. We note that the same inequality will hold if f and g are both nonincreasing.

Negative association of random variables $x_1, \ldots, X_n$ allows to write inequality

$$\mathbb{E}\left(\exp\left(h \sum_{i=1}^n X_i\right)\right) \le \prod_{i=1}^n \mathbb{E}(\exp(hX_i)),$$

which gives the Chernoff bound (Chapter I). Thus it is useful to establish the negative association of random variables.

8. A Variety of Number Theoretic Inequalities

Using inequality (59) of Lecture 15 we can derive other conclusions than that in Theorem 49.

For arbitrary $A, B \subset \mathbb{N}$,

(i) $\underline{D}A \cdot \overline{D}B \le \underline{D}[A, B] \cdot \overline{D}(A, B)$
(ii) $\overline{D}A \cdot \overline{D}B \le \overline{D}[A, B] \cdot \overline{D}(A, B)$
(iii) $\overline{D}A \cdot \underline{D}B \le \overline{D}[A, B] \cdot \underline{D}(A, B)$.

It is a wide field of research to investigate for which sets the various densities exist.

Several equivalent and also nonequivalent inequalities can be derived from Theorem 49 and Theorem 50 in §6. They deal with sets of multiples and nonmultiples. For the example $A = \{1\}, B = \mathbb{N}$ we have

$$1 = d(A, B) \not\le dA \cdot dB = 0,$$

which shows that for the inequality in Theorem 50, it is essential to work with sets of multiples.

We list now first elementary relations between such sets.

$$M(A), M(B) \supset M[A, B],$$
$$M(A) \cap M(B) = M[A, B],$$
$$M(A) \cap M(B) = [M(A), M(B)],$$
$$M(A) \cap M(B) \supset M(A \cap B),$$
$$M(A) \cap M(B) \supset M(A \times B),$$
$$M(A, B) \supset M(A \cup B),$$

$$M(A,B) = (M(A),M(B)),$$
$$N(A \cup B) = (N(A),N(B)),$$
$$N(A \cup B) = N(A) \cap N(B),$$
$$N(A) \cup N(B) = [N(A),N(B)].$$

Useful are also these two identities.

$$dM(A \cup B) = dM(A) + dM(B) - dM[A,B],$$
$$dN(A \cup B) = dN(A) + dN(B) - d(N(A) \cup N(B)).$$

Now we deduce directly from Theorem 49

$$DN(A)DN(B) \leq D[N(A),N(B)]D(N(A),N(B)).$$

Since $(N(A),N(B)) = N(A \cup B)$ we thus arrive at the following inequality. For finite $A,B \subset \mathbb{N}$

$$dN(A) \cdot dN(B) \leq dN(A \cup B) \cdot d[N(A),N(B)]. \tag{30}$$

Notice that this is by the factor $d[N(A),N(B)]$ better than Behrend's inequality.

The inequality (50) in Lecture 15 is by the summand $dN(A,B) \cdot (1 - \mathbb{N}[A,B])$ better than Behrend's inequality. Quite surprisingly these two inequalities are different and none implies the other!

Example Let $A = \{3,4\}$, $B = \{6\}$. Then $dN(A) = 1 - \left(\frac{1}{3} + \frac{1}{4} - \frac{1}{12}\right) = \frac{1}{2}$, $dN(B) = \frac{5}{6}$, $dN(A \cup B) = dN(A) = \frac{1}{2}$, $d[N(A),N(B)] = \frac{11}{12}$, $(A,B) = \{2,3\}$, $dN(A,B) = \frac{1}{3}$, $[A,B] = \{6,12\}$, $dN[A,B] = dN(\{6\}) = dN(B) = \frac{5}{6}$.

We have therefore $dN(A \cup B) - dN(A,B)(1 - dN[A,B]) = \frac{1}{2} - \frac{1}{3}\left(1 - \frac{5}{6}\right) = \frac{4}{9} < \frac{11}{24} = \frac{1}{2} \cdot \frac{11}{12} = dN(A \cup B) \cdot d[N(A),N(B)]$.

Example Let $A = \{2\}$, $B = \{2,3\}$. Then $dN(A) = \frac{1}{2}$, $dN(B) = 1 - \frac{1}{2} - \frac{1}{3} + \frac{1}{6} = \frac{1}{3}$, $dN(A \cup B) = dN(B) = \frac{1}{3}$, by R_{10} $[N(A),N(B)] = N(A) \cup N(B) = N(A)$ and thus $d[N(A),N(B)] = \frac{1}{2}$, $dN(A \cup B) = dN(B) = \frac{1}{3}$, $dN(A,B) = 0$.

Theorem 93 *For finite $A,B \subset \mathbb{N}$*

$$dN[A,B]dN(A \cup B) \leq dN(A)dN(B).$$

Equality holds exactly if $N(A) \supset N(B)$ or $N(B) \supset N(A)$.

Proof. Since $dM(A \cup B) = dM(A) + dM(B) - dM[A,B]$, an equivalent inequality is

$$(1 - dM(A)) \cdot (1 - dM(B)) \geq (1 - dM[A,B]) \cdot (1 - dM(A) - dM(B) + dM[A,B]).$$

This is equivalent to

$$(dM(A) - dM[A,B])(dM(B) - dM[A,B]) \geq 0.$$

Since $M(A), M(B) \supset M[A,B] = M(A) \cap M(B)$, this inequality holds. Furthermore, if neither $M(A) \subset M(B)$ nor $M(B) \subset M(A)$ holds, then we have strict inequality. Otherwise we have equality, because one factor vanishes. □

Corollary 8 *For finite* $A, B \subset \mathbb{N}$

$$dN[A,B]dN(A,B) \leq dN(A)dN(B).$$

Proof. Just notice that $N(A,B) \subset N(A \cup B)$ and apply Theorem 93. □

Theorem 94 *For finite* $A, B \subset \mathbb{N}$

$$d(M(A) \cap M(B)) \cdot d(M(A) \cup M(B)) \leq dM(A) \cdot dM(B).$$

Equality holds exactly if

$$M(A) \supset M(B) \text{ or } M(B) \supset M(A).$$

Proof. From

$$d(M(A) \cup M(B)) + d(M(A) \cap M(B)) = dM(A) + dM(B)$$

and

$$d(M(A) \cup M(B)) \geq dM(A), \quad dM(B) \geq d(M(A) \cap M(B))$$

the inequality follows, because if $dM(A) = x_1$, $dM(B) = x_2$, $d(M(A) \cap M(B)) = y_1$, $d(M(A) \cup M(B)) = y_2$ we have $(y_2 - y_1)^2 \geq (x_2 - x_1)^2$ or equivalently $(y_2 + y_1)^2 - 4y_2y_1 \geq (x_2 + x_1)^2 - 4x_1x_2$ and $x_1x_2 \geq y_1y_2$, since $x_1 + x_2 = y_1 + y_2$.

Equality holds exactly if

$$d(M(A) \cup M(B)) = dM(A) \text{ resp. } dM(B)$$

and

$$d(M(A) \cap M(B)) = dM(B) \text{ resp. } dM(A)$$

This means that $M(A) \supset M(B)$ (resp. $M(B) \supset M(A)$). □

Corollary 9 *For finite* $A, B \subset \mathbb{N}$

$$dM(A \cap B) \cdot dM(A \cup B) \leq dM(A) \cdot dM(B).$$

Equality holds exactly if $B \subset M(A \cap B)$ *or* $A \subset M(A \cap B)$.

Proof. Use that $M(A) \cup M(B) = M(A \cup B)$ and that $M(A \cap B) \subset M(A) \cap M(B)$. The equality characterization starts now with

$$dM(A \cup B) + dM(A \cap B) \leq dM(A) + dM(B)$$

and $dM(A \cup B) \geq dM(A), dM(B) \geq dM(A \cap B)$ and proceeds as before. □

Combining the last theorem with (51) in Lecture 15 we get now

$$\max(d(M(A) \cup M(B)) \cdot d(M(A) \cap M(B)), dM(A \cdot B)) \leq dM(A) \cdot dM(B).$$

This is truly better than any one of the two inequalities.

Example Let $A = \{2\}$, $B = \{3\}$. Then $d(M(A) \cup M(B)) = \frac{2}{3}$, $d(M(A) \cap M(B)) = \frac{1}{6}$, and $dM(A \times B) = \frac{1}{6}$ and $\frac{2}{3} \cdot \frac{1}{6} = \frac{1}{9} < \frac{1}{6}$. On the other hand for $A = B = \{2,3\}$ we have $d(M(A) \cup M(B)) = d(M(A) \cap M(B)) = \frac{2}{3}$, $dM(A \times B) = dM(\{4,6,9\}) = \frac{1}{4} + \frac{1}{6} + \frac{1}{9} - \frac{1}{12} - \frac{1}{36} - \frac{1}{18} + \frac{1}{36} = \frac{14}{36}$ and $\frac{2}{3} \cdot \frac{2}{3} = \frac{16}{36} > \frac{14}{36}$. In this case $\frac{16}{36}$ is tight and so is $\frac{1}{6}$ in the former case.

Another example shows that the two bounds can be close.

Example For $A = \{4,6\}$, $B = \{9,10\}$ we have $d(M(A) \cup M(B)) = \frac{19}{45}$, $d(M(A) \cap M(B)) = \frac{1}{9}$ and $dM(A \times B) = \frac{7}{108}$.

Exercises

1. Derive from RBK Harris' inequality [H60]

$$P(A \cap B) \geq P(A)P(B), \qquad \text{if} \quad A \uparrow, B \uparrow$$

   or, equivalently

$$P(A \cap B) \leq P(A)P(B), \qquad \text{if} \quad A \uparrow, B \downarrow.$$

   This inequality, which is one basic tool in percolation theory, is now a special case of FKG (ad 4).

2. Show that RBK implies the van den Berg/Kesten inequality

$$P(A * B) \leq P(A)P(B), \qquad \text{if} \quad A \uparrow, B \uparrow. \tag{31}$$

   (ad 4)

3. For $A, D \downarrow \subset I^n$, show that
$$A * D \supset A \vee D.$$

   (ad 4)

4. Suppose that for $\alpha, \beta, \gamma, \delta : \quad I^n \to \mathbb{R}_+$

$$\alpha(a)\beta(d) \leq \gamma(a \vee d)\delta(a \wedge d) \quad \forall a, d \in I^n, \qquad \text{then}$$
$$\alpha(A)\beta(D) \leq \gamma(A * D)\delta(A \wedge D) \quad \forall A, D \downarrow \subset I^n.$$

   Hint: Use Birkhoff's Theorem that $(I^n, \vee, \wedge)$ can isomorphically be embedded into $(\{0,1\}^n, \vee, \wedge)$ or verify directly that $(\wedge, \vee)$ is $\mathcal{M}$-expansive on I. (ad 4)

5. For $P = \prod\limits_{t=1}^{n} p_t$ on I^n, show that

$$P(A)P(D) \leq P(A*D)P(A \wedge D); \quad A, D \downarrow \subset I^n.$$

(ad 4)

6. Negative association is a stronger property than negative correlation. Prove that two binary random variables (taking values in $\{0,1\}$) are negatively associated iff they are negatively correlated (ad 7).

7. Here is a result useful for probabilists:
   There are many ways of describing positive dependence, for example the strong FKG inequalities and association. It is known that for Bernoulli random variables the strong FKG inequalities are equivalent to all the conditional distributions being associated, which is in turn equivalent to all the conditional distributions having positively correlated marginals. These and similar definitions are extended to point processes on $\mathbb{R}^d$. Construct examples to show that, unlike the analogous Bernoulli random variable case, these conditions are no longer equivalent, although some are implied by others. Consult [BF90] if necessary (ad 7).

8. Let $m_i \in \{0,1\}$, $i \in [n]$, $\sum_{m_i} = m \leq n$,

$$P(B_1 = m_1, \ldots, B_n = m_n) = \frac{1}{\binom{n}{m}}.$$

Prove using FKG that random variables are negative associated.
Hint: Let $f : \mathbb{R}^m \to \mathbb{R}$, $g : \mathbb{R}^{n-k} \to \mathbb{R}$ be two nondecreasing functions. For $S \in \binom{[n]}{m}, \bar{S} = [n] \setminus S$, define functions

$$f'(S) = \frac{1}{m!} \sum_{\tau} f(\tau(S)), \quad g' = \frac{1}{(n-m)!} \sum_{\rho} g(\rho(\bar{S})),$$

where τ and ρ range over all permutations of S and $\bar{S}$, correspondingly. Under natural partial orders on $\mathbb{R}^k$ and $\mathbb{R}^{n-k}$ function f' is nondecreasing and g' nonincreasing. For measure $\mu = \binom{n}{m}^{-1}$

$$\mu(S)\mu(S') \leq \mu(S \vee S')\mu(S \wedge S').$$

At last show that

$$\sum_{S} f'(S)\mu(S) = \mathbb{E}(f(B_i, i \in I)),$$

$$\sum_{S} g'(S)\mu(S) = \mathbb{E}(g(B_i, i \in [n] \setminus I)),$$

$$\sum_{S} f'(S)g'(S)\mu(S) = \mathbb{E}(f(B_i, i \in I)g(B_i, i \in [n] \setminus I))$$

and use FKG (ad 7).

Research Problems

1. For $M, N \leq 2^n$ characterize the optimal pairs (A', B')

$$|A' \square B'| = \max_{\substack{A, B \subset 2^{[n]} \\ |A| = M, |B| = N}} |A \square B|$$

   in terms of lex and colex orders (ad 4).
2. A characterization of all cases of equality was started for Marica/Schönheim in [AH93], was approached for more general correlation inequalities in [AK95c], and should be continued or even completed.

Notes to Lecture 16

1. Density Properties

Considering the relevance of the work reported in Combinatorial Number Theory, we draw attention to a documentation of its popularity, especially among Hungarian mathematicians in [F93]. For another account see also [PS95]. In the section "Extremal problems under divisibility constrains for finite sets" of [E] it is mentioned how the natural correspondence between divisibility properties for numbers and intersection properties for sets led to a fruitful advancement of methods in both, Number Theory and Combinatorics. This correspondence also led to establishing old and new number theoretic inequalities as consequence of correlation inequalities; such examples are reported in Lecture 15. Finally Section 2.3 "Densities for primitive, prefix-free, quotient, and square-free sets" in [E] reports on works [AK96e], [AKS99], [AKS00].

By a primitive set here we mean a set $A \subset \mathbb{N}$ such that for $a, b \in A$, $a \nmid b$. P. Erdös in [E35] proved that if A is a primitive set, then for some constant c

$$\sum_{a \in A} \frac{1}{a \log a} < c.$$

This easily implies (proving by contradiction and using partial summation) that

$$A(x) = |A \cap [1, x]| < \frac{x}{\log \log x \log \log \log x}$$

is valid for infinitely many $x \in \mathbb{N}$. In [AKS99] the following lower bound on $A(x)$ is proved: for all $\varepsilon > 0$ there exists a primitive set A such that for $x > x(\varepsilon)$ we have

$$A(x) > \frac{x}{\log \log x (\log \log \log x)^{1+\varepsilon}}.$$

Recently, a common generalization of Theorems 56 and 60 asked for in [AK96b] was established by Ahlswede and Blinovsky in [AB06b]. In Theorem 95 below the analog of $f(n,1)$ in the case of algebraic number fields is considered.

In [AKS04] Ahlswede, Khachatrian, and Sárközy consider a variety of problems concerning the density of primitive sets, under various weightings.

Given a weighting $f(m) \geq 0$ supported on the natural numbers, define

$$S(f,A) = \sum_{a \in A} f(a).$$

Then, if $A \subseteq \{1,\ldots,n\}$, and if $f(m) \neq 0$ for some $m \leq n$, define the density of A relative to f to be

$$\delta(f,A,n) = \frac{S(f,A)}{S(f,\{1,2,\ldots,n\})}.$$

Then, one considers the question of how dense a primitive set A can be relative to f, and the natural function for measuring this is

$$F(f,n) = \max_{\substack{A \subset [n] \text{ primitive}}} \delta(f,A,n).$$

The authors prove several bounds for the function $F(f,n)$ when f is a *smooth weighting*, a *multiplicative weighting*, and when it is a *combinatorial weighting* (a special case of multiplicative weightings): A weighting $f(m)$ is said to be *smooth* if it is not eventually 0, is a nonincreasing function for all $m > m_0$, and satisfies $0 \leq f(m) \leq 1$. A weighting $f(m)$ is multiplicative if f is a multiplicative function satisfying $f(m) \geq 0$ and $f(1) = 1$. Finally, a weighting f is combinatorial if it is multiplicative and satisfies $f(p) = 0$ or 1 for every prime p, and satisfies $f(p^\alpha) = 0$ for all $\alpha \geq 2$ and all primes p.

Most of the results boil down to clever applications of Sperner's lemma, as well as applications of the fact that any dyadic integer interval $\{n+1,\ldots,2n\}$ forms a primitive set.

Used are also results on a combinatorial conjecture by Frankl [F88]. Improvements or even a complete solution would improve also the number theoretic results.

We propose to continue the studies of [AKS04] on densities for primitive sets in the following directions.

Research Program C

1. *Densities for primitive sets*

We already know

(i) for $f_1(m) \equiv 1$ $F(f,n) = \frac{1}{2}(1 + o(1))$

(ii) for $f_2(m) = \frac{1}{m}$ $F(f,n) \sim \frac{1}{\sqrt{2\pi}\sqrt{\log\log n}}$

(iii) for $f_3(m) = \begin{cases} 0 & \text{for } m = 1 \\ \frac{1}{m\log m} & \text{for } m > 1 \end{cases}$ $F(f,n) = O\left(\frac{1}{\log\log n}\right).$

For a fixed weighting we are interested in the growth of $F(f,n)$ as $n \to \infty$. In (i) we have $\lim_{n \to \infty} F(f,n) = \frac{1}{2}$ while in the other cases the limit equals 0.

Problem 1 How fast must f decrease to guarantee that the limit equals 0? Is it true that for f_σ with $f_\sigma(m) = \frac{1}{m^\sigma}$ $\liminf_{n \to \infty} F(f_\sigma, n) > 0$ for $0 \le \sigma < 1$, but $\lim_{n \to \infty} F(f,n) = 0$ if $\frac{1}{m^{1-\varepsilon_m}} > f(m) \ge \frac{1}{m}$ for large m with $\varepsilon_m \to 0$?

Problem 2 Is (iii) optimal or is $F(f,n) = o(\frac{1}{\log\log n})$ possible? What happens for instance if $f(m) = \frac{(\log\log m)^c}{m \log m}$ $(c > 0)$?

Problem 3 Is $F(f,n)$ monotone in f for $1 \ge f(m) \ge \frac{1}{m\log m}$? In case not, is it then at least true that for $0 \le \sigma_1 < \sigma_2 \le 1$ one has $F(f_{\sigma_1}, n) > F(f_{\sigma_2}, n)$ (for $n > n_0(\sigma_1, \sigma_2)$)?

2. *Size of elements in extremal primitive sets*

The elements in extremal sets tend to be large in average $(> n/2)$.

Problem 4 Is it true that

(i) $\sum_{a \in A} a > (1-\varepsilon)|A|\frac{n}{2}$ (notice that $\sum_{a \in A} a > (1+\varepsilon)|A|\frac{n}{2}$ is possible)
(ii) and
$$|A \cap (n/2, n]| > (1-\varepsilon)|A \cap [1, n/2]|$$
always hold?

Problem 5 We already know that the previous properties hold for the asymptotic density. Does it hold also for, say, $f_\sigma(m)$, if $\sigma > 0$ is smaller?

3. *The smallest element of an extremal set*

The smallest element of an extremal set cannot be $O(1)$ for ordinary density.

Problem 6 Can this be shown for $f_\sigma(m)$ with $\sigma > 0$ "small"? Presumably this is true for $f(m) = \frac{1}{m}$. If $f(m) = \frac{1}{m^\sigma}$, σ large, then $2 \in A$ must hold.

Problem 7 If $\sigma \ge 1$, the weighting is $f_\sigma(m)$, and $k = k(\sigma)$ is large enough, does it then hold for $n > n_0(\sigma, k)$ that a_1 in a primitive A is smaller than k or $S(f,A) < \sum_{p(\le n)} \frac{1}{p \log p}$?

4. *Uniqueness of extremal sets*

For $f(m) \equiv 1$ there are many extremal sets.

Problem 8 Is it true that almost every $m \in [n]$ (or even almost every k-element subset of $[n]$, k fixed) can be extended to an extremal set? We conjecture that in every other case the extremal set is unique. For $f_\sigma(m)$, σ large, we know this.

5. *The structure of extremal sets*

Define $N(k,n) = \{m : m \leq n, \Omega(m) = k\}$. We call $A \subset [n]$ $N(k,n)$ typical if for fixed f and $n \to \infty$ $\delta(f, A \triangle N(k,n), n) = o(\delta(f, A, n))$.

Problem 9 Is it true that for "small" f an extremal set $N(k,n)$ is typical? For how fast decreasing f is this the case?

6. *Existence of many disjoint "larger" (almost extremal) primitive subsets*

For $f(m) = \frac{1}{m}$ there are "many" primitive disjoint subsets $A_1, \ldots, A_k$ of $[n]$ with $\delta(f, A_i, n) > (1 - \varepsilon) F(f, n)$, for $f(m) = 1$, $f(m) = \frac{1}{m \log m}$ there are no $A_1, \ldots, A_k$ with these properties. Where is the limit?

We are grateful to A. Granville for having communicated to us in March 2004 a problem B. Poonen had once asked him, and which might be solvable using some of the results in [AKS04].

Suppose that $S \subset \mathbb{N}$ and $S(n) = S \cap [n]$. Consider the asymptotic density (known to exist) $dM(S(n))$ and the ratio $r(n) = \frac{|M(S(n)) \cap [n]|}{n}$.
Must

$$\lim_{n \to \infty} dM(S(n)) = \lim_{n \to \infty} r(n)? \tag{32}$$

One can prove that the answer is no by constructing sets S where the limit on the RHS does not exist; but, the limit on the LHS always exists because $dM(S(n))$ increases as $n \to \infty$, and is bounded from above by 1. An example of a set S for which the limit on the RHS does not exist is a union of dyadic integer intervals $\{x_i + 1, \ldots, 2x_i\}$, where the x_i's are chosen to be very far apart. When $n = 2x_i$, for some i, there will be a higher proportion of integers $m \leq n$ divisible by some element of S than when $n = x_i$.
So, one can modify the question in the following way.

Problem 10 If we have that $\lim_{n \to \infty} r(n)$ exists, must it follow then that (32) holds?
Finally, we mention a conceivable sharpening of Theorem 56 of Lecture 16.

Conjecture (Ahlswede/Khachatrian; also Erdös) In Theorem 56 one can choose for every k $n(k) = cp_k^2$ for a suitable constant c. Presently, we have only $n(k) = \prod_{p \leq (p_{c_1 k})} p_{c_2 k}$.

Towards Combinatorial Algebraic Number Theory

After all these contributions to Combinatorial (Elementary) Number Theory, which in particular widens the area treated in [HT88], we open a new area of research by considering now seemingly basic extremal properties for algebraic number fields. We present with complete proofs our recent work. For its understanding knowledge about number fields is required.

We prove that for all sufficiently large N_0 a maximal set of ideals of the maximal order of an algebraic number field, such that any pair of ideals from this set is not coprime and the norm of each ideal does not exceed N_0, is of the form $E(N_0) = \{\theta : N(\theta) \leq N_0, \ \theta = \eta_1 u\}$, where $\{\eta_1, \eta_2, \ldots\}$ is the set of prime ideals of the maximal order and $N(\eta_3) > 2$.

In the paper [AK95b] the authors investigated the problem of finding the maximal sets of integers bounded from above by some number N_0 without $k+1$ coprimes. There it was proved that for all sufficiently large N_0 the unique maximal set is

$$E(N_0, k) = \{n \leq N_0, \ n = p_i u, \ i = 1, \ldots, k\}, \tag{33}$$

where $p_1 < p_2 < \ldots$ is the sequence of prime numbers. Shortly before it was proved in [AK94b] that this assertion is not valid for all N_0, i.e., for small values of N_0 the set $E(N_0, k)$ is not maximal. These facts completely solved the problem of Erdős of determining the maximal sets of integers without $k+1$ coprimes.

It is natural to extend this problem to the case of algebraic numbers. Here we concentrate our attention on the problem when $k = 1$. It is a straightforward result that in the ring of integers the maximal set (for arbitrary N_0) is $E(N_0, 1)$. The answer is not so obvious in the case of ideals in the maximal order of an algebraic number field. Moreover, we can prove that the analogous result is true only for large enough N_0 and only when the norm satisfies $N(\eta_3) > 2$. We consider the maximal order $\mathcal{B}$ of the algebraic number field K, which is a finite extension of the rationals [1] R and $(K : R) = n$. Denote the set of integer ideals of the maximal order by Θ and the set of ideals whose norms do not exceed N_0 by $\Theta(N_0)$.

Let $\Omega = \{\eta_1, \eta_2, \ldots\}$ be the set of prime ideals of the order $\mathcal{B}$, which are ordered in such a way that their norms do not decrease, i.e., $N(\eta_i) \leq N(\eta_{i+1})$. Recall that for an arbitrary $\eta \in \Omega$, $N(\eta) = p^f$ for some prime p and positive integer f. We say that two ideals $\theta_1, \theta_2 \in \Theta$ are coprime if they do not have any common multiple in their prime ideal decomposition. The problem we are going to solve here is to determine for all sufficiently large N_0 the maximal set of ideals from $\Theta(N_0)$ such that it does not contain a pair of coprime ideals. The main problem here in comparison with the ring of integers, which was considered in [AK96a], is that the norm of prime ideals is not a strictly increasing function. We find that the solution of this problem is an interesting interaction between the methods of the work [AK96a] and a diametric problem. This interaction is based on the special properties of intersecting antichains, which we establish here.

Here is the main result.

Theorem 95 (Ahlswede and Blinovsky) *If $N(\eta_2) > 2$, then for sufficiently large N_0 any maximal set of ideals $O(N_0)$ without coprimes and with a norm not exceeding N_0 is one from*

$$E(N_0, \eta_i) \overset{\Delta}{=} \{\theta \in \Theta(N_0) : \theta = \eta_i u\}, i = 1, \ldots, k,$$

where k is the maximal number such that $N(\eta_i) = N(\eta_1)$, $i \leq k$.

---

[1] Here R denotes the field of rational numbers whereas the usually used letter Q denotes in this paper the alphabet $\{0, 1, \ldots, q-1\}$.

If $N(\eta_2) = 2$, $N(\eta_3) > 2$, then the maximal set is one from $E(N_0, \eta_1)$, $E(N_0, \eta_2)$.

Note that in this theorem we still have the condition (as in the ring of integers from [AK95b]), that N_0 must be sufficiently large, and one additional condition, that $N(\eta_3) > 2$. In the case $N(\eta_3) = 2$ we do not even have a conjecture what the maximal set of ideals with restricted norm and without coprimes is and we will show that the maximal density of such a set can be achieved on several sets of ideals.

Define $\mathcal{O}(N_0)$ as the family of maximal sets of ideals from Θ without coprime pairs whose norm does not exceed N_0. Next we assume that $N(\eta_3) > 2$. We say that two ideals $\theta_1, \theta_2 \in \Theta$ intersect in the ith position if $\eta_i | \theta_1$, $\eta_i | \theta_2$. We need the notion of left compressedness of $D \subset \Theta(N_0)$. We say that D is left compressed if for all $d \in D$ such that

$$d = \eta_\ell^i u, \ \eta_\ell \nmid u, \ i \geq 1,$$

and all $\eta_k : \ k < \ell$, we have

$$\bar{d} = \eta_k^i u \in D.$$

Denote by $\mathcal{C}(N_0)$ the family of sets, which belong to the family $\mathcal{S}(\mathcal{N}_r)$ of sets of ideals without coprimes and with a norm not exceeding N_0 and which has the additional property that each set from this family is left compressed. Next we consider a set of ideals from $\mathcal{C}(N_0)$. It is easy to show (and it was done for example in Lemma 1 of [AK95b]) that

$$\mathcal{O}(N_0) \bigcap \mathcal{C}(N_0) \neq \emptyset.$$

Note that $D \in \mathcal{O}(N_0)$ is a downset, i.e., if $d = \eta_{i_1}^{\alpha_1} \ldots \eta_{i_\ell}^{\alpha_\ell} \in D$, then $\bar{d} = \eta_{i_1} \ldots \eta_{i_\ell} \in D$ and D is also an upset in the sense that

$$D = M(D) \bigcap \Theta(N_0),$$

where $M(\mathcal{A})$ is the set of multiples of $\mathcal{A} \subset \Theta$. For $D \subset \Theta$ we denote by $P(D) \subset \Theta$ the set of ideals such that for $\theta_1, \theta_2 \in P(D), \theta_1 \nmid \theta_2$ and $D \subset M(P(D))$. It is easy to see that, if $D \in \mathcal{O}(N_0)$, then

$$D = M(P(D)) \bigcap \Theta(N_0)$$

and $P(D)$ is the set of square-free ideals.

Lemma 55 *For all sufficiently large N_0 there exists a t (which does not depend on N_0 and depends only on K) such that any two $\theta_1, \theta_2 \in \mathcal{A} \in \mathcal{O}(N_0) \bigcap \mathcal{C}(N_0)$ are i-intersecting for some $i \leq t$.*

Lemma 56 *The density of $\mathcal{A} \in \mathcal{O}(N_0)$ equals $1/N(\eta_1)$.*

The proof of Lemma 56 uses some results about intersecting antichains, which we introduce later.

Lemma 57 *If $N(\eta_2) > 2$, then the density $1/N(\eta_1)$ is achieved on one of the sets $E(N_0, \eta_i)$; $i = 1, \ldots, k$. If $N(\eta_2) = 2$, then the maximal density $1/2$ is achieved on two sets, $E(N_0, \eta_1)$ and $E(N_0, \eta_2)$.*

The statement of the next lemma is well known ([N74]).

Lemma 58 (Prime ideal theorem) *The following relation is valid*

$$\#\{\eta \in \Omega : N(\eta) \le z\} = \frac{z}{\log z}(1 + o(1)), \ z \to \infty.$$

Proof of Lemma 55. Let

$$\pi(z) = \{\eta \in \Omega : N(\eta) \le z\}$$

be the number of prime ideals with norm not exceeding z. Our proof is based on the following statement, which was proved in [AK96a].

Proposition 24 *For all $\mathcal{A} \in \mathcal{O}(N_0) \cap \mathcal{C}(N_0)$ no $a \in P(\mathcal{A})$ has divisor η_i, $i \ge s$, where $s \ge 2$ is the minimal number, such that for $z \in R_+$ the following inequality is valid*

$$2\pi(z) \le \pi(\eta_s z). \tag{34}$$

Notice that this statement looks different from Lemma 4 in [AK96a] but the essential parts of the proofs coincide.

Now it is easy to see that for a given field K inequality (34) is always true for $s > s_0$, where s_0 is sufficiently large. Indeed, let us choose z_0 such that

$$\frac{1}{2}\frac{z}{\log z} \le \pi(z) \le 2\frac{z}{\log z}, \ z > z_0.$$

The possibility of such a choice follows from the mentioned Lemma 58. Then also

$$\frac{1}{2}\frac{p_s z}{\log(p_s z)} \le \pi(p_s z) \le 2\frac{p_s z}{\log(p_s z)}, \ z > z_0.$$

Now we choose s_0 such that for $s \ge s_0$

$$2\frac{2z}{\log z} \le \frac{p_s z}{\log(p_s z)}. \tag{35}$$

If $z < 2$, then $\pi(z) = 0$ and (34) is valid. If $2 \le z < z_0$, then we choose s_1 such that

$$\pi(p_s z) \ge \pi(p_s 2) \ge 2\pi(z_0) \ge 2\pi(z), \ s \ge s_1. \tag{36}$$

At last if $t = \max(s_0, s_1)$, then (35) and (36) imply (34). Lemma 55 is proved. $\square$

Thus for some t, which is independent of N_0, each ideal from $P(\mathcal{A})$ has no divisors η_j, $j \ge t$. Hence we should consider only $P(\mathcal{A})$ such that $\theta = \eta_{i_1} \dots \eta_{i_r} \in P(\mathcal{A})$, $i_1 < \dots < i_r \le t$ for some t, which depends only on K. We assume that the square-free ideals $a_1, a_2, \dots \in P(\mathcal{A})$ are ordered colexicographically. Hence there exists a natural one-to-one correspondence between ideals from $P(\mathcal{A})$ and binary t-tuples. The set of t-tuples, which correspond to $P(\mathcal{A})$, is an intersecting antichain. Now we are going to investigate some properties of intersecting antichains. First of all note that a maximal $\mathcal{A}(N_0)$ must have maximal asymptotic density as $N_0 \to \infty$. The density $d\mathcal{A}(N_0)$ is equal to

$$d\mathcal{A}(N_0) = \sum_i dB^i, \tag{37}$$

where dB^i is the density of the set of ideals B^i from $\mathcal{A}$, which are divisible by $a_i \in P(\mathcal{A})$ and are not divisible by $a_j \in P(\mathcal{A})$, $j < i$. By left compressedness of the set $\mathcal{A}$ it follows that if $a_i = \eta_{j_1} \dots \eta_{j_{r_i}}$ ($j_1 < \dots < j_{r_i}$) and $N(\eta_j) = q_j$, then

$$B^i = \left\{ \theta \in \Theta(N_0) : \theta = \eta_{j_1}^{\alpha_1} \dots \eta_{j_{r_i}}^{\alpha_{r_i}} u, \; \alpha_j \geq 1, \; \left(u, \prod_{j \leq j_{r_i}} \eta_j \right) = 1 \right\}$$

and hence

$$dB^i = \sum_{\alpha_{j_p} \geq 1} \frac{1}{q_{j_1}^{\alpha_1} \dots q_{j_{r_i}}^{\alpha_{r_i}}} \prod_{j \leq j_{r_i}} \left(1 - \frac{1}{q_j} \right) = \prod_{j \leq j_{r_i}, \; j \neq j_p; \; p=1,\dots,r_i} (q_j - 1) \left(\prod_{j=1}^{j_{r_i}} q_j \right)^{-1}$$

and

$$d\mathcal{A}(N_0) = \sum_i \prod_{j \neq j_p; \; p=1,\dots,r_i} (q_j - 1) \left(\prod_{j=1}^{j_{r_i}} q_j \right)^{-1}. \tag{38}$$

We now consider the t-tuples, whose jth element is chosen from the alphabet $\{0, 1, \dots, q_j - 1\}$, and consider sets $\mathcal{A}(t)$ of t-tuples such that every pair of t-tuples has a common unit in some position (possibly different for different pairs). As it was shown in [AK98], the cardinality of a maximal set of t-tuples from $Q^t = \{0, 1, \dots, q - 1\}^t$ such that its diameter is d coincides with the maximal cardinality of a set of t-tuples from Q^t such that every pair of t-tuples from this set has $t - d$ common ones. The same is true for $Q^t = \prod_{i=1}^t Q_i = \{0, \dots, q_1 - 1\} \times \dots \times \{0, \dots, q_t - 1\}$. The characterization of all such maximal sets constitutes a diametric problem. In our case $d = t - 1$. In [L79a] was proved (and it also follows from [AK98]) that maximal subsets from Q^t with diameter $d = t - 1$ are of the form

$$\mathcal{A}_{ij} = \{(a_1, \dots, a_t) \in Q^t : a_i = j\}, \; i = 1, \dots, t, \; j = 0, \dots, q - 1.$$

Their cardinality is q^{t-1}. We use this result to show the validity of the following

Proposition 25 *Any maximal set from Q^t with diameter $d = t - 1$ is one of the form*

$$\mathcal{A}_{ij} = \{(a_1, \dots, a_t) \in Q^t : a_i = j\}, \; i = 1, \dots, k, \; j = 0, \dots, q - 1$$

if $q_1 = \dots = q_k = q < q_{k+1}$, $q_2 > 2$
and

$$\mathcal{A}_{ij} = \{(a_1, \dots, a_t) \in Q^t : a_i = j\}, \; i = 1, 2, \; j = 0, 1$$

if $q_1 = q_2 = 2 < q_3$.

Proof. For $t = 1$ or $t = 2, q_1 = q_2 = 2$ the statement is obvious. Next we suppose that $t > 1$ and if $t > 2$, then $q_3 > 2$. The proof will use induction on t.

Suppose that $\mathcal{A} \subset Q^t$ is a maximal intersecting set. It can be easily seen that $|\mathcal{A}| = \prod_{i=2}^{t} q_i$. We set $\mathcal{A} = \bigcup_{j=0}^{q_t-1} \mathcal{A}_j$, where $\mathcal{A}_j = \{a \in \mathcal{A} : a = (a_1, \ldots, a_{t-1}, j)\}$. Denote $\mathcal{A}'_j = \{x \in \mathcal{A}_j : \bar{x}^i \text{ intersects with all } \bar{y}^i, \ y \in \mathcal{A}\}$, where $\bar{x}^i = (x_1, \ldots, x_{i-1}, x_{i+1}, \ldots, x_t)$. Denote also $\mathcal{T} = \bigcup_{j=0}^{q_t-1} \left(\mathcal{A}_j \setminus \mathcal{A}'_j \right)$.

We assume that $q_t > q_1$. Otherwise the proof of the lemma reduces to the proof of Theorem 2 from [L79a], which states the result for the case $Q_i = \{0, 1, \ldots, q-1\}$ for $i = 1, \ldots, t$.

Consider two cases:

Case $\mathcal{T} = \mathcal{A}$. It is easy to see that for each $(a_1, \ldots, a_{t-1}) \in \{0, \ldots, q_1 - 1\} \times \ldots \times \{0, \ldots, q_{t-1} - 1\}$ there exists not more than one $a_t \in \{0, \ldots, q_t - 1\}$ such that $a = (a_1, \ldots, a_t) \in \mathcal{A}$. But in this case $|\mathcal{A}| \leq \prod_{j=1}^{t-1} q_j$, which contradicts to the maximality of $\mathcal{A}$.

Case $\mathcal{T} \neq \mathcal{A}$. It can be easily seen that if $q_1 < q_t$, then $\mathcal{T} = \emptyset$. Indeed, consider the decomposition $\mathcal{T} = \bigcup_{j=0}^{q_1-1} \mathcal{T}_j$, where $\mathcal{T}_j = \{a = (a_1, \ldots, a_t) \in \mathcal{T} : a_1 = j\}$. With the pigeon-hole principle follows the existence of an $i \in \{0, \ldots q_1 - 1\}$ such that $|\mathcal{T}_i| \geq |\mathcal{T}|/q_1$. Then the set

$$\mathcal{A}' = \bigcup_{j=0}^{q_t-1} \mathcal{A}'_j \bigcup \{(a_1, \ldots, a_{t-1}, m), \ m \in \{0, \ldots, q_t - 1\}$$

$$\text{and } (a_1, \ldots, a_{t-1}) = \bar{a}^t \text{ for some } a \in \mathcal{T}_i \}.$$

is intersecting and $|\mathcal{A}'| > |\mathcal{A}|$, which is a contradiction.

Next, if $\mathcal{T} = \emptyset$, then $\mathcal{A} = \bigcup_{j=0}^{q_t-1} \mathcal{A}'_j$ and $\mathcal{B} = \{(a_1, \ldots, a_{t-1}) : (a_1, \ldots, a_t) \in \mathcal{A}\}$ is an intersecting set. By maximality of $\mathcal{A}$ we have $\mathcal{A} = \{(a_1, \ldots, a_{t-1}, m), \ m \in \{0, \ldots, q_t - 1\}, \ (a_1, \ldots, a_{t-1}) \in \mathcal{B}\}$. Hence to maximize $|\mathcal{A}|$ we should maximize the intersecting set $\mathcal{B}$, but this set consists of $(t-1)$-tuples and we can use induction. This completes the proof. $\qquad\square$

Now we turn to some facts about intersecting antichains. We introduce several relations that have independent interest; however, for our proofs we only need Proposition 28.

Intersecting Antichains

Let us have an antichain $\mathcal{A} \subset 2^{[t]}$ satisfying at the same time for arbitrary $A_1, A_2 \in \mathcal{A}$, $A_1 \cap A_2 \neq \emptyset$. Such a set we call an intersecting antichain. Denote by $\mathcal{A}_i \subset \mathcal{A}$ the set of binary t-tuples such that i is the last position where every $A \in \mathcal{A}_i$ has a one. We start with a simple but interesting inequality.

Proposition 26 *If $\mathcal{A}$ is an intersecting antichain, then*

$$\sum_{i=1}^{t} \frac{|\mathcal{A}_i|}{2^i} \leq \frac{1}{2}. \tag{39}$$

(It is easy to see that bound (39) is tight; for example there is equality, when $|\mathcal{A}_i| = 0$ for $i \geq 2$, and $|\mathcal{A}_1| = 1$. There is equality also in many other cases, as we will show later).

Proof. Denote by $\mathcal{B}_i$ the set of vectors obtained from $\mathcal{A}_i$ by deleting the last $t - i$ zeros. The vectors from $\mathcal{B}_i$ have i components. $\mathcal{B} = \bigcup_{i=1}^{t} \mathcal{B}_i$ is a prefix-free code and $|\mathcal{B}_i| = |\mathcal{A}_i|$. Hence by the Kraft inequality we have

$$\sum_{i=1}^{t} \frac{|\mathcal{B}_i|}{2^i} \leq 1$$

and hence

$$\sum_{i=1}^{t} \frac{|\mathcal{A}_i|}{2^i} \leq 1.$$

Now for every i and every $b \in \mathcal{B}_i$ consider all possible continuations of i-tuple b to the length t. The number of such continuations is 2^{t-i}. This way we obtain a set of different t-tuples $\mathcal{C}$,

$$|\mathcal{C}| = \sum_{i=1}^{t} |\mathcal{B}_i| 2^{t-i} = \sum_{i=1}^{t} |\mathcal{A}_i| 2^{t-i}. \tag{40}$$

At the same time the set $\mathcal{C}$ is intersecting and hence

$$|\mathcal{C}| \leq 2^{t-1}. \tag{41}$$

Therefore

$$\sum_{i=1}^{t} |\mathcal{A}_i| 2^{t-i} \leq 2^{t-1}$$

and we obtain (39). Equality in (39) is achieved also on the intersecting antichain

$$\mathcal{A} = \begin{cases} \binom{[t]}{\frac{t+1}{2}}, & \text{if } 2 \nmid t, \\ \left\{ A \in \binom{[t]}{\frac{t}{2}+1} : 1 \notin A \right\} \cup \left\{ A \in \binom{[t]}{\frac{t}{2}} : 1 \in A \right\}, & \text{if } 2 \mid t. \end{cases} \tag{42}$$

This can be easily seen by the fact that the set $\mathcal{A}$ is an intersecting antichain whose sets $\mathcal{A}_i$ generate the sets $\mathcal{B}_i$ such that all possible continuations of the sets $\mathcal{B}_i$ to the length t form the intersecting set $\mathcal{C}$:

$$\mathcal{C} = \begin{cases} \bigcup_{j=(t+1)/2}^{t} \binom{[t]}{j}, & \text{if } 2 \nmid t, \\ \left\{ A \in \bigcup_{j=t/2+1}^{t} \binom{[t]}{j} : 1 \notin A \right\} \cup \left\{ A \in \bigcup_{j=t/2}^{t} \binom{[t]}{j} : 1 \in A \right\}, & \text{if } 2 \mid t \end{cases}$$

and this intersecting set has cardinality 2^{t-1}. Another proof of this fact can be done by induction (by proving relation (43) below). Consider for example the case $2 \nmid t$. We have

$$|\mathcal{A}_i| = a_i = \binom{i-1}{\frac{t-1}{2}}.$$

Hence

$$\sum_{i=1}^{t} \frac{a_i}{2^i} = \sum_{i=(t-1)/2}^{t-1} \frac{\binom{i}{\frac{t-1}{2}}}{2^{i+1}} = \frac{1}{2} \sum_{i=(t-1)/2}^{t-1} \frac{\binom{i}{\frac{t-1}{2}}}{2^i}.$$

We are done if we can show that

$$g(c) = \sum_{i=c}^{2c} \frac{\binom{i}{c}}{2^i} = 1. \tag{43}$$

We prove (43) by induction. For $c = 0, 1$ it is true. Then

$$g(c+1) = \sum_{i=c+1}^{2c+2} \frac{\binom{i}{c+1}}{2^i} = \sum_{i=c+1}^{2c+2} \frac{\binom{i-1}{c}}{2^i} + \sum_{i=c+2}^{2c+2} \frac{\binom{i-1}{c+1}}{2^i}$$

$$= \frac{1}{2} \left(g(c) + \frac{\binom{2c+1}{c}}{2^{2c+1}} + g(c+1) - \frac{\binom{2c+2}{c+1}}{2^{2c+2}} \right) = \frac{1}{2}(1 + g(c+1)).$$

We can generalize inequality (39) to the case of r-intersecting antichains $\mathcal{A}$, i.e., when $|A_1 \cap A_2| \geq r$ for all $A_1, A_2 \in \mathcal{A}$. We use Katona's Lemma: if $\mathcal{C} \subset 2^{[t]}$ is an r-intersecting set, then

$$|\mathcal{C}| \leq K(t,r) = \begin{cases} \sum_{i=(t+r)/2}^{t} \binom{t}{i}, & 2|(t+r), \\ 2\sum_{i=(t+r-1)/2}^{t-1} \binom{t-1}{i}, & 2 \nmid (t+r). \end{cases}$$

and instead of inequality (41) we obtain

Lemma 59 *If $\mathcal{A}$ is an r-intersecting antichain, then*

$$\sum_{i=1}^{t} \frac{|\mathcal{A}_i|}{2^i} \leq \frac{K(t,r)}{2^t}. \tag{44}$$

Note that everywhere instead of the antichain condition we can consider the weaker condition that $\bigcup_{i=1}^{t} \mathcal{B}_i$ is a prefix-free code. However, when $r > 1$, equality in (44) is achieved only on the antichain $\mathcal{A}$ consisting of the minimal elements of Katona's set (about Katona's set see for example [AK05]), i.e., when

$$\mathcal{A} = \begin{cases} \binom{[t]}{\frac{t+r}{2}}, & \text{if } 2|(t+r), \\ \left\{ A \in \binom{[t]}{\frac{t+r-1}{2}} : 1 \notin A \right\} \cup \left\{ A \in \binom{[t]}{\frac{t+r+1}{2}} : 1 \in A \right\}, & \text{if } 2 \nmid (t+r) \end{cases}$$

We can find further generalizations of inequality (39), for example when the ground alphabet is q−ary. Note that the maximal number of intersecting t−tuples from

$Q^t = \{0, 1, \ldots, q-1\}^t$ is q^{t-1}. Hence if we consider $\mathcal{A}_i \subset \mathcal{A} \subset Q^t$ as the set of t-tuples such that i is the position of their rightmost nonzero symbol, we can write (39) with q instead of 2. However, more useful for our purpose will be the model, when we take into account only positions of t-tuples from $\mathcal{A}$, which contain ones. For $a^t = (a_1, \ldots, a_t) \in Q^t$ define

$$B(a^t) = \{j : a_j = 1\}$$

and for $\mathcal{A} \subset Q^t$ denote $\mathcal{B}(\mathcal{A}) = \{B(a^t), a^t \in \mathcal{A}\}$. Let also $L(\mathcal{A})$ be the set of minimal elements of $\mathcal{B}(\mathcal{A})$. Denote by $\mathcal{A}_{i,\omega} \subset L(\mathcal{A})$ the set of t-tuples each having its last one in position i with the whole number of ones equal to ω. Then the following relation is valid:

$$\sum_{i=1}^t \sum_{\omega=1}^i \frac{|\mathcal{A}_{i,\omega}|(q-1)^{i-\omega}}{q^i} \leq \frac{1}{q}. \tag{45}$$

The proof of this inequality involves similar counting arguments as the proof of (39).

To find a generalization of (45) for the case of r-intersecting sets we should know the formula for the maximal cardinality of a q–ary set $\mathcal{A}$ such that for every $A_1, A_2 \in \mathcal{A}$, $|A_1 \cap A_2| \geq r$, where intersection means the set of positions, where both A_1 and A_2 have ones.

Proposition 27 *If for $\mathcal{A} \subset Q^t, L(\mathcal{A})$ is an r-intersecting antichain, then*

$$\sum_{i=1}^t \frac{1}{q^i} \sum_{\omega=1}^i |\mathcal{A}_{i,\omega}|(q-1)^{i-\omega} \leq \frac{N_q(t,r)}{q^t},$$

where $N_q(t,r)$ is the maximal cardinality of a set from $[q]^t$ whose diameter does not exceed $t - r$.

At last we need one, the most general case, when $\mathcal{A} \subset \prod_{i=1}^t Q_i = \{0, \ldots, q_1 - 1\} \times \cdots \times \{0, \ldots, q_t - 1\}$. In this case, we have the following generalization of (45) (and correspondingly (39)), the proof of which we leave to the reader.

Proposition 28 *The following relation is valid:*

$$\sum_{i=1}^t \sum_{C \in L(C): \, s^+(C)=i} \prod_{j \in [i] \setminus C} (q_j - 1) \prod_{j=i+1}^t q_j \leq \prod_{j=2}^t q_j \tag{46}$$

where $s^+(C)$ is the position of the rightmost one of C.

Proof of Theorem 95. Now we summarize the facts that we have obtained and prove Theorem 95. Note that the expression on the LHS of (46) is equal to the number of t-tuples in some set $C \subset \prod_{i=1}^t Q_i$ with intersecting $L(C)$ and if $L(C) = P(\mathcal{A}(N_0))$, then it is proportional up to $\prod_{i=1}^t q_i$ to the density (37) of $\mathcal{A}(N_0) \subset \Theta(N_0)$ where we use the one-to-one correspondence between binary t-tuples and square-free ideals $\theta \in \Theta$, such that $\eta_\tau \nmid \theta$ when $\tau > t$. Solving the diametric problem in this case, we see that the maximum of the LHS of (46) for left compressed sets and hence the maximum of the density of $A(N_0)$ is achieved (only) when $L(C) = \{(1,0,\dots,0)\}$. As the number of possible $P(A(N_0))$ such that for $\theta \in P(A(N_0))$ we have $\eta_\tau \not| \theta$ when $\tau > t$ is bounded from above independently of N_0, there exists N' such that when $N_0 > N'$ we have for the maximal $A(N_0)$:

$$A(N_0) = \{\theta \in \Theta(N_0): \ \theta = \eta_1 u\}. \tag{47}$$

This maximal set is unique among left compressed sets. This proves Lemma 56. To prove Lemma 57 and Theorem 95 note that for not left compressed sets, in the case $N(\eta_2) > 2$ we have additional to (47) possibilities $\{\theta \in \Theta(N_0): \ \theta = \eta_i u\}$, $i = 2,\dots,k$ each of which is a maximal set and in the case $N(\eta_2) = 2$, $N(\eta_3) > 2$ we have one additional to (47) maximal set $\{\theta \in \Theta(N_0): \ \theta = \eta_2 u\}$. This proves Theorem 95. $\qquad\qquad\qquad\qquad\qquad\qquad\qquad\qquad\qquad\qquad\qquad\qquad\qquad \square$

Remark In the case, when $N(\eta_3) = 2$, the density $1/2$ is achieved besides the set (47) also on the set

$$A''(N_0) = \{\theta \in \Theta(N_0): \ \theta = \eta_1 \eta_2 u, \ = \eta_1 \eta_3 u, \ = \eta_2 \eta_3 u\} \tag{48}$$

and at the present we are not able to determine in the general case when $N(\eta_3) = 2$ which set of ideals is maximal.

Note that the results that do not use the strict increase of the norm along the set of ideals $\eta_1, \eta_2, \dots$ are still valid for the set of ideals as for the set of positive integers. Let us give an example. Write

$$\zeta(A,s) = \sum_{\eta \in A} \frac{1}{N(\eta)^s}, \ s > 1,$$

where A is some set of ideals. The lower Dirichlet density $\underline{D}(A)$ of the set A is defined as follows:

$$\underline{D}(A) \overset{\Delta}{=} \liminf_{s \to 1^+} \zeta(A,s).$$

For an arbitrary pair of divisors η_1, η_2 denote by (η_1, η_2) $([\eta_1, \eta_2])$ their greatest common divisor (least common multiple) and for two sets of ideals A, B let

$$(A,B) = \{(\eta_1, \eta_2); \ \eta_1 \in A, \ \eta_2 \in B\},$$

$$[A,B] = \{[\eta_1, \eta_2]; \ \eta_1 \in A, \ \eta_2 \in B\}.$$

The following inequality is valid:

$$\underline{D}(A)\underline{D}(B) \leq \underline{D}([A,B])\underline{D}((A,B)).$$

This inequality is from the class of correlation inequalities. The proof of this inequality is literally the same as when the sets A and B are sets of positive integers and this proof can be found in [AK97a].

Research Perspectives with Informational Aspects: Three Research Programs

Here we try to stimulate readers to try to explore three new directions in Combinatorial Extremal Theory, which are not covered by the lectures. They all relate to Information Theory.

A Direction in Extremal Theory of Sequences: Creating Order with Simple Machines

In [AZ89] and [AYZ90] a new field of research, creating order in sequence spaces with simple machines, was introduced. People spend a large amount of time creating order in various circumstances. We contribute to a theory of ordering. In particular we try to understand how much "order" can be created in a "system" under constraints on our "knowledge about the system" and on the "actions we can perform in the system."

We have a box that contains β objects at time t labeled with numbers from $\mathcal{X} = \{0, \ldots, \alpha - 1\}$. The state of the box is $s_t = (s_t(1), \ldots, s_t(\alpha))$, where $s_t(i)$ denotes the number of balls at time t labeled by i.

Assume now that an arbitrary sequence $x^n = (x_1, \ldots, x_n) \in \mathcal{X}^n$ enters the box iteratively. At time t an organizer $\mathcal{O}$ outputs an object y_t and then x_t enters the box. $x^n = (x_1, \ldots, x_n)$ is called an input and $y^n = (y_1, \ldots, y_n)$ an output sequence. The organizer's behavior must obey the following rules.

Constraints on matter. The organizer can output only objects from the box. At each time t he must output exactly one object.

Constraints on mind. The organizer's strategy depends on the following

(a) His knowledge about the time t. The cases where $\mathcal{O}$ has a timer and has no timer are denoted by T^+ and T^-, respectively.
(b) His knowledge about the content of the box. O^- indicates that the organizer knows at time t only the state s_t of the box. If he also knows the order of entrance times of the objects, we write O^+.
(c) The passive memory (π, β, φ). At time t the organizer remembers the output letters $y_{t-\pi}, \ldots, y_{t-1}$ and can see the incoming letters $x_{t+1}, \ldots, x_{t+\varphi}$.
    Let $\mathcal{F}_n(\pi, \beta, \varphi, T^-, O^-)$ be the set of all strategies for (T^-, O^-), length n and a given memory (π, β, φ) and $\mathcal{S}$ be the set of all states. A strategy $f_n : \mathcal{X}^n \times \mathcal{S} \to \mathcal{X}^n$ assigns to each pair (x^n, s_1) an output y^n. Denote $\mathcal{Y}(f_n)$ the image of $\mathcal{X}^n \times \mathcal{S}$ under f_n. Also denote $\|\mathcal{Y}(f_n)\|$ the cardinality of $\mathcal{Y}(f_n)$.
    Now we define the *size*

$$N_\alpha^n(\pi, \beta, \varphi) = \min\{\|\mathcal{Y}(f_n)\| : f_n \in \mathcal{F}_n(\pi, \beta, \varphi, T^-, O^-)\}$$

and the *rate*

$$v_\alpha(\pi,\beta,\varphi) = \lim_{n\to\infty} \frac{1}{n}\log N_\alpha^n(\pi,\beta,\varphi).$$

Analogously, we define in the case (T^-,O^+) the quantities $O_\alpha^n(\pi,\beta,\varphi)$, $\omega_\alpha(\pi,\beta,\varphi)$, in the case (T^+,O^-) the quantities $T_\alpha^n(\pi,\beta,\varphi)$, $\tau_\alpha(\pi,\beta,\varphi)$ and in the case (T^+,O^+) the quantities $G_\alpha^n(\pi,\beta,\varphi)$, $\gamma_\alpha(\pi,\beta,\varphi)$.

(d) The active memory. Now the organizer has additional memory of size m, where he is free to delete or store any relevant information at any time. Here we are led to study the quantities $N_\alpha^n(\pi,\beta,\varphi,m)$, $v_\alpha(\pi,\beta,\varphi,m)$, etc.

Survey of the Results

$\pi,\quad \phi$	$v_2(\pi,\beta,\varphi)$
0, 0	1
0, 1	1
1, 0	$\sup_\delta (1-(\beta-1)\delta)h\left(\frac{\delta}{1-(\beta-1)\delta}\right)$
$\pi,\quad \infty$	$1/\beta$
$\infty,\quad \leq \beta-1$	$\log\lambda^*$, where λ^* is the largest root of $\lambda^{\beta+1+\varphi} = \lambda^{\lceil(\beta+1+\varphi)/2\rceil} + \lambda^{\lfloor(\beta+1+\varphi)/2\rfloor}$
$\infty,\quad \geq \beta-1$	$1/\beta$

Furthermore, the following relations hold.

$$\omega_2(\infty,\beta,\varphi) = v_2(\infty,\beta,\varphi), \qquad\qquad \omega_2(\pi,\beta,\infty) = v_2(\pi,\beta,\infty),$$
$$\tau_2(\pi,\beta,\varphi) = v_2(\infty,\beta,\varphi) \text{ for } \pi \geq 1, \quad \tau_2(0,2,0) = \log((\sqrt{5}+1)/2).$$
$$\lim_{\beta\to\infty} v_3(0,\beta,0) = 1.$$

In the model of active memory we have for the memory size $m = 2$ that $v_2(0,\beta,0,2) = v_2(1,\beta,0) = \log\lambda_\beta$, where λ_β is the positive root of $\lambda^\beta - \lambda^{\beta-1} - 1 = 0$.

The general case, where the size α of the set $\mathcal{X}$, the size β of the box, and the memory parameters π,φ, and m are arbitrary, has not been solved yet. This is the cardinal goal for our research to aim at within this field. We have the following **Conjectures**

1. $\lim\limits_{\varphi\to\infty} v_2(\pi,\beta,\varphi) \neq v_2(\pi,\beta,\infty)$ (in the analogous case for $\pi\to\infty$ equality holds)
2. $\lim_{\beta\to\infty} v_\alpha(0,\beta,0) = \log_2\lceil(\alpha+1)/2\rceil$ (for $\alpha = 2$ and $\alpha = 3$ this is true)
3. $\omega_2(0,\beta,0) = v_2(1,\beta-1,0)$

In a probabilistic model, the objects or letters are produced by a stochastic process, which in the simplest case is a sequence $(X_t)_{t=1}^\infty$ of i.i.d. RV's with values in $\mathcal{X} = \{0,1,\ldots,\alpha-1\}$ and generic distribution P_X. In Information Theory, this is also called a discrete, memoryless source. For a strategy f_n, which depends on the triple

(π, β, φ), let $Y^n = Y_1 \ldots Y_n$ be the output sequence corresponding to $X^n = X_1 \ldots X_n$. Let $F_\alpha^n(\pi, \beta, \varphi, P_X)$ be the set of strategies restricted to block length n.

We use the "per letter" entropy $\frac{1}{n} H(Y^n)$ as performance criterion and define

$$\eta_\alpha(\pi, \beta, \varphi, P_X) = \lim_{n \to \infty} \min_{f_n \in F_\alpha^n(\pi, \beta, \varphi, P_X)} \frac{1}{n} H(Y^n).$$

This is the smallest mean entropy of the output process, which can be achieved by $\mathcal{O}$ with strategies based on his knowledge. It corresponds to the optimal rate $v_\alpha(\pi, \beta, \varphi)$ in the nonprobabilistic model. Our new quantity is much harder to analyze.

In the first nontrivial case $\beta = 2$ and $\pi = \infty$, $\varphi = 0$ only the simplest nontrivial source, namely the binary symmetric source defined by $P_X(0) = P_X(1) = 1/2$, could be analyzed.

Theorem 96 *The strategy that is locally optimal for every $t = 1, 2, \ldots$ is optimal. Moreover for the disjoint events $D_k = E_k \setminus E_{k+1}$, where $E_k = \{Y^k = 01010\ldots\}$, $q(k) = Prob(D_k)$ satisfies $\sum_{n=1}^{\infty} q(k) = 1$ and*

$$\eta_2(\infty, 2, 0, P_X) = \frac{H(q)}{\sum\limits_{k=1}^{\infty} k q(k)} = 0,5989\ldots \tag{49}$$

The formula (49) has a nice structure. It suggests a general principle for arbitrary sources. However, already the binary nonsymmetric source is difficult to solve. Finally, we mention the survey of Vanroose, pages 603–613 in [N].

Directions of Developments of Our Basic Model for Sequences

Multiple in- and outputs: s inputs and s outputs, varying number of outputs, merging, splitting, correlation

Objects with special features: Varying-length objects, death-birth, idle objects, box with exclusion rule

Compound objects: Box with reaction rules, representatives, objects with many properties, exchanging parts of objects

Errors: Probabilistic, confusion rule, frequency rule, receiver can distinguish only certain objects

Applications

Production of goods, arrival of goods and documents, garbage collection

Extensions of the Basic Model

A combined theory of ordering and source coding
Ordering, sorting, and Maxwell's demon
A calculus of machines: comparisons of machines, commutativity

We want to emphasize that the subject discussed falls into the large area of combinatorial problems concerning sequence–subsequence relations, which includes genetic studies. Actually several results of this book have extensions in this context. Intensively studied have been shadows and isoperimetry under the sequence–subsequence relation, briefly reported in Section 6.3 of [E].

Information Flows in Networks

We continue now with the subject whose origin is generally attributed to [ACLY00]. The founder of Information Theory Claude E. Shannon, who set the standards for efficient transmission of channels with noise by introducing the idea of coding, also wrote together with Peter Elias and Amiel Feinstein a basic paper on networks ([SEF56]) discussing algorithmic aspects of the Min Cut – Max Flow Theorem ([FF56]), saying that for flows of physical commodities like electric currents or water, satisfying Kirchhoff's laws, the maximal flow equals the minimal cut.

With the stormy development of Computer Science, there is an ever increasing demand for designing and optimizing information flows over networks – for instance in the Internet.

Data, that is strings of symbols, are to be sent from sources $s_1,\ldots,s_n$ to their destinations, sets of node sinks $D_1,\ldots,D_n$.

Computer Scientists quickly realized that it is beneficial to copy incoming strings at processors sitting at nodes of the network and to forward copies to adjacent nodes. This task is called multicasting.

However, quite surprisingly *they did not consider coding*, which means here to produce not only copies, but, more generally, new output strings as deterministic functions of incoming strings.

A *Min–Max-Theorem was discovered and proved for information flows* by Ahlswede, Cai, Li, and Yeung in [ACLY00].

Its statement can be simply explained. For one source only, that is $n = 1$, in the notation above, and $D_1 = \{d_{11},d_{12},\ldots,d_{1t}\}$ let F_{1j} denote the max-flow value, which can go for any commodity like water in case of Ford/Fulkerson from s_i to d_{1i}. The same water cannot go to several sinks. However, the amount of $\min_{1 \le j \le t} F_{1j}$ bits can go *simultaneously* to $d_{11},d_{12},\ldots$ and d_{1t}. Obviously, this is best possible. It has been referred to as ACLY-Min-Max-Theorem. To the individual F_{1j} Ford/Fulkerson's Min-Cut-Max-Flow Theorem applies.

It is very important that in the starting model there is no noise, and it is amazing for how long Computer Scientists did the inferior multicasting allowing only copies. (It is perhaps surprising that Shannon seems not to have realized the consequences of the basic difference between classical and information flows.)

Network flows with *more than one source* are much harder to analyze and lead to a wealth of old and new combinatorial extremal problems.

Even nicely characterized classes of *error-correcting codes* come up as being isomorphic to a complete set of solutions of flow problems *without errors*!

Also *optimal anticodes* (see Theorem 5) arise in such a role!

On the classical side for instance orthogonal *Latin Squares* arise.

With NetCod 2005 – the first workshop on Network Coding Theory and Applications, April 7, 2005, Riva, Italy the *new subject Network Coding* was put to start.

It is known that classical network flows have many connections to combinatorial extremal problems like Baranyai's factorization theorem ([B75]) or especially for matching problems. Information flows promise more such connections as for example in [WJK06]. There may be a great challenge not only coming to *Combinatorics* but also to *Algebraic Geometry* and its present foundations.

We draw attention to the chapter on Network Coding in [G], pages 858–897.

Information Theory and the Regularity Lemma

Next we introduce an example of how the knowledge of Information Theory helps to solve combinatorial problems. Tao in [T06] approached Szemerédi's Regularity Lemma from the perspectives of Probability Theory and Information Theory instead of Graph Theory and as a technical tool he proved the following

Lemma 60 *Let Y, and X, X' be discrete random variables taking values in $\mathcal{Y}$ and $\mathcal{X}$, respectively, where $\mathcal{Y} \subset [-1, 1]$, and with $X' = f(X)$ for a (deterministic) function f. Then we have*

$$\mathbb{E}(|\mathbb{E}(Y|X') - \mathbb{E}(Y|X)|) \leq 2I^{1/2}(X; Y|X').$$

Readers familiar with Information Theory immediately find out that the inequality in the Lemma is a Pinsker-type inequality [P64] between the variational distance in the LHS and the divergence of random variables in the RHS and this helps to prove this inequality in a regular way [A07] with constant $\sqrt{2 \ln 2}$ instead of 2 in the RHS, and this constant is best possible.

References

[AH93]   R. Aharoni and R. Holzman, Two and a half remarks on the Marica/Schönheim inequality, J. Lond. Math. Soc., (2), Vol. 48, No. 3, 385–395, 1993

[AK96]   R. Aharoni and U. Keich, A generalization of the Ahlswede-Daykin inequality, Discrete Math., Vol. 152, No. 1–3, 1–12, 1996

[A79]   R. Ahlswede, Coloring hypergraphs: A new approach to multi-user source coding I, J. Combin. Inform. System Sci., Vol. 4, No. 1, 76–115, 1979

[A80]   R. Ahlswede, Coloring hypergraphs: A new approach to multi-user source coding II, J. Combin. Inform. System Sci., Vol. 5, No. 3, 220–268, 1980

[A87]   R. Ahlswede, On code pairs with specified Hamming distances, Colloq. Math. Soc. János Bolyai, Vol. 52, Combinatorics, Eger, 9–47, 1987

[A96]   R. Ahlswede, Report on work in progress in combinatorial extremal theory: Shadows, AZ-identity, matching, Ergnzungsreihe des SFB 343 "Diskrete Strukturen in der Mathematik", Universität Bielefeld, No. 95–004, 1996

[A99]   R. Ahlswede, Asymptotical isoperimetric problem, Proceedings 1999 IEEE ITW, Krüger National Park, South Africa, June 20–25, 85–87, 1999

[A01]   R. Ahlswede, Advances on extremal problems in number theory and combinatorics, European Congress of Mathematics, Barcelona 2000, Vol. I, 147–175, Carles Casacuberta, Rosa Maria Miró-Roig, Joan Verdera, and Sebastiá Xambó-Descamps, editors, Progress in Mathematics, Vol. 201, Birkhäuser Verlag, Basel, 2001

[A06a]   R. Ahlswede, Solution of Burnashev's Problem and a sharpening of the Erdos/Ko/Rado Theorem, General Theory of Information Transfer and Combinatorics, Lecture Notes in Computer Science, Vol. 4123, Springer, Berlin Heidelberg New York, 1006–1009, 2006

[A06b]   R. Ahlswede, Another diametric theorem in Hamming spaces: optimal group anticodes, Proceedings of IEEE Information Theory Workshop, Punta del Este, Uruguay, March 13–17, 212–216, 2006

[A06c]   R. Ahlswede, On set coverings in Cartesian product spaces, General Theory of Information Transfer and Combinatorics, Lecture Notes in Computer Science, Vol. 4123, Springer, Berlin Heidelberg New York, 926–937, 2006

[A07]   R. Ahlswede, The final form of Tao's inequality relating conditional expectation and conditional mutual information, Advances in Mathematics of Communications, Vol. 1, No. 2, 239–242, 2007

[AAERS96]   R. Ahlswede, N. Alon, P.L. Erdös, M. Ruszinko, and L.A. Székely, Intersecting Systems, Comb. Probab. Comput., Vol. 6, No. 2, 127–137, 1996

[AA94]   R. Ahlswede and I. Althöfer, The asymptotic behaviour of diameters in the average, J. Comb. Theory B, Vol. 61, No. 2, 167–177, 1994

[AA07]   R. Ahlswede and H. Aydinian, On security of statistical databases, SIAM
              J. Discrete Math., submitted

[AAK98]   R. Ahlswede, H. Aydinian, and L. Khachatrian, The intersection theorem for
              direct products, Eur. J. Comb., Vol. 19, No. 6, 649–661, 1998

[AAK03a]   R. Ahlswede, H. Aydinian, and L. Khachatrian, Maximum number of constant
              weight vertices of the unit n-cube contained in a k-dimensional subspace, Com-
              binatorica, Vol. 23, No. 1, 1–18, 2003

[AAK03b]   R. Ahlswede, H. Aydinian, and L. Khachatrian, Forbidden $(0,1)$-vectors in hy-
              perplanes of $\mathbb{R}^n$: the restricted case, Des. Codes Cryptogr., Vol. 29, No. 1–3,
              17–18, 2003

[AAK03c]   R. Ahlswede, H. Aydinian, and L. Khachatrian, Maximal antichains under di-
              mension constrains, Discrete Math., Vol. 273, No. 1–3, 23–29, 2003

[AAK03d]   R. Ahlswede, H. Aydinian, and L. Khachatrian, More about shifting tech-
              niques, Eur. J. Comb., Vol. 24, No. 5, 551–556, 2003

[AAK03e]   R. Ahlswede, H. Aydinian, and L. Khachatrian, Extremal problems under di-
              mension constraints, Discrete Math., Special issue: EuroComb'01 - J. Nesetril,
              M. Noy, and O. Serra, editors, Vol. 273, No. 1–3, 9–21, 2003

[AAK04a]   R. Ahlswede, H. Aydinian, and L. Khachatrian, On shadows of intersecting
              families, Combinatorica, Vol. 24, No. 4, 555–566, 2004

[AAK04b]   R. Ahlswede, H. Aydinian, and L. Khachatrian, On Bohman's conjecture re-
              lated to a sum packing problem of Erdos, Proc. Amer. Math. Soc., Vol. 132,
              No. 5, 1257–1265, 2004

[AAK05]   R. Ahlswede, H. Aydinian, and L. Khachatrian, Forbidden $(0,1)$-vectors in Hy-
              perplanes of $\mathbb{R}^n$: the unrestricted case, Des. Codes Cryptogr., Vol. 37, No. 1,
              151–167, 2005

[ABK96]   R. Ahlswede, B. Balkenhol, and L.H. Khachatrian, Some properties of fix-free
              codes, Proceedings First INTAS International Seminar on Coding Theory and
              Combinatorics, Thahkadzor, Armenia, 20–33, 1996

[ABCABDM06]   R. Ahlswede, L. Bäumer, N. Cai, H. Aydinian, V. Blinovsky, C. Deppe, and
              H. Mashurian, editors, General Theory of Information Transfer and Com-
              binatorics, Lecture Notes in Computer Science, vol. 4123, Springer, Berlin
              Heidelberg New York, 2006

[ABEK02]   R. Ahlswede, C. Bey, K. Engel, and L. Khachatrian, The t-intersection problem
              in the truncated Boolean lattice, Eur. J. Comb., Vol. 23, No. 5, 471–487, 2002

[AB95]   R. Ahlswede and S.L. Bezrukov, Edge isoperimetric theorems for integer point
              arrays, Appl. Math. Lett., Vol. 8, No. 2, 75–80, 1995

[ABBMM93]   R. Ahlswede, S.L. Bezrukov, A. Blokhuis, K. Metsch, and G.E. Moorhouse,
              On partitioning the n-cube into sets with mutual distance 1, Appl. Math. Lett.,
              Vol. 6, No. 4, 17–19, 1993

[AB06b]   R. Ahlswede and V. Blinovsky, Maximal sets of integers not containing $k+1$
              pairwise coprimes and having divisors from a specified set of primes, Special
              Issue in Honor of Jacobus H. van Lint of J. Combinatorial Theory Series A,
              Vol. 113, No. 8, 1621–1628, 2006

[AB08]   R. Ahlswede and V. Blinovsky, Multiple packing in sum-type metric spaces,
              General Theory of Information Transfer and Combinatorics, Special Issue of
              Discrete Applied Math., Vol. 156, 1469–1477, 2008

[AC91]   R. Ahlswede and N. Cai, On sets of words with pairwise common letter in
              different positions, Extremal Problems for Finite Sets, Visográd (Hungary),
              25–38, 1991

[AC93a]   R. Ahlswede and N. Cai, A generalization of the AZ identity, Combinatorica,
              Vol. 13, No. 3, 241–247, 1993

[AC93b]   R. Ahlswede and N. Cai, On extremal set partitions in Cartesian product spaces,
              Comb. Probab. Comput., Vol. 2, No. 3, 211–220, 1993

[AC94]   R. Ahlswede and N. Cai, On partitioning and packing products with rectangles,
              Comb. Probab. Comput., Vol. 3, No. 4, 429–434, 1994

[AC96]   R. Ahlswede and N. Cai, Cross-disjoint pairs of clouds in the interval lattice, The Mathematics of Paul Erdos, Vol. I; R.L. Graham, and J. Nesetril, editors, Algorithms and Combinatorics B, Springer, Berlin Heidelberg New York, 155–164, 1996

[AC97a]   R. Ahlswede and N. Cai, Models of multi-user write-efficient memories and general diametric theorems, Inf. Comput., Vol. 135, No. 1, 37–67, 1997

[AC97b]   R. Ahlswede and N. Cai, General edge-isoperimetric inequalities, Part 1: Information theoretical methods, Eur. J. Comb., Vol. 18, No. 4, 355–372, 1997

[AC97c]   R. Ahlswede and N. Cai, General edge-isoperimetric inequalities, Part 2: A local-global principle for lexicographical solutions, Eur. J. Comb., Vol. 18, No. 5, 479–489, 1997

[AC97d]   R. Ahlswede and N. Cai, Shadows and isoperimetry under the sequence-subsequence relation, Combinatorica, Vol. 17, No. 1, 11–29, 1997

[ACLY00]   R. Ahlswede, N. Cai, S.Y.R. Li, and R.W. Yeung, Network information flow, Preprint 98-033, SFB 343 "Diskrete Strukturen in der Mathematik", Universitt Bielefeld, IEEE Trans. Inf. Theory, Vol. 46, No. 4, 1204-1216, 2000

[ACZ89]   R Ahlswede, N. Cai, and Z. Zhang, A general 4-words inequality with consequences for 2-way communication complexity, Adv. Appl. Math., Vol. 10, No. 1, 75–94, 1989

[ACZ92a]   R. Ahlswede, N. Cai, and Z. Zhang, Diametric theorems in sequence spaces, Combinatorica, Vol. 12, No. 1, 1–17, 1992

[ACZ92b]   R. Ahlswede, N. Cai, and Z. Zhang, Rich colorings with local constraints, J. Comb. Inf. Syst. Sci., Vol. 17, No. 3–4, 203–216, 1992

[ACZ94]   R. Ahlswede, N. Cai, and Z. Zhang, A new direction in extremal theory for graphs, Inf. Syst. Sci., Vol. 19, No. 3–4, 269–280, 1994

[ACZ96]   R. Ahlswede, N. Cai, and Z. Zhang, Higher level extremal problems, Comb. Inf. Syst. Sci., Vol. 21, No. 3–4, 185–210, 1996

[AC98]   R. Ahlswede and I. Csiszár, Common randomness in Information Theory and Cryptography, Part II: CR capacity, IEEE Trans. Inf. Theory, Vol. 44, No. 1, 225–240, 1998

[AD78]   R. Ahlswede and D. Daykin, An inequality for the weights of two families of sets, their unions and intersections, Z. Wahrscheinlichkeitstheorie und verw. Gebiete, Vol. 43, No. 3, 183–185, 1978

[AD79a]   R. Ahlswede and D. Daykin, The number of values of combinatorial functions, Bull. Lond. Math. Soc., Vol. 11, No. 1, 49–51, 1979

[AD79b]   R. Ahlswede and D.E. Daykin, Inequalities for a pair of maps $S \times S \to S$ with S a finite set, Math. Zeitschrift, Vol. 165, No. 3, 267–289, 1979

[AEP84]   R. Ahlswede, A. El Gamal, and K.F. Pang, A two family extremal problem in Hamming space, Discrete Math. Vol. 49, No. 1, 1–5, 1984

[AEG95]   R. Ahlswede, P.L. Erdös, and N. Graham, A splitting property of maximal antichains, Combinatorica, Vol. 15, No. 4, 475–480, 1995

[AGK76]   R. Ahlswede, P. Gács, and J. Körner, Bounds on conditional probabilities with applications in multi-user communication, Z. Wahrscheinlichkeitstheorie und verw. Geb., Vol. 34, No. 2, 157–177, 1976

[AK77]   R. Ahlswede and G. Katona, Contributions to the geometry of Hamming spaces, Discrete Math., Vol. 17, No. 1, 1–22, 1977

[AK78]   R. Ahlswede and G. Katona, Graphs with maximal number of adjacent pairs of edges, Acta Math. Acad. Sci. Hung., Vol. 32, No. 1–2, 97–120, 1978

[AK94a]   R. Ahlswede and L. Khachatrian, The maximal length of cloud-antichains, Discrete Math., Vol. 131, No. 1–3, 9–15, 1994

[AK94b]   R. Ahlswede and L. Khachatrian, On extremal sets without coprimes, Acta Arith., Vol. 66, No. 1, 89–99, 1994

[AK95a]   R. Ahlswede and L. Khachatrian, Density inequalities for sets of multiples, J. Number Theory, Vol. 55, No. 2, 170–180, 1995

[AK95b]   R. Ahlswede and L. Khachatrian, Maximal sets of numbers not containing $k+1$ pairwise coprime integers, Acta Arith., Vol. 72, No. 1, 77–99, 1995

[AK95c]   R. Ahlswede and L. Khachatrian, Towards characterising equality in correlation inequalities, Eur. J. Comb., Vol. 16, No. 4, 315–328, 1995

[AK96a]   R. Ahlswede and L. Khachatrian, Sets of integers and quasi-integers with pairwise common divisor, Acta Arith., Vol. 74, No. 2, 141–153, 1996

[AK96b]   R. Ahlswede and L. Khachatrian, Sets of integers with pairwise common divisor and a factor from a specified set of primes, Acta Arith., Vol. 75, No. 3, 259–276, 1996

[AK96c]   R. Ahlswede and L. Khachatrian, Optimal pairs of incomparable clouds in multisets, Graphs and Combinatorics, Vol. 12, No. 2, 97–137, 1996

[AK96d]   R. Ahlswede and L. Khachatrian, The complete nontrivial-intersection theorem for systems of finite sets, J. Comb. Theory Ser. A, Vol. 76, No. 1, 121–138, 1996

[AK96e]   R. Ahlswede and L. Khachatrian, Classical results on primitive and recent results on cross-primitive sequences, The Mathematics of P. Erdos, Vol. I; R.L. Graham, and J. Nesetril, editors, Algorithms and Combinatorics B, Springer, Berlin Heidelberg New York, 104–116, 1996

[AK97a]   R. Ahlswede and L. Khachatrian, Number-theoretic correlation inequalities for Dirichlet densities, J. Number Theory, Vol. 63, No. 1, 34–46, 1997

[AK97b]   R. Ahlswede and L. Khachatrian, The complete intersection theorem for systems of finite sets, Eur. J. Comb., Vol. 18, No. 2, 125–136, 1997

[AK97c]   R. Ahlswede and L. Khachatrian, Counterexample to the Frankl/Pach conjecture for uniform dense families, Combinatorica, Vol. 17, No. 2, 299–301, 1997

[AK98]   R. Ahlswede and L. Khachatrian, The diametric theorem in Hamming spaces – optimal anticodes, Adv. Appl. Math., Vol. 20, No. 4, 429–449, 1998

[AK99]   R. Ahlswede and L. Khachatrian, A pushing-pulling method: New proofs of intersection theorems, Combinatorica, Vol. 19, No. 1, 1–15, 1999

[AK00a]   R. Ahlswede and L. Khachatrian, Splitting properties in partially ordered sets and set systems, Numbers, Information and Complexity, Kluwer, Boston, 29–44, 2000

[AK00b]   R. Ahlswede and L. Khachatrian, A diametric theorem for edges, J. Comb. Theory Ser. A, Vol. 92, No. 1, 1–16, 2000

[AK05]   R. Ahlswede and L. Khachatrian, Katona's intersection theorem: four proofs, Combinatorica, Vol. 25, No. 1, 105–110, 2005

[AKMS03]   R. Ahlswede, L. Khachatrian, C. Mauduit, and A. Sárközy, A complexity measure for families of binary sequences, Periodica Math. Hung., Vol. 46, No. 2, 107–118, 2003

[AKS99]   R. Ahlswede, L. Khachatrian, and A. Sárközy, On the counting function of primitive sets of integers, J. Number Theory, Vol. 79, No. 2, 330–344, 1999

[AKS00]   R. Ahlswede, L. Khachatrian, and A. Sárközy, On prefix-free and suffix-free sequences of integers, Numbers, Information and Complexity, Special volume in honour of R. Ahlswede on occasion of his 60th birthday, I. Althfer, N. Cai, G. Dueck, L.H. Khachatrian, M. Pinsker, A. Srkzy, I. Wegener, and Z. Zhang, editors, Kluwer, Boston, 1–16, 2000

[AKS04]   R. Ahlswede, L. Khachatrian, and A. Sárközy, On the density of primitive sets, J. Number Theory, Vol. 109, No. 2, 319–361, 2004

[AK75]   R. Ahlswede and J. Körner, Source coding with side information and a converse for degraded broadcast channels, IEEE Trans. Inf. Theory, Vol. 21, No. 6, 629–637, 1975

[AMS06a]   R. Ahlswede, C. Mauduit, and A. Sárközy, Large families of pseudorandom sequences of k symbols and their complexity, Part I, General Theory of Information Transfer and Combinatorics, Lecture Notes in Computer Science, Vol. 4123, Springer, Berlin Heidelberg New York, 293–307, 2006

[AMS06b]   R. Ahlswede, C. Mauduit, and A. Sárközy, Large families of pseudorandom sequences of k symbols and their complexity, Part II, General Theory of Information Transfer and Combinatorics, Lecture Notes in Computer Science, Vol. 4123, Springer, Berlin Heidelberg New York, 308–325, 2006

[AM88]   R. Ahlswede and M. Mörs, Inequalities for code pairs, Eur. J. Comb., Vol. 9, No. 2, 175–181, 1988

[AW02]   R. Ahlswede and A. Winter, Strong converse for identification via quantum channels, IEEE Trans. Inf. Theory, Vol. 48, No. 3, 569–579, 2002

[AYZ97]   R. Ahlswede, E. Yang, and Z. Zhang, Identification via compressed data, IEEE Trans. Inf. Theory, Vol. 43, No. 1, 48–70, 1997

[AYZ90]   R. Ahlswede, J. Ye, and Z. Zhang, Creating order in sequence spaces with simple mashines, Inf. Comput., Vol. 89, No. 1, 47–94, 1990

[AZ89]   R. Ahlswede and Z. Zhang, Contributions to a theory of ordering for sequence spaces, Probl. Control Inf. Theory, Vol. 18, No. 4, 197–221, 1989

[AZ90a]   R. Ahlswede and Z. Zhang, An identity in combinatorial extremal theory, Adv. Math., Vol. 80, No. 2, 137–151, 1990

[AZ90b]   R. Ahlswede and Z. Zhang, On cloud-antichains and related configurations, Discrete Math., Vol. 85, No. 3, 225–245, 1990

[AZ98]   M. Aigner and G. Ziegler, Proofs from the Book, Springer, Berlin Heidelberg New York, 1998

[ACNS82]   M. Ajtai, V. Chvátal, M. Newborn, and E. Szemerédi, Crossing-free subgraphs, Annals Discrete Math., Vol. 12, 9–12, 1982

[AKS80]   M. Ajtai, J. Komlós, and E. Szemerédi, A note on Ramsey numbers, J. Comb. Theory Ser. A, Vol. 29, No. 3, 354–360, 1980

[AKS81]   M. Ajtai, J. Komlós, and E. Szemerédi, A dense infinite Sidon sequence, Eur. J. Comb., Vol. 2, No. 1, 1–11, 1981

[A74]   V. Alekseev, The number of monotone k-valued functions, Problemy Kibernet., Vol. 28, 5–24, 1974 (in Russian)

[A98]   N. Alon, The Shannon capacity of a union, Combinatorica, Vol. 18, No. 3, 301–310, 1998

[AL06]   N. Alon and E. Lubetzky, Uniformly cross intersecting families, preprint, arXiv:math.CO/0608173, 2006

[AS92]   N. Alon and J. Spencer, The Probabilistic Method, Wiley, New York, 1992

[AS95]   N. Alon and B. Sudakov, Disjoint systems, Random Structures Algorithms, Vol. 6, No. 1, 13–20, 1995

[ACDKPSWZ00] I. Althöfer, N. Cai, G. Dueck, L. Khachatrian, M. Pinsker, A. Sarközy, I. Wegener, and Z. Zhang, editors, Numbers, Information and Complexity, Special volume in honour of R. Ahlswede on the occasion of his 60th birthday, Kluwer, Boston, 501–516, 2000

[A76]   T. Apostol, Intoduction to Analytic Number Theory, Springer, Berlin Heidelberg New York, 1976

[A88]   H. Aydinian, A T-family extremal problem in Hamming space, Probl. Control Inf. Theory, Vol. 17, No. 1, 33–36, 1988

[A67]   K. Azuma, Weighted sums of certain dependent random variables, Tôhoku Math. J. (2), Vol. 19, 357–367, 1967

[BS96]   R. Balasubramanian and K. Soundararajan, On a conjecture of R. L. Graham, Acta Arith., Vol. 75, No. 1, 1–38, 1996

[B75]   Z. Baranyai, On the factorization of the complete uniform hypergraph. Infinite and finite sets, in Infinite and Finite Sets, Vol. 1, Proceedings a Colloquium held at Keszthely, June 25-July 1, 1973, Dedicated to Paul Erdös on his 60th Birthday, A. Hajnal, R. Rado, and V. T. Sós, editors, 91–108, 1975

[B81]   V. Baston, Inequalities involving maps of finite sets, Mathematische Zeitschrift, Vol. 177, 207–210, 1981

[B48]   F.A. Behrend, Generalization of an inequality of Heilbronn and Rohrbach, Bull. Amer. Math. Soc., Vol. 54, 681–684, 1948

[BF87]   J. van den Berg and U. Fiebig, On a combinatorial conjecture concerning disjoint occurences of events, Ann. Probab., Vol. 15, No. 1, 354–374, 1987

[BK85]   J. van den Berg and H. Kesten, Inequalities with applications to percolation and reliability, J. Appl. Probab., Vol. 22, No. 3, 556–569, 1985

[B67]   A. Bernstein, Maximally connected arrays on the n-cube, SIAM J. Appl. Math., Vol. 15, 1485–1489, 1967

[B05a]   C. Bey, Polynomial LYM inequalities, Combinatorica, Vol. 25, No. 1, 19–38, 2005

[B94]   S.L. Bezrukov, Isoperimetric problems in discrete spaces, Extremal problems for finite sets (Visegrd, 1991), 59–91, Bolyai Soc. Math. Stud., Vol. 3, Jnos Bolyai Math. Soc., Budapest, 1994

[BS02]   S.L. Bezrukov and O. Serra, A local-global principle for vertex-isoperimetric problems, Discrete Math., Vol. 257, No. 2–3, 285–309, 2002

[B73a]   G. Birkhoff, Lattice Theory, Amer. Math. Soc., Providence, 1973

[B86a]   V. Blinovsky, Bounds for codes in the case of list decoding of finite volume, Probl. Inf. Transm., Vol. 22, No. 1, 7–19, 1986

[B97]   V. Blinovsky, Asymptotic Combinatorial Coding Theory, Kluwer, Boston, 1997

[B00]   V. Blinovsky, Lower bound for the linear multiple packing of the binary Hamming space, J. Comb. Theory Ser. A, Vol. 92, No. 1, 95–101, 2000

[B01a]   V. Blinovsky, Error probability exponent of list-of-L decoding at zero rate, unpublished manuscript, 2001

[B01b]   V. Blinovsky, Error probability exponent of list decoding at low rates, Probl. Inf. Transm., Vol. 37, No. 4, 277–287, 2001

[B05b]   V. Blinovsky, Code bounds for multiple packing over a non-binary finite alphabet, Probl. Inf. Transm., Vol. 41, No. 1, 23–32, 2005

[B96]   T. Bohman, A sum packing problem of Erdös and the Conway-Guy sequence, Proc. Amer. Math. Soc., Vol. 124, No. 12, 3627–3636, 1996

[B80]   H.-W. Bollmann, Ungleichungen für verbandsgeordnete Maßräume, diploma thesis, University of Bielefeld, 1980

[B65]   B. Bollobás, On generalized graphs, Acta Math. Acad. Sci. Hung., Vol. 16, 447–452, 1965

[B73b]   B. Bollobás, Sperner systems consisting of pairs of complementary subsets, J. Comb. Theory Ser. A, Vol. 15, 363–366, 1973

[B88]   B. Bollobás, Combinatorics, Cambridge University Press, Cambridge, 1988

[B98]   B. Bollobás, Modern Graph Theory, Springer, Berlin Heidelberg New York, 1998

[BB90]   B. Bollobás and G. Brightwell, Parallel selection with high probability, SIAM J. Discrete Math., Vol. 3, No. 1, 21–31, 1990

[BL90]   B. Bollobás and I. Leader, An isoperimetric inequality on the discrete torus, SIAM J. Discrete Math., Vol. 3, No. 1, 32–37, 1990

[BL91]   B. Bollobás and I. Leader, Edge-isoperimetric inequalities in the grid, Combinatorica, Vol. 11, No. 4, 299–314, 1991

[BL93]   B. Bollobás and I. Leader, Maximal sets of given diameter in the grid and the torus, Discrete Math., Vol. 122, No. 1–3, 15–35, 1993

[BL04]   B. Bollobás and I. Leader, Isoperimetric problems for r-sets, Comb. Probab. Comput., Vol. 13, No. 2, 277–279, 2004

[BK07]   A. de Bonis and G. Katona, Largest families without an r-fork, Order, Vol. 24, 181–191, 2007

[BKS05]   A. de Bonis, G. Katona, and K. Swanepoel, Largest family without $A \cup B \subseteq C \cup D$, J. Comb. Theory, Ser. A, Vol. 111, No. 2, 331–336, 2005

[B86b]   J. Borden, Coding for write-undirectional memories, preprint, 1986

[BCR99]   C. Borgs, J. Chayes, and D. Randall, The van den Berg-Kesten-Riemer inequality: a review, Perplexing problems in probability, Progr. Probab., Vol. 44, Birkhuser, Boston, MA, 195–173, 1999

[BHM00]   L. Branković, P. Horak, and M. Miller, An optimization problem in statistical databases, SIAM J. Discrete Math., Vol. 13, No. 3, 346–353, 2000

[B85]   G. Brightwell, Universal correlations in finite posets, Order, Vol. 2, No. 4, 129–144, 1985

[B86c]   G. Brightwell, Some correlation inequalities in finite posets, Order, Vol. 2, No. 4, 387–402, 1986

[BF90]   R. Burton,Jr and M. Franzosa, Positive dependence properties of point processes, Ann. Probab., Vol. 18, No. 1, 359–377, 1990

[C86]   N. Cai, A bound of sizes of code pairs satisfying the strong 4-words property for Lee distances, J. System Sci. Math. Sci., Vol. 6, No. 2, 129–135, 1986

[CF92]   A.R. Calderbank and P. Frankl, Improved upper bounds concerning the Erdös-Ko-Rado theorem, Comb. Probab. Comput., Vol. 1, No. 2, 115–122, 1992

[C99]   Y.G. Chen, On sums and products of integers, Proc. Amer. Math. Soc., Vol. 127, No. 7, 1927–1933, 1999

[CO82]   F. Chin and G. Ozsoyoglu, Auditing and inference control in statistical databases, IEEE Trans. Softw. Eng., Vol. 8, No. 6, 574–582, 1982

[CG98]   F. Chung and R. Graham, Erdös on graphs, his legacy of unsolved problems, A K Peters, Ltd., Wellesley, 1998

[C72]   V. Chvátal, Unsolved Problem No. 7, Hypergraph Seminar, Proceedings of First Working Sem., Ohio State Univ., Columbus, Ohio, 1972, Lecture Notes in Mathematics, Vol. 411, Springer, Berlin Heidelberg, New York, 1974

[C71]   G. Clements, Sets of lattice points which contain a maximal number of edges, Proc. Amer. Math. Soc., Vol. 27, 13–15, 1971

[CL69]   G. Clements and B. Lindström, A generalization of a combinatorial theorem of Macaulay, J. Comb. Theory, Vol. 7, 230–238, 1969

[CG68]   J.H. Conway and R.K. Guy, Sets of natural numbers with distinct sums, Notices Amer. Math. Soc., Vol. 15, 345, 1968

[CG87]   T. Cover and B. Gopinath, Open Problems in Communication and Computation, Springer, Berlin Heidelberg New York, 1987

[DD97a]   T.N. Danh and D.E. Daykin, Ordering integer vectors for coordinate deletions, J. Lond. Math. Soc. (2), Vol. 55, No. 3, 417–426, 1997

[DD97b]   T.N. Danh and D.E. Daykin, Sets of 0,1 vectors with minima sets of subvectors, Rostock Math. Kolloq. Vol. 50, 47–52, 1997

[DE36]   H. Davenport and P. Erdös, On sequences of positive integers, Acta. Arith., Vol. 2, 147–151, 1937

[D74]   D.E. Daykin, Erdös-Ko-Rado from Kruskal-Katona, J. Comb. Theory Ser. A, Vol. 17, 254–255, 1974

[D77]   D.E. Daykin, A lattice is distributive iff $|A||B| \leq |A \vee B||A \wedge B|$, Nanta Math., Vol. 10, No. 1, 58–60, 1977

[D80]   D.E. Daykin, A hierarchy of inequalities, Stud. Appl. Math., Vol. 63, No. 3, 263–264, 1980

[DKW79]   D.E. Daykin, D. Kleitman, and D. West, On the number of meets between two subsets of a lattice, J. Comb. Theory Ser. A, Vol. 26, No. 2, 135–155, 1979

[DL76]   D.E. Daykin and L. Lovasz, The number of values of a Boolean function, J. Lond. Math. Soc. (2), Vol. 12, No. 2, 225–230, 1976

[DT94]   D.E. Daykin and T. Thu, The dual of the Ahlswede-Zhang identity, J. Comb. Theory Ser. A, 68, No. 1, 246–249, 1994

[DP85]   P. Delsarte and P. Piret, An extension of an inequality by Ahlswede, El Gamal, and Pang for pairs of binary codes, Discrete Math., Vol. 55, No. 3, 313–315, 1985

[DKM04]   J. Demetrovich, G.O.H. Katona, and D. Miklos, On the security of individual data, Lecture Notes in Computer Science, Springer, Vol. 2942, 49–58, 2004

[D82]   D.E.R. Denning, Cryptology and Data Security, Addison-Wesley, Reading, 1982

[DS06]   C. Deppe and H. Schnettler, On the 3/4-conjecture for fix-free codes, European Conference on Combinatorics, Graph Theory and Applications, S. Felsner, editor, DMTCS Proceedings Volume AE, 111–116, 2005

[DS89]   J. Deuschel and D. Strook, Large Deviations, Academic, Boston, 1989

[DF83]   M. Deza and P. Frankl, Erdös-Ko-Rado theorem – 22 years later, SIAM J. Alg. Disc. Meth., Vol. 4, 419–431, 1983

[DS05]   I. Dinur and S. Safra, On the hardness of approximating minimum vertex cover, Ann. Math. (2), Vol. 162, No. 1, 439–485, 2005

[DD63]   P.G.L. Dirichlet and R. Dedekind, Lectures on Number Theory, Translated from the German original "Vorlesungen über Zahlentheorie", AMS, 1999

[D-F02]   J. Domingo-Ferrer, Inference Control in Statistical Databases, Springer, Berlin Heidelberg New York, 2002

[DK90]   D. Du and D. Kleitman, Diameter and radius in the Manhattan metric, Discrete Comput. Geom., Vol. 5, No. 4, 351–356, 1990

[DPR96]   D. Dubhashi, V. Priebe, and D. Ranjan, Negative dependence through the FKG inequality, BRICS Report Series, RS-96-27, 1996

[DR98]   D. Dubhashi and D. Ranjan, Balls and bins: A study in negative dependence, Random Struct. Algor., Vol. 13, No. 2, 99–124, 1998

[E58]   H. Eggleston, Convexity, Cambridge University Press, Cambridge, 1958

[E05]   O. Einstein, Properties of intersecting families of ordered set, preprint, 2005

[E97a]   G. Elekes, On the number of sums and products, Acta Arith., Vol. 81, No. 4, 365–367, 1997

[ENR00]   G. Elekes, M. Nathanson, and I. Ruzsa, Convexity and sumsets, J. Number Theory, Vol. 83, No. 2, 194–201, 2000

[E57]   P. Elias, List decoding for noisy channels, Wescon Convention Rec., Pt.2, Inst. Radio Eng., 94–104, 1957

[E91]   P. Elias, Error-correcting codes and list decoding, IEEE Trans. Inf. Theory, Vol. 37, No. 1, 5–12, 1991

[E84]   K. Engel, Optimal representations, LYM posets, Peck posets, and the Ahlswede-Daykin inequality, Rostock Math. Kolloq., Vol. 26, 63–68, 1984

[E97b]   K. Engel, Sperner Theory, Encyclopedia of Mathematics and its Applications, Vol. 65, Cambridge University Press, Cambridge, 1997

[E35]   P. Erdös, On sequences of integers no one of which is divisible by any other, J. Lond. Math. Soc., Vol. 10, 126–128, 1935

[E45]   P. Erdös, On a lemma of Littlewood and Offord, Bull. Amer. Math. Soc., Vol. 51, 898–902, 1945

[E56]   P. Erdös, Problems and results from additive number theory, Colloq. Théorie des Nombres, Bruxelles, 1955, Liege & Paris, 1956

[E62]   P. Erdös, Remarks in number theory, IV, Mat. Lapok 13, 228–255, 1962

[E65]   P. Erdös, Extremal problems in number theory, in Theory of Numbers, A.L. Whiteman, editor, Amer. Math. Soc., Providence, 181–189, 1965

[E73]   P. Erdös, Problems and results in combinatorial number theory, Chapt. 12 in: A Survey of Combinatorial Theory, J.Srivastava, et al., editors, North-Holand, Amsterdam, 1973

[E80]   P. Erdös, A survey of problems in combinatorial number theory, Ann. Discrete Math., Vol. 6, 89–115, 1980

[E87]   P. Erdös, My joint work with Richard Rado, Surveys in Combinatorics, London Math. Soc. Lecture Note Ser., Vol. 123, 53–80, Cambridge University Press, Cambridge, 1987

[E90]   P. Erdös, Some of my favorite unsolved problems, in: A Tribute to Paul Erdös, A. Baker, B. Bollob'as, and A.Hajnal, editors, Cambridge University Press, Cambridge, 1990

[EH63]   P. Erdös and H. Hanani, On a limit theorem in combinatorial analysis, Publ. Math. Debrecen, Vol. 10, 10–13, 1963

[EK86]   P.L. Erdös and G. Katona, Convex hulls of more-part Sperner families, Graphs
              Comb., Vol. 2, No. 2, 123–134, 1986
[EKR61]   P. Erdös, C. Ko, and R. Rado, Intersection theorems for systems of finite sets,
              Quart. J. Math. Oxford Ser. (2), Vol. 12, 313–320, 1961
[ES92]   P. Erdös and A. Sárközy, On sets of coprime integers in intervals, Mathematical
              Institute of the Hungarian Academy of Sciences, 1992
[ESS69]   P. Erdös, A. Sárközy, and E. Szemerédi, On some extremal properties of se-
              quences of integers, Ann. Univ. Sci. Budapest, Eötvös, Vol. 12, 131–135, 1969
[ESS80]   P. Erdös, A. Sárközy, and E. Szemerédi, On some extremal properties of se-
              quences of integers, II, Publ. Math., Vol. 27, No. 1–2, 117–125, 1980
[ES69]   P. Erdös and J. Schönheim, On the set of non-pairwise coprime division of a
              number, Comb. Theory Appl., Vol. I, 369–376, 1969
[ESS94]   P. L. Erdös, A. Seress, and L. A. Szekely, On intersecting chains in Boolean
              algebras, Comb. Probab. Comput., Vol. 3, 57–62, 1994
[ES83]   P. Erdös and E. Szemerédi, On sums and products of integers, Studies in pure
              mathematics, Birkhäuser, Basel, 213–218, 1983
[F70]   P.G. Farrell, Linear binary anticodes, Electron. Lett., Vol. 6, 419–421, 1970
[F68]   W. Feller, An Introduction to Probability Theory and its Applications, Vol 1,2,
              Wiley, New York, 1968
[F84]   P. Fishburn, A correlation inequality for linear extensions of posets, Order,
              Vol. 1, No. 2, 127–137, 1984
[F86]   P. Fishburn, Maximizing a correlational ratio for linear extensions of posets,
              Order, Vol. 3, No. 2, 159–167, 1986
[F91]   P. Fishburn, A note on linear extensions and incomparable pairs, J. Comb.
              Theory Ser. A, Vol. 56, No. 2, 290–296, 1991
[F92]   P. Fishburn, Correlation in partially ordered sets, Discrete Appl. Math., Vol. 39,
              No. 2, 173–191, 1992
[FDS88]   P. Fishburn, P. Doyle, and L. Shepp, The match set of a random permutation
              has the FKG property, Ann. Probab., Vol. 16, No. 3, 1194–1214, 1988
[FS91]   P. Fishburn and L. Shepp, On FKB conjecture for disjoint intersections, Dis-
              crete Math., Vol. 98, No. 2, 105–122, 1991
[F98]   K. Ford, Sums and products from a finite set of real numbers, Ramanujan J.,
              Vol. 2, No. 1–2, 59–66, 1998
[FF56]   L. R. Ford and D. R. Fulkerson, Maximal flow through a network, Canad. J.
              Math., Vol. 8, 399–404, 1956
[FKG71]   C. Fortuin, P. Kasteleyn, and J. Ginibre, Correlation inequalities for some par-
              tially ordered sets, Comm. Math. Phys., Vol. 22, 89–103, 1971
[F77]   P. Frankl, On the minimum number of disjoint pairs in a family of finite sets,
              J. Comb. Theory Ser. A, Vol. 22, No. 2, 249–251, 1977 .
[F78]   P. Frankl, The Erdös-Ko-Rado theorem is true for $n = ckt$, Colloq. Math. Soc.
              Math. Bolyai, Vol. 18, 365–375, 1978
[F87]   P. Frankl, The shifting technique in extremal set theory, Surveys in combina-
              torics, Lond. Math. Soc. Lecture Note Ser., Vol. 123, Cambridge University
              Press, Cambridge, 81–110, 1987
[F88]   P. Frankl, Old and new problems on finite sets, Congressus Numerantium,
              Vol. 67, 243–256, 1988
[FF80]   P. Frankl and Z. Füredi, The Erdös-Ko-Rado theorem for integer sequences,
              SIAM J. Algebraic Discrete Methods, Vol. 1, No. 4, 316–381, 1980
[FF87]   P. Frankl and Z. Füredi, Colored packing of sets, North-Holland Math. Stud.,
              Vol. 149, 165–177, 1987
[FP84]   P. Frankl and J. Pach, On disjointly representable sets, Combinatorica, Vol. 4,
              No. 1, 39–45, 1984
[FR85]   P. Frankl and V. Rödl, Near perfect coverings in graphs and hypergraphs, Eur.
              J. Comb., Vol. 6, No. 4, 317–326, 1985

[FR87]   P. Frankl and V. Rödl, Forbidden intersections, Trans. Amer. Math. Soc., Vol. 300, No. 1, 259–286, 1987

[FW81]   P. Frankl and R. M. Wilson, Intersection theorems with geometric consequences, Combinatorica, Vol. 1, No. 4, 357–368, 1981

[FLS07]   Y. Frein, B. Lévêque, and A. Sebö, Optimizing diversity, Electron. Notes Discrete Math., Vol. 29, 73–77, 2007

[F93]   R. Freud, Paul Erdös 80 – A personal account, Period. Math. Hung., Vol. 26, No. 2, 87–93, 1993

[FK06]   Z. Füredi and G. Katona, 2-bases of quadruples, Comb. Probab. Comput., Vol. 15, No. 1–2, 131–141, 2006

[FR05]   Z. Füredi and M. Ruszinkó, Large convex cones in hypercubes, Electron. Notes Discrete Math., Vol. 21, 283–284, 2005

[FK77]   M. Furlmeier and N. Kalus, Isopermetrische Probleme in endlichen Rumen, diploma thesis, University of Bielefeld, 1977

[G68]   R. Gallager, Information Theory and Reliable Communication, Wiley, New York, 1968

[G63]   T. Gallai, Neuer Beweis eines Tutte'schen Satzes, Magyar Tud. Akad. Mat. Kutató Int. Közl, Vol. 8, 135–139, 1963

[G70]   R. Graham, Unsolved problem 5749, Amer. Math. Mon., Vol. 77, 775, 1970

[G82]   R. Graham, Linear extensions of partial orders and the FKG inequality, NATO Adv. Study Inst Ser. C: Math. Phys. Sci., Vol. 83, 1982

[GRS90]   R. Graham, B. Rothschild, and J.H. Spencer, Ramsey Theory, Wiley, New York, 1990

[GYY80]   R. Graham, A. Yao, and F. Yao, Some monotonicity properties of partrial orders, SIAM J. Algebraic Discrete Methods, Vol. 1, No. 4, 251–258, 1980

[G71]   G. Grätzer, Lattice Theory, W.H. Freeman and Company, San Francisco, 1971

[G99]   J. Griggs, Database security and the distribution of subset sums in $\mathbb{R}^n$, Bolyai Soc. Math. Studs., Vol. 7, 223–252, 1999

[GK08]   J. Griggs and G. Katona, No four sets forming an N, J. Comb. Theory, Ser. A, Vol. 115, No. 4, 677–685, 2008

[GST84]   J. Griggs, J. Stahl, and W. Trotter, A Sperner theorem on unrelated chains of subsets, J. Comb. Theory Ser. A, Vol. 36, No. 1, 124–127, 1984

[G66]   R. Gupta, The chromatic index and the degree of a graph, Notices Amer. Math. Soc., Vol. 13, 719, 1966

[GR99]   J. Gupta and B. Rao, Van den Berg-Kesten inequality for the Poisson Boolean model for continuum percolation, Sankhyā Ser. A, Vol. 61, No. 3, 337–346, 1999

[H79]   W. Haemers, On some problems of Lovasz concerning the Shannon capacity of a graph, IEEE Trans. Inf. Theory, Vol. 25, No. 2, 231–232, 1979

[HR73]   A. Hajnal and B. Rothschild, A generalization of the Erdös-Ko-Rado theorem on finite set systems, J. Comb. Theory Ser. A, Vol. 15, 359–362, 1973

[HKG08]   H. Halabian, M. Khosravifard, and T.A. Gulliver, Some sufficient conditions for the existence for the existence of d-ary fix-free codes, Proceedings of Internantional ITG Conference on Source and Channel Coding, Ulm, Germany, 2008

[HR66]   H. Halberstam and K. Roth, Sequences, Clarendon Press, Oxford, 1966

[HL67]   M. Hall Jr. and J. van Lint, Constant distance code pairs, Nederl. Akad. Wetensch. Indag. Math., Vol. 47, No. 1, 41–45, 1985

[HT88]   R.R. Hall and G. Tenenbaum, Divisors, Cambridge Tracts in Mathematics, Vol. 90, 1988

[HK99]   K. Harada and K. Kobayashi, A note on the fix-free property, IEICE Trans. Fundam., Vol. E82-A, No. 10, 2121–2128, 1999

[H64]   L. Harper, Optimal assignment of numbers to vertices, J. Soc. Ind. Appl. Math., Vol. 12, No. 1, 131–135, 1964

[H67b]   L. Harper, A necessary condition on minimal cube numberings, J. Appl. Probab., Vol. 4, 397–401, 1967

[H04]   L. Harper, Global Methods for Combinatorial Isoperimetric Problems, Cambridge Studies in Advanced Math., Vol. 90, 2004

[H60]   T.E. Harris, A lower bound for the critical probability in a certain percolation process, Proc. Cambr. Phil. Soc., Vol. 56, 13–20, 1960

[H76]   S. Hart, A note on the edges of the n-cube. Discrete Math., Vol. 14, No. 2, 157–163, 1976

[H37]   H. Heilbronn, On an inequality in the elementary theory of numbers, Proc. Cambr. Phil. Soc., Vol. 33, 207–209, 1937

[HM67b]   A.J.W. Hilton and E.C. Milner, Some intersection theorems for systems of finite sets, Quart. J. Math. Oxford Ser. (2), Vol. 18, 369–384, 1967

[H63]   W. Hoeffding, Probability inequalities for sums of bounded random variables, J. Amer. Statist. Assoc., Vol. 58, 13–30, 1963

[H74]   R. Holley, Remarks on the FKG inequalities, Comm. Math. Phys., Vol. 36, 227–231, 1974

[HBM99]   P. Horak, L. Branković, and M. Miller, A combinatorial problem in database security, Discrete Appl. Math., Vol. 91, No. 1–3, 119–126, 1999

[J-DSV84] K. Joag-Dev, L. Shepp, and R. Vitale, Remarks and open problems in the area of the FKG inequality, IMS Lecture Notes-Monograph Series, Vol. 5, 121–126, 1984

[KKS95]   J. Kahn, J. Komlós, and E. Szemerédi, On the probability that a random ± 1-matrix is singular, J. Amer. Math. Soc., Vol.8, No. 1, 223–240, 1995

[KR80]   S. Karlin and Y. Rinott, Classes of ordering measures and related correlation inequalities, I. Multivariate totally positive distributions, J. Multivariate Analysis, Vol. 10, No. 4, 467–498, 1980

[K64]   G. Katona, Intersection theorems for systems of finite sets, Acta Math. Acad. Sci. Hung., Vol. 15, 329–337, 1964

[K68]   G. Katona, A theorem of finite sets, Theory of graphs (Proc. Colloq., Tihany, 1966), 187–207, Academic, New York, 1968

[K72]   G. Katona, A simple proof of the Erdös-Chao Ko-Rado theorem, J. Comb. Theory Ser. B, Vol. 13, 183–184, 1972

[K75]   G. Katona, The Hamming-sphere has minimum boundary, Stud. Sci. Math. Hung., Vol. 10, No. 1–2, 131–140, 1975

[K00]   G. Katona, The cycle method and its limits, Numbers, Information and Complexity (Bielefeld, 1998), Kluwer, Boston, 129–141, 2000

[KT83]   G. Katona and T. Tarján, Extremal problems with excluded subgraphs in the n-cube, Lecture Notes in Mathematics, Vol. 1018, 84–93, 1983

[K77]   J. Kemperman, On the FKG inequality for measures on a partially ordered space, Indag. Math., Vol. 39, No. 4, 313–331, 1977

[K95]   J. Kim, The Ramsey number $R(3,t)$ has order of magnitude $t^2/\log t$, Random Struct. Algor., Vol. 7, No. 3, 173–207, 1995

[K66a]   D. Kleitman, Families of non-disjoint subsets, J. Comb. Theory, Vol. 1, 153–155, 1966

[K66b]   D. Kleitman, On a combinatorial conjecture of Erdös, J. Comb. Theory, Vol. 1, 209–214, 1966

[KF88]   D. Kleitman and M. Fellows, Radius and diameter in Manhattan lattices, Discrete Math., Vol. 73, No. 1–2, 119–125, 1988

[KKR71]   D. Kleitman, M. Krieger, amd B. Rothschild, Configurations maximizing the number of pairs of Hamming-adjacent lattice points, Stud. Appl. Math, Vol. 50, 115–119, 1971

[KS81]   D. Kleitman and J. Shearer, A monotonicity property of partial orders, Stud. Appl. Math., Vol. 65, No. 1, 81–83, 1981

[KF61]   A. Kolmogorov and S. Fomin, Elements of the Theory of Functions and Func-
            tional Analysis, Vol.2; Measure, the Lebesgue integral, Hilbert space, Graylock
            Press, 1961
[K90]   J. Komlós, A strange pigeonhole principle, Order, Vol. 7, No. 2, 107–113, 1990
[K91]   E. Koppenrade, Durchmesserprobleme in der Lee-Metrik, Diploma Thesis,
            University of Bielefeld, 1991
[KS92]   J. Körner and G. Simonyi, A Sperner-type theorem and qualitative indepen-
            dence, J. Comb. Theory Ser. A, Vol. 59, No. 1, 90–103, 1992
[K00]   D.S. Krotov, $\mathbb{Z}_4$-linear perfect codes, Diskr. Analiz Issled. Oper., Ser. 1, Vol. 7,
            No. 4, 78–90, 2000 (in Russian)
[K01]   D.S. Krotov, $\mathbb{Z}_4$-linear Hadamard and extended perfect codes, Electron. Notes
            Discrete Math., Vol. 6, 107–112, 2001
[K07]   D.S. Krotov, $\mathbb{Z}_{2^k}$-duality, $\mathbb{Z}_{2^k}$-linear Hadamard codes, and co-$\mathbb{Z}_{2^k}$-linear
            1-perfect codes, IEEE Trans. Inf. Theory, Vol. 53, No. 4, 1532–1537, 2007
[K63]   J.B. Kruskal, The number of simplicies in a complex, in: Mathematical Op-
            timization Techniques, University of California Press, Berkely, California,
            251–278, 1963
[L69]   P. Lancaster, Theory of Matrices, Academic, New York, 1969
[L83]   T. Leighton, Complexity Issues in VLSI, MIT, Cambridge, 1983
[L96]   Z. Lengvárszky, The Marica/Schönheim inequality in lattices, Bull. Lond.
            Math. Soc., Vol. 28, No. 5, 449–454, 1996
[L64]   J. Lindsey, Assignment of numbers to vertices, Amer. Math. Monthly, Vol. 71,
            508–516, 1964
[LZ67]   B.A. Lindström and H.O. Zetterström, A combinatorial problem in the k-adic
            number systems, Proc. Amer. Math. Soc., Vol. 18, 166–170, 1967
[L98]   J. van Lint, Introduction to Coding Theory, Springer, Berlin Heidelberg New
            York, 1998
[L79a]   M. Livingston, An ordered version of the Erdös-Ko-Rado theorem, J. Comb.
            Theory Ser. A, Vol. 26, No. 2, 162–165, 1979
[L79b]   L. Lovász, On the Shannon capacity of a graph, IEEE Trans. Inf. Theory,
            Vol. 25, No. 1, 1–7, 1979
[LP86]   L. Lovász and M. Plummer, Matching Theory, North-Holland Mathematics
            Studies, 121, North-Holland, Amsterdam, 1986
[L88]   W.F. Lunnon, Integer sets with distinct subset sums, Math. Comp., Vol. 50, No.
            181, 297–320, 1988
[MS77]   F.J. MacWilliams and N.J. Sloane, The Theory of Error-Correcting Codes,
            I. North-Holland Math. Libr., Vol. 16. North-Holland, Amsterdam, 1977
[M95]   F. Maire, On the shadow of squashed families of k-sets, Electron. J. Comb.
            Vol. 2, R16, 1995
[M74]   G.A. Margulis, Probabilistic characteristics of graphs with large connectivity,
            Probl. Pererdav ci Informacii, Vol. 10, No. 2, 101–108, 1974
[MS69]   J. Marica and J. Schönheim, Differences of sets and a problem of Graham,
            Canad. Math. Bull., Vol. 12, 635–637, 1969
[MO79]   A. Marshall and I. Olkin, Inequalities: Theory of Majorization and its Applica-
            tions, Academic, New York, 1979
[MS82]   A. Marshall and M. Shaked, A class of multivariate new better than used dis-
            tributions, Ann. Probab., Vol. 10, No. 1, 259–264, 1982
[M96]   K. Marton, A measure concentration inequality for contracting Markov chains,
            Geom. Funct. Anal., Vol. 6, No. 3, 556–571, 1996
[M98]   K. Marton, Measure concentration for a class of random processes, Probab.
            Theory Relat. Fields, Vol. 110, No. 3, 427–439, 1998
[MRRW77]   R. McEliece, E. Rodemich, H. Rumsey, and L. Welch, New upper bounds on
            the rate of a code via the Delsarte-MacWilliams inequalities, IEEE Trans. Inf.
            Theory, Vol. IT-23, No. 2, 157–166, 1977

[MRS91]   M. Miller, I. Roberts, and I. Simpson, Application of symmetric chains to an optimization problem in the security of statistical databases, Bull. Inst. Combin. Appl., Vol. 2, 47–58, 1991

[MS94]   M. Miller and J. Seberry, Relative compromise of statistical databases, Austral. Comput. J., Vol. 21, No. 2, 51–62, 1994

[M82]   A. Moon, An analogue of the Erdös-Ko-Rado theorem for the Hamming schemes $H(n, q)$, J. Comb. Theory Ser. A, Vol. 32, No. 3, 386–390, 1982

[M1903]   R. Muirhead, Some methods applicable to identities and inequalities of symmetric algebraic functions of n letters, Proc. Edinburgh Math. Soc., Vol. 21, 144–157, 1903

[N74]   W. Narkiewicz, Elementary and Analytic Theory of Algebraic Numbers, Polish Scientific Publishers, Warszawa, 1974

[N97]   M. Nathanson, On sums and products of integers, Proc. Amer. Math. Soc., Vol. 125, No. 1, 9–16, 1997

[N00]   M. Nathanson, Elementary Methods in Number Theory, Graduate Texts in Mathematics, Vol. 195, Springer, Berlin Heidelberg New York, 2000

[NT99]   M. Nathanson and G. Tenenbaum, Inverse theorems and the number of sums and products. Structure theory of set addition. Astérisque, No. 258, xiii, 195–204, 1999

[O88]   A. Odlyzko, On subspaces spanned by random selections of ± 1 vectors, J. Comb. Theory Ser. A, Vol. 47, No. 1, 124–133, 1988

[O56]   H. H. Ostmann, Additive Zahlentheorie I,II, Springer, Berlin Heidelberg New York, 1956

[P05]   B. Patkos, AZ-type identities, CEU Math Preprint Series, 2005

[P64]   M. Pinsker, Information and information stability of random variables and processes, Holden-Day, San Francisco, 1964

[PS95]   C. Pomerance and A. Sárközy, Combinatorial number theory, in Handbook of Combinatorics, R.Graham, M.Grötschel, and L.Lovász, editors, MIT, Cambridge, MA, 967–1018, 1995

[P74]   C. Preston, A generalization of the FKG inequalities, Comm. Math. Phys., Vol. 36, 233–241, 1974

[R30]   F.P. Ramsey, On a problem of formal logic, Proc. London Math. Soc. Ser. 2, Vol. 30, 264–286, 1930

[R91]   D. Reimer, Proof of the van den Berg-Kesten conjecture, Comb. Probab. Comput., Vol. 9, No. 1, 27–32, 1991

[R69]   A. Rényi, Lectures on the theory of search, Institute of Statistics Mimeo Ser., No. 600.7, University of North Carolina, Chapel Hill, 1969

[RS93]   Y. Rinott and M. Saks, Correlation inequalities and a conjecture for permanents, Combinatorica, Vol. 13, No. 3, 269–277, 1993

[R85]   V. Rödl, On packing and covering problems, Eur. J. Comb., Vol. 6, No. 1, 69–78, 1985

[R37]   H. Rohrbach, Beweis einer zahlentheoretischen Ungleichung, J. Reine U. Angew. Math., Vol. 177, 193–196, 1937

[RS62]   J. Rosser and L. Schoenfeld, Approximate formulas for some functions of prime numbers, Ill. J. Math., Vol. 6, 64–89, 1962

[R65]   D. Rutherford, Introduction to Lattice Theory, Oliver & Boyd, Edinburg, 1965

[S48a]   E. Schmidt, Der Brunn-Minkowskische Satze und sein Spiegeltheorem sowie die isoperimetrische Eigenschaft der Kugel in der euklidischen und hyperbolischen Geometrie, Math. Ann., Vol. 120, 307–422, 1948

[S48b]   E. Schmidt, Die Brunn-Minkowskische Ungleichung und ihr Spiegelbild sowie die isoperimetrische Eigenschaft der Kugel in der euklidischen und nichteuklidischen Geometrie I, Math. Nachr., Vol. 1, 81–157, 1948

[S49]   E. Schmidt, Die Brunn-Minkowskische Ungleichung und ihr Spiegelbild sowie die isoperimetrische Eigenschaft der Kugel in der euklidischen und nichteuklidischen Geometrie II, Math. Nachr., Vol. 2, 171–244, 1949

[S05]   H. Schnettler, On the 3/4-conjecture for fix-free codes, diploma thesis, University of Bielefeld, arXiv:0709.2598, 2005

[S69]   J. Schönheim, Unsolved problems, W.T. Tutte, editor, Recent Progress in Combinatorics, Proceedings of the Third Waterloo Conference on Combinatorics, Academic, New York, 1969

[S59]   M.P. Schützenberger, A characteristic property of certain polynomials of E.F. Moore and C.E. Shannon, in : RLE Quarterly Progress Report No. 55, Research Laboratory of Electronics, MIT, 117–118, 1959

[S73]   P. Seymour, On incomparable collections of sets, Mathematika, Vol. 20, 208–209, 1973

[S56]   C. Shannon, The zero error capacity of a noisy channel, Institute of Radio Engineers, Trans. Inf. Theory, Vol. IT-2, 8–19, 1956

[SEF56]   C. Shannon, P. Elias, and A. Feinstein, A note on the maximum flow through a network, IEEE Trans. Inf. Theory, Vol. 2, No. 4, 117–119, 1956

[SGB67]   C. Shannon, R. Gallager, and E. Berlekamp, Lower bounds to error probability for coding in discrete memoryless channels, I, II, Inf. Control, Vol.10, No. 1, 65–103, Vol.5, 522–552, 1967

[S83]   J. Shearer, A note on the independence number of triangle-free graphs, Discrete Math., Vol. 46, No. 1, 83–87, 1983

[S80]   L. Shepp, The FKG inequality and some monotonicity properties of partial orders, SIAM J. Algebraic Discrete Methods, Vol. 1, No. 3, 295–299, 1980

[S82]   L. Shepp, The XYZ conjecture and the FKG inequality, Ann. Probab., Vol. 10, No. 3, 824–827, 1982

[S89]   G. Simonyi, On write-undirectional memory codes, IEEE Trans. Inf. Theory, Vol. 35, No. 3, 663–669, 1989

[S90]   W. Suen, A correlation inequality and Poisson limit theorem for non-overlapping balanced subgraphs of random graph, Random Struct. Alg., Vol. 1, No. 2, 231–242, 1990

[ST85]   C. Szabó and G. Tóth, Maximal sequences not containing 4 pairwise coprime integers, Mat. Lapok, Vol. 32, 253–257, 1985 (in Hungarian)

[S97]   L. Székely, Crossing numbers and hard Erdös problems in discrete geometry, Comb. Probab. Comput., Vol. 6, No. 3, 353–358, 1997

[T91]   M. Talagrand, A new isoperimetric inequality and the concentration of measure phenomenon, Geometric aspects of functional analysis (1989–90), Lecture Notes in Mathematics, Springer, Berlin Heidelberg New York, Vol. 1469, 94–124, 1991

[T95]   M. Talagrand, Concentration of measure and isoperimetric inequalities in product spaces, Publications Math., Inst. des Hautes Études Sci., No. 81, 73–205, 1995

[T06]   T. Tao, Szemerédi's regularity lemma revisited, Contrib. Discrete Math., Vol. 1, No. 1, 8–28, 2006

[TV05]   T. Tao and V. Vu, On the singularity probability of random Bernoulli matrices, arXiv:math.CO/0501313v1, 20 Jan., 2005

[T93]   T. Thu, An induction proof of the Ahlswede-Zhang identity, J. Comb. Theory Ser. A, Vol. 62, No. 1, 168–169, 1993

[T97]   T. Thu, Identities for combinatorial extremal theory, Bull. London Math. Soc., Vol. 29, No. 6, 693–696, 1997

[T07]   T. Thu, An AZ-style identity and Bollobás deficiency, J. Comb. Theory Ser. A, Vol. 114, No. 8, 1504–1514, 2007

[T08]   T. Thu, On half-way AZ-style identities, J. Comb. Theory, submitted

[T98]   H.T. Thanh, An extremal problem with excluded sub-posets in the Boolean lattice, Order, Vol. 15, No. 1, 51–57, 1998

[T49]   M. Tomić, Théorème de Gauss relatif au centre de gravité et son application (Serbian, Russian, and French summaries), Bull. Soc. Math. Phys. Serbie, Vol. 1, 31–40, 1949

[TV91]   M. Tsfasman and S. Vladut, Algebraic Geometric Codes, Kluwer, Boston, 1991

[T89]   Z. Tuza, Intersection properties and extremal problems for set systems, in: G.Halsz and V.Sós, editors, Irregularities of Partitions, Vol.8 of Algorithms and Combinatorics, Springer, Berlin Heidelberg New York, 141–151, 1989

[V64]   V. Vizing, On an estimate of the chromatic class of a p-graph, Discret Analiz, Vol. 3, 25–30, 1964 (in Russian)

[WW77]   D.L. Wang and P. Wang, Discrete isoperimetric problems, SIAM J. Appl. Math., Vol. 32, No. 4, 860–870, 1977

[W84]   R. Wilson, The exact bound on the Erdös-Ko-Rado theorem, Combinatorica, Vol. 4, No. 2–3, 247–257, 1984

[W83]   P. Winkler, Correlation among partial orders, SIAM J. Algebraic Discrete Methods, Vol. 4, No. 1, 1–7, 1983

[W86]   P. Winkler, Correlation and order, Contemp. Math., Vol. 57, 151–174, 1986

[WJK06]   Y. Wu, K. Jain, and S.-Y. Kung, A unification of Edmonds' routing theorem and Ahlswede et al's network coding theorem, Joint Special Issue of IEEE Trans. Inf. Theory and IEEE/ACM Trans. on Networking and Information Theory, 2006

[Y79]   A. Yao, Some complexity questions related to distributive computing, Proceedings of 11th Annual ACM Symposium on Theory of Computing, 209–219, 1979

[Y01]   S. Yekhanin, Sufficient conditions of existence of fix-free codes, Proceedings of International Symposium on Information Theory, Washington, D.C., 284, 2001

[Y04]   S. Yekhanin, Improved upper bound for the redundancy of fix-free codes, IEEE Trans. Inf. Theory, Vol. 50, No. 11, 2815–2818, 2004

[YY01]   C. Ye and R. W. Yeung, Some basic properties of fix-free codes, IEEE Trans. Inf. Theory, Vol. 47, No. 1, 72–87, 2001

Index

List of Symbols

Universitext

Nikulin, V. V.; Shafarevich, I. R.: Geometries and Groups

Oden, J. J.; Reddy, J. N.: Variational Methods in Theoretical Mechanics

Øksendal, B.: Stochastic Differential Equations

Øksendal, B.; Sulem, A.: Applied Stochastic Control of Jump Diffusions. 2nd edition

Orlik, P.; Welker, V.: Algebraic Combinatorics

Perrin, D.: Algebraic Geometry

Poizat, B.: A Course in Model Theory

Polster, B.: A Geometrical Picture Book

Porter, J. R.; Woods, R. G.: Extensions and Absolutes of Hausdorff Spaces

Procesi, C.: Lie Groups

Radjavi, H.; Rosenthal, P.: Simultaneous Triangularization

Ramsay, A.; Richtmeyer, R. D.: Introduction to Hyperbolic Geometry

Rautenberg, W.: A concise Introduction to Mathematical Logic

Rees, E. G.: Notes on Geometry

Reisel, R. B.: Elementary Theory of Metric Spaces

Rey, W. J. J.: Introduction to Robust and Quasi-Robust Statistical Methods

Ribenboim, P.: Classical Theory of Algebraic Numbers

Rickart, C. E.: Natural Function Algebras

Rotman, J. J.: Galois Theory

Rubel, L. A.: Entire and Meromorphic Functions

Ruiz-Tolosa, J. R.; Castillo E.: From Vectors to Tensors

Runde, V.: A Taste of Topology

Rybakowski, K. P.: The Homotopy Index and Partial Differential Equations

Sabbah, C.: Isomonodromic Deformations and Frobenius Manifolds

Sagan, H.: Space-Filling Curves

Salsa, S.: Partial Differential Equations in Action

Samelson, H.: Notes on Lie Algebras

Sauvigny, F.: Partial Differential Equations I

Sauvigny, F.: Partial Differential Equations II

Schiff, J. L.: Normal Families

Schirotzek, W.: Nonsmooth Analysis

Sengupta, J. K.: Optimal Decisions under Uncertainty

Séroul, R.: Programming for Mathematicians

Seydel, R.: Tools for Computational Finance

Shafarevich, I. R.: Discourses on Algebra

Shapiro, J. H.: Composition Operators and Classical Function Theory

Simonnet, M.: Measures and Probabilities

Smith, K. E.; Kahanpää, L.; Kekäläinen, P.; Traves, W.: An Invitation to Algebraic Geometry

Smith, K. T.: Power Series from a Computational Point of View

Smorynski, C.: Self-Reference and Modal Logic

Smoryński, C.: Logical Number Theory I. An Introduction

Srivastava: A Course on Mathematical Logic

Stichtenoth, H.: Algebraic Function Fields and Codes

Stillwell, J.: Geometry of Surfaces

Stroock, D. W.: An Introduction to the Theory of Large Deviations

Sunder, V. S.: An Invitation to von Neumann Algebras

Tamme, G.: Introduction to Étale Cohomology

Tondeur, P.: Foliations on Riemannian Manifolds

Toth, G.: Finite Möbius Groups, Minimal Immersions of Spheres, and Moduli

Tu, L. W.: An Introduction to Manifolds

Verhulst, F.: Nonlinear Differential Equations and Dynamical Systems

Weintraub, S. H.: Galois Theory

Wong, M. W.: Weyl Transforms

Xambó-Descamps, S.: Block Error-Correcting Codes

Zaanen, A.C.: Continuity, Integration and Fourier Theory

Zhang, F.: Matrix Theory

Zong, C.: Sphere Packings

Zong, C.: Strange Phenomena in Convex and Discrete Geometry

Zorich, V. A.: Mathematical Analysis I

Zorich, V. A.: Mathematical Analysis II